AF443742

Technical Writer's Handbook

Technical Writer's Handbook

HARRY E. CHANDLER

AMERICAN SOCIETY FOR METALS
Metals Park, Ohio 44073

Library of Congress Catalog Card No. 82-73610

ISBN: 0-87170-151-0

SAN 204-7586

To
Mary Lou, Susan and Mikey, Mike and Teresa,
Teresa, Steve, and Julie

Preface____

This handbook is actually three books in one:

Part I, *Writing for Busy Readers*: an audience-related format for technical writing designed by the author. Includes how-to needed to implement the format.

Part II, *Customs, Practices, and Standards for Technical Writing*: guidelines for writing and editing 13 different standard formats plus chapters on libel law, patent and trademark law, and copyright law. The formats: technical reports, abstracts and summaries, technical papers, preparing and presenting technical talks, technical articles, technical books, standards and specifications, proposals and idea submissions, technical directions, committee reports, memos and letters, technical news releases, and theses. Information is directly from the source, as opposed to a review and interpretation of what has been published on these subjects. Sources include readers/writers of technical literature who were asked for opinions and recommendations in an extensive survey. The United States, Canada, and England are represented.

Part III, *An Anthology of Stylebooks*: a collection of style rules from a variety of sources. The aim is to present a cross section of practices.

In all three parts, my underlying motive is to present a composite, not a consensus, of answers to "how are others doing this?"

In all three parts, I intend to champion a personal standard of excellence that refutes, "What do you expect—it's only technical writing?" I take the position there is no reason technical writing can't be on a par with all other types of writing. Such a standard, I believe, should be adhered to whether one is writing only one sentence or a technical treatise. As a part of this standard, I also suggest the need for enthusiasm for what one is writing. Enthusiasm, I submit, has a direct bearing on quality, and comes from being totally immersed in one's subject, and from being totally convinced one has something worthwhile to say, and that it can be said in an interesting—not dull—manner.

To the same end, I believe that taking oneself too seriously in writing should be avoided. Going off the deep end breeds preaching, pomposity, pontificating, overwriting, verbosity, and other afflictions. The opposite condition, being relaxed and keeping a low profile, can open new doors for the writer. He writes for his readers, rather than for himself. He sees the virtues of being direct and straightforward, being concise, being clear, and being able to stop at an appropriate place.

Further, I feel no useful purpose is served by putting down technical writers and technical writing in general. I contest the implication that all technical people have certain unspecified writing disabilities. Such charges are irresponsible and counter-productive. For one thing, self-esteem is an important part of the effective writer's makeup. For another, if there is a basic problem, it stems from a source that is seldom, if ever, talked about. There is a language problem. I describe it as a standard, first-draft language of approximation. Until one can escape the influence of this language, I contend, one cannot graduate to a higher plateau of writing.

Part of the problem is vocabulary. Part of the problem is how we use our vocabularies in thinking as writers. Technical people, in my experience, usually have separate writing, speaking, and reading vocabularies. They seldom use the last two in writing because of the influence of standard language. The writing process is a trial and error business. You experiment, evaluate, reject, try again, re-evaluate, etc. Standard language encourages us to shortcut this disorganized process. We accept the first word that comes to mind, usually the standard word, and do not go through the tortuous process of finding out what we have to say and developing an acceptable way of saying it.

I do not suggest that standard language should be outlawed. I don't think it can be avoided. But I propose: Don't settle for standard language, the first word that comes to mind, until you have investigated and experimented with alternatives.

For example, I once received a request for instructions to authors writing for *Metal Progress* magazine. We had just changed our editorial format, which would necessitate some minor changes in our instructions. I chose to send an unexpurgated version.

I wrote in part:

"A copy of our instructions to authors is attached. Even though we have changed our format, the instructions are still good."

I was dissatisfied with the second sentence. It didn't say exactly what I had in mind, but I couldn't put my finger on the trouble immediately.

I put the letter aside and handled another piece of business before going back to the errant sentence. As is often the case, I thought maybe the problem was in choice of words. "Still good" caught my eye.

I tried a second time:

"The brief statement of recommendations is still on target even though we recently changed format."

Another way of doing it; "still good" is eliminated, but the sentence is wordy. At this point, I realized my problem was content, rather than language. It took some time to reach this point, but all I needed to say was:

"A copy of our instructions to authors is attached. Our recent change in editorial format does not call for any major changes in the instructions."

Not perfect, but closer than "still good." Getting closer and closer and closer in this sense is one of the secrets of good writing, including good technical writing.

Use of language and thinking like a writer are topics in Part I, *Writing for Busy Readers*.

I want to make clear that writing for busy readers is a format that stands on its own. It is not held out as an alternative to or substitute for the 13 formats discussed in Part II, *Customs, Practices, and Standards in Technical Writing*.

In fact, I wrote Part I before I wrote Part II because I did not want to be influenced in any manner by the latter. Neither did I want to appear to compete with or disagree with what is said in Part II.

In fact, I disagree with only one thing in Part II: the widely held notion that one can write effectively to two or more audiences at the same time. I feel one must pick an audience, technical or nontechnical, and write to it exclusively.

The practice I do not buy prescribes introductions and general use of background materials for the nontechnical reader of texts written for technical readers. In my experience, technical people are annoyed with tutorial information they already know. The nontechnical reader, on the other hand, understands, partly at least, only the introductions and background materials. The bulk of the text, all figures, and all tables are Greek to him.

Contents___________________________

Part III: An Anthology of Stylebooks

Part I

Writing for Busy Readers

CHAPTER 1

Introduction to Part I

At this point, you, the reader, have already started to look for a more complete statement of my subject, "Writing for Busy Readers," than I gave you in the title of Part I of this handbook.

However, I, the writer, have a natural reluctance to accommodate you immediately. I feel an introduction is needed to give you background that will help you understand my statement of the subject when I unveil it. I feel secure in taking a slow, indirect approach because the practice has the sanction of universal usage in technical writing.

The example illustrates conflicts of interest between readers and writers of technical literature that are the basis of my thesis.

I propose: Look at common technical writing practices such as the paragraphs above (call it delaying the statement of the subject with an introduction) the way a reader does from the viewpoints of what he expects from this literature and how he reads it. Writing for busy readers (writing for a defined audience, if you will) is the product of such analysis. It is a style of technical writing that runs against the dictates of certain common technical writing practices—11 in all. Practice No. 1 (delaying statement of subject) has been cited. I name the remainder following brief statements of my understanding of why and how technical literature is read—key factors in this format.

First, why it is read.

We need to remind ourselves that reading is still the most common way a professional has to keep up with current and emerging developments and trends in his field or fields. Reading is a part of the job and a never-ending chore. The reader never catches up, and couldn't, even if he spent all of his time reading. And he can't do this because he always has a number of other important things to do. He is a discerning and grouchy reader because he is busy and because he is looking for original material, information, and ideas that are new to him. He does not like to be put in a position where he is forced to read something he already knows in the hope that sooner or later the writer will get around to what he is looking for. He is particularly irked if he has to waste time reading a piece that does not give him a decent return on his investment in reading time.

How does the busy reader read?

My conception of the process is touched upon in the opening example. I skipped a necessary detail. The reader goes from the title to the beginning of the text hoping to start picking up further evidence of what the writer's subject is. We did not mention that the reader exercises an option before reaching this point. He may decide not to read a piece after sampling its title. Since he is busy, finding he should not read a piece is as important as finding he should. He can always go on to another piece of technical literature or to other activities.

The reader still has two more "stay in" or "get out" options—one after he feels he has found the statement of the subject in text and one after he has checked out a portion of the writer's documentation of the subject.

The reader would prefer the writer to state his subject fully in the first sentence or first or second paragraph of text, then proceed immediately to documentation —primary documentation, not a traffic jam of secondary documentation before the author gets around to what he is supposed to be writing about. The busy reader is certainly not seeking an indiscriminate mixture of secondary and primary documentation that forces him to prospect for what he wants. Obviously, he would like to be a selective or "scan" reader. This is not always possible because of the influence of one or more of the common writing practices that follows:

Practice No. 1:   Delaying the statement of the subject early on in the text with an introduction or background information. The impact is on how we read.

Practice No. 2:   Following the statement of the subject with information that does not start to document the subject in a direct manner. A practice that conflicts with how a busy reader reads.

Practice No. 3:   Overusing devices, such as connectives that aren't needed, that retard the flow or progress of a piece. The reader feels as if he were on a treadmill.

Practice No. 4:   Overusing generalities and claims unsupported, or not adequately supported, by specifics. In this case, and the four practices that follow, the problem is in the "why we read" category. The reader does not get the information, the specifics, he is looking for.

Practice No. 5:   Overusing standard information—tutorial information the reader already knows, reads over and over and over involuntarily, and is secondary documentation in terms of the stated subject.

Practice No. 6:   Overusing standard language. An impoverished writing vocabulary that is unequal to the task at hand.

Practice No. 7:   Overusing jargon, clichés, buzzwords, truisms, sophistry. Jargon can be a private language unknown to some readers. Clichés have many of the limitations of generalities and standard language. Truisms package the obvious in an overabundance of words. Sophistry is a kind of word con game.

Practice No. 8:   Using various forms of unnecessary repetition. The reader is forced to read words that do not inform. Flow is impeded.

Practice No. 9:   Using stilted language—words that are not appropriate for the task at hand.

Practice No. 10:   Overusing certain sentence construction practices. What is of secondary importance is put at the front of a sentence; what is of primary importance is in the back seat. Piling up adjectives obstructs readability. Passive construction is an unnecessarily long, indirect means of expression.

Practice No. 11:   Overusing commercialism. The main problem is not giving the reader the information he is looking for.

It should be said that you can't be critical of a technical writer for observing these practices because they have the sanction of universal usage. They are part of the common law of technical writing. What the busy reader wants doesn't come naturally to the the technical writer. In brief, the busy reader wants the most possible new and useful information and ideas in the shortest possible reading time and in a readable manner.

Practice No. 1—Delaying the Statement of the Subject With an Introduction

Title:
"A New Tin Coating Process."
First sentence of text:
"Tin and its alloys have been widely used for centuries to produce corrosion-resistant, solderable finishes."
Second sentence:
"Now, thanks to work performed by the International Institute, another new electroless method of applying tin coatings has been developed which promises to create even wider applications for this versatile metal."

The two sentences, in miniature, illustrate what we are talking about. The first sentence is background, an introduction. It delays the statement of the subject, which is in the second half of the second sentence.

What we learn, in addition to what was said in the title, is that the "electroless method promises to create even wider applications for this versatile metal." This is where the writer should have started. The first sentence, as far as the reader experienced in this field is concerned, does not contain any original material. It is standard, handbook, review information. The reader, it is assumed, is knowledgeable in the subject. Otherwise, he would not opt to read the piece. He wants an update. What's new. What's new to him.

What can be done with the example?

Drop the first sentence. Start with the second. A rewrite is necessary, but first a couple of other common practices should be pointed out.

Reread the second sentence:

"Now, thanks to work performed by the International Institute, another new electroless method of applying tin coatings has been developed which promises to create even wider applications for this versatile metal."

"Performed" is both a standard and stilted word, and the first nine words do not belong in the front of the sentence. What the reader is looking for is at the back. Upside-down construction promotes slow, indirect expression.

"Which promises to create even wider applications for this versatile metal" is a claim or generality that tells the reader little unless it is backed by specific information and/or data—direct documentation that should be started immediately.

A second example adds realism to the introduction problem.

Title:

"Interpretation of Acoustic Emission Data to Indicate Flaw Significance."

Introduction:

"Acoustic emission (AE) is the basis for a nondestructive, passive technique to monitor structures for flaw growth. AE is an elastic wave generated by energy released during deformation and/or fracture of a solid material. This monitoring technique has been evolving over the last 15 years.

"To evaluate the integrity of a structure such as a nuclear reactor pressure boundary continuously, three functions must be performed:

"1.   Detect the occurrence of flaw formation and/or growth (failure event).

"2.   Determine the location of the failure event in the structure.

"3.   Evaluate the significance of the failure event to continued structural integrity.

"The AE method has been demonstrated to be capable of performing the first two functions under a variety of conditions. Capability to perform the third function by AE, however, still remains largely to be developed. This paper concerns such development."

The introduction goes on for 318 more words.

Going back to the title, the reader, who, we must assume, knows AE and is looking for an update, was promised new information on the "interpretation of acoustic emission data to indicate flaw significance."

After 122 words of introduction, one shoe is dropped. Ability to interpret the significance of AE data "still remains largely to be developed." After 305 more words, the writer concludes, "The program is in its early stages. Limited results to date are discussed." Obviously, the writer has little to write about. Chances are the discerning reader would have departed after the first warning at the 122-word mark.

In summary, the busy reader style conflicts with the technical writer's natural urge to explain, to amplify, to orient, to teach, to expound.

The alternative: place the statement of subject at the front end of the text.

Practice No. 2—Not Starting To Document the Subject Immediately After It Is Stated

Back to the last example, in which the writer stated his subject in two stages: after 122 words and after 318 words. Moving what is left over behind the statement of the subject doesn't solve anything. The writer has reached his next obligation to the busy reader: start to document "slight progress has been made . . ." immediately.

Secondary documentation should not even be considered before primary documentation has been completed. At this time, you may conclude secondary documentation would serve no useful purpose, or you may find you do not have room for more words.

A piece is typically short of primary documentation once it has been weeded. This generally does not pose an insurmountable problem. Technical writers normally have what they need at hand or close by.

It is worth repeating that secondary documentation is typically standard information found in texts, handbooks, technical books, and other references, or it involves a development or trend being overkilled in current journal papers and technical talks.

The next example illustrates the point.

The subject is presented in a single sentence: the first.

"The use of rapid solidification (RS) of liquid steels will continue to grow in volume."

The next two sentences are secondary documentation. One would be hard pressed to pick up an appropriate journal today and not find a piece on the subject. The following has been repeated *ad nauseam* in the current literature:

"RS can be achieved by splat cooling of atomized streams by self or mass quenching of either laser or electron beam melted surfaces. Cooling rates in both instances can reach 10^3 to 10^7 C per second."

What the reader is looking for is decent development of why "use will continue to grow."

The writer obliges in sentence No. 4. Use will continue to grow, he writes, because "RS produces an extremely homogeneous and very fine grained material. . . ." Then he expands on how properties are improved.

Here is a longer example. A piece with a title and five paragraphs: 1, statement of subject, where it should be. 2, a mix of primary and secondary documentation —the former is in italics. 3, secondary documentation. 4 and 5, primary documentation.

Title:

"T-250, a New Maraging Steel."

No. 1, Subject:

"A cobalt-free, precipitation hardening maraging steel with mechanical properties similar to those of 18% nickel maraging 250 type steel containing 7½% cobalt is expected to gain widespread acceptance as a replacement for the cobalt-bearing alloy. Its lower cost due to the elimination of cobalt probably will result in a greater number of applications."

No. 2:

"Cobalt is a strategic material for which the United States and many other countries are dependent upon foreign sources. Shortages and interruptions in cobalt supplies have temporarily disrupted production of cobalt maraging steels in recent years. Further, the cost of the cobalt maraging steels is linked to the price of cobalt, which has risen drastically in recent years and tends to fluctuate. *In addition to eliminating the need for cobalt, the new steel employs less molybdenum, another strategic material, than the cobalt-bearing grade.*"

No. 3:

"The new maraging steel was a joint research and development project of ABC Company with XYZ Research & Development Center, which holds the patent rights for the new material. ABC introduced the new titanium-strengthened grade at the Paris Air Show."

No. 4:

"To differentiate the new titanium-strengthened (cobalt-free) maraging steel from the cobalt-strengthened maraging steel, ABC Company employs the prefix 'C' for cobalt-strengthened grades and the prefix 'T' for the titanium-strengthened grade. The nominal compositions of the two steels are listed below for comparative purposes.

	New T-250	Previous C-250
"Nickel	18.50%	18.50%
Cobalt	None	7.50
Molybdenum	3.00	4.80
Titanium	1.40	0.40
Aluminum	0.10	0.10."

No. 5:

"The new steel has been melted in full-scale production heats and manufactured into a variety of mill product forms, including billet, bar, forgings, plate, and sheet. Its processing is similar to that of the cobalt-bearing 18% nickel 250 grade maraging steel. Both annealed and fully aged hardness properties are identical to those of the cobalt grade. T-250 may be fabricated using standard machining, grinding, and aging (heat treatment) techniques."

Analyze the piece the way a busy reader would.

The title gives him a shorthand statement of the subject. He reads 54 words in the first paragraph and finds that the new alloy is a cobalt-free grade of a maraging steel.

Direct documentation—including further information about the new composition and reasons why it is expected to catch on—is expected to come next. Instead, the next 62 words are tutorial information, broken by 22 words of primary information (italicized) in the last sentence of the paragraph.

The next 42 words are secondary documentation. What the reader has been looking for thus far is sandwiched between 104 words.

The last two paragraphs, which add up to 132 words, are primary documentation. The bottom line: the piece contains 312 words, not counting the title. The statement of the subject requires 54 words. Of the remaining 258 words, the reader is forced to read 104 words that do not help him and tend to irritate him. His option to stop reading the piece is delayed because secondary and primary documentation are mixed. What he is looking for is the opening 54 words and 154 words of primary documentation scattered over the remainder of the piece.

Here is a short piece with a maximum return on reading time.

Title:

"New Way to Compute Tubing Tolerances."

Subject:

"A new technique for calculating seamless steel tubing tolerances and stock removal has been developed by the Steel Division of ABC Company."

Primary documentation:

"This involves the use of linear equations to express the exact tolerances and stock removal requirements for the desired tube sizes. The technique eliminates the present step table method and is ideal for programmable calculators; consequently, clean-up tube sizes can be calculated directly in field offices and/or customer plants."

A total of 71 words—22 for the subject, 49 for primary documentation. No secondary documentation. Not a word wasted.

Once more, the technical writer is advised to resist the temptation to develop his piece in a slow, rambling, leisurely manner.

Practice No. 3—Overusing Devices That Retard Flow

Flow, or continuity, is grounded in readability. The feeling a reader gets when a piece reads smoothly from sentence to sentence and paragraph to paragraph and the writer makes steady progress in delivering his message. No stopping, or slowing down, or frequent and unnecessary references to something in the last sentence, or a sentence or paragraph before it.

Go back to the previous example titled, "New Way to Compute Tubing Tolerances."

Sentence No. 1:

"A new technique for calculating seamless steel tubing tolerances and stock removal has been developed by the Steel Division of ABC Company."

Sentence No. 2:

"*This* involves the use of linear equations to express the exact tolerances and stock removal requirements for the desired tube sizes."

Sentence No. 3:

"*The technique* eliminates the present step table method and is ideal for programmable calculators."

Sentence No. 4:

"*Consequently*, clean-up tube sizes can be calculated directly in field offices and/or customer plants."

Note that "this" and "the technique," which are italicized, refer to the first sentence. You take a step back before taking one forward.

"Consequently" is an unnecessary connector between the third and fourth sentences. The aim is to tie the two sentences together, a grossly overused, tedious practice.

A rewrite is in order.

In the second sentence, drop the first five words, "This involves the use of," and start with "linear equations." The reconstructed sentence reads, "Linear equations express the exact tolerances and stock removal requirements for the desired tube sizes."

In the third sentence, "the technique" can be dropped. My new version has two sentences: "The present step table method is eliminated. Usage of programmable calculators is ideal."

"Consequently," the connector, is used here in this sense: "taking all of the foregoing, you reach this conclusion." Connectors are also used as bridges or transitions between succeeding sentences or between the last sentence in a paragraph and the first sentence in the new paragraph. The idea is to show a relationship. Restraint is advised. Connectors and transitions are often used where they aren't needed. In this case, "consequently" should be pruned.

Notice how much more smoothly the rewrite reads than the original:

"A new technique for calculating seamless steel tubing tolerances and stock removal has been developed by the Steel Division of ABC Company. Linear equations express exact tolerances and stock removal requirements for desired tube sizes. The present step table method is eliminated. Usage of programmable calculators is ideal. Clean-up tube sizes can be calculated directly in field offices and/or customer plants."

An occasional "this" or "that" is tolerable, but unnecessary use of connectives should be avoided.

Practice No. 4—Overusing Generalities and Claims

When we write, "ABC Company has developed a new, fine-grain steel bar offering significant machinability improvements," and do not follow with supporting specifics, we are observing a common practice that shortchanges the busy reader.

Generalities and claims raise questions directly or indirectly. Answer them. In the example, evidence is needed to back up "significant machinability improvements." Generalities are widely used by all technical writers. When a writer has something to sell, his generalities are usually in the form of claims.

Pick up any piece of technical literature. Search for examples: "Improved internal cleanliness," "improved surface quality," "improved formability," "superior transverse impact toughness," "reduced tooling costs," "improved productivity," "ideal for use as fittings," "have successfully passed all tests," "shows great promise," "has exceptional corrosion resistance," "outstanding impact strength," "can eliminate most of the problems," "is the leading flux," "can deliver the lowest hydrogen content in the industry."

In each example, the writer stopped after stating his generality or claim. The reader must take what is said at face value or, more likely, go elsewhere to get the information he was promised but did not get.

The technical writer is not attempting to withhold information, and probably isn't aware that he is leaving the reader holding the bag, when he writes, "While most of the equipment is not revolutionary in the area of near net shape technology, it does represent significant incremental steps in the efficiency of material utilization," and does not follow with a reasonable explanation of "significant incremental steps."

An example worthy of emulation:

Subject:

"The high cost of heat treatment provided the stimulus for the development of X steel, which acquires its strength properties on cooling from the forging temperature."

Data are presented in the next sentence:

"The medium carbon steel (0.35 to 0.45 C) contains 0.10 to 0.15 vanadium, which provides precipitation strengthening."

A comparison follows:

"The resulting tensile properties are comparable to those of quenched and tempered steels; impact properties are lower."

Then an explanation supplemented by data:

Minor modifications in steel chemistry (e.g., increasing silicon content to 0.7) contribute to improved toughness."

In conclusion, representative applications:

"The steel is being used in critical applications like crankshafts and connecting rods."

The piece is only 88 words long (26 for the subject, 62 documentation). Not a word is wasted. The reader gets an attractive return for his investment in reading time. Once more, shortness per se is not a virtue, but unnecessary length is a sin.

The cardinal rule is: never leave an unsupported claim or generality.

Practice No. 5—Overusing Standard Information

This practice was touched upon in some detail in prior discussions of primary and secondary documentation. The subject is re-introduced because within the context of generalities and claims I can begin to show the cumulative negative effect of approved practices on content.

In frequency, standard information is as common as generalities and claims.

A short example:

Generality: "Titanium and its alloys have a solid future in aerospace."

Standard information follows:

"Titanium's elevated temperature properties make for substantial use in fuel efficient engines. Its corrosion resistance, high strength, and low density are attractive for general airframe use . . ."

Neither sentence has any surprises for anyone in this field. For one thing, it is textbook information. For another, it is repeated in practically every piece on the subject the reader sees.

Now we have a basis for distinguishing between primary and secondary information. Secondary information is known by the reader, and/or it is not the information he has been promised by the title and statement of the subject.

What about the student or inexperienced reader?

As stated before, secondary documentation may be primary documentation to them because they do not know it. However, the writer should not attempt to cater to both sets of readers. The busy reader style is for the reader who knows the fundamentals and is looking for an update.

Writing at the tutorial level requires a different approach to content. It's for a different audience. Title, subject, and documentation must be selected for the neophyte. If these things are done, the busy reader style can be applied here too. When the experienced busy reader encounters such a piece, he will skip it; but you should not make either reader sift through documentation for scraps of information he is looking for.

The tailoring required in both instances is called writing for an audience. I repeat this because of the importance of the point.

Practice No. 6—Overusing Standard Language

In writing this section, I experimented with two different ways of presenting my material: start with a series of general statements not supported by examples, or start with examples and use them as the foundation for general statements. I tried a number of combinations before I was satisfied.

What does all this have to do with standard language, the announced subject?

A key point I want to make is that a user of standard language never gets around to a second way of writing anything.

Say I am a user of standard language and want to express "becoming smaller." I settle for the first word that comes to mind—probably the standard word "reduced." This is as far as I will ever go as a user of standard language. I won't stop and worry, "Is 'reduced' the right word?" I don't even get a chance to use my much richer reading and talking vocabularies. I'm limited to my much smaller and less powerful writing vocabulary.

Further, it would not occur to me to go beyond the process of word selection to the next plateau and ask, "Do I really have a grasp of what I am trying to say?" and "Am I expressing myself in the most effective way I know?"

In other words, standard language is a first-draft language and a product of first-draft thinking. The normal painful writing process is cut short.

Why does the practice persist?

Because we see this language in all the technical literature we read and hear. The language looks right and sounds right because of familiarity. It is the standard. No one questions it. It has the sanction of universal usage.

Now for some examples.

A writer wants to express the idea "improved."

He starts a paragraph with "While many new products are being offered, it is worth noting that many of the workhorses of previous years are being continually *improved.*"

He closes the paragraph with "The result of these *improvements* is a heavier gage steel with *improved* resistance to temper embrittlement and *improved* low temperature toughness."

Among other things, the example provides an insight into why readers complain about the boring sameness of this literature. Variety is not the answer, though. One should go beyond hunting for a synonym.

A frequent problem with standard language is that it conveys only approximate meanings and is seldom the right word in this sense: the reader is looking for content, for meanings, for precise, specific information. He also hopes the writer will convey his message in an understandable manner. Now we are no longer solely in the realm of word selection. We must rethink what we are trying to do. To have second thoughts, third thoughts. To refine and re-refine. To feel the hopeless, miserable feeling of a writer who is stuck. After all this, you may find the standard word is the right one after all; but the only way you can find out is to struggle through the entire process.

Developing an awareness of standard language is the necessary first step in a self-help program. You will be amazed. In the page of source material I am working from, I see—in addition to the aforementioned "improved"—the following:

"Manifested," "increasingly," "requires," "obtain," "tailored," "offers," "found," "provide," "involving," "reducing," and "achieved."

Note that "manifested" and "achieved" are also stilted language which is out of place in this literature.

A few words of advice: at the very least, be critical of the first word that comes to mind. Remove your speaking and reading vocabularies from cold storage. New horizons will appear. You will start to make use of your ingenuity, your ability to invent, to imagine. Things standard language prevents us from doing.

Practice No. 7—Overusing Jargon, Clichés, Buzzwords, Truisms, Sophistry

"We prefer C-Rad, but if it isn't possible, we'll settle for I-Rad," the speaker told a peer group. He paused. Looked about his audience, expecting comment, which came just as he restarted.

"What is C-Rad?"

A murmur of common concern followed. Another question was asked.

"And while you are about it, what is I-Rad?"

The speaker was using jargon known only within his corporation. As is often the case, he had to be reminded that not everyone knew it.

"C-Rad," by the way, stands for "contract research and development," and "I-Rad" is "independent research and development."

The incident illustrates a weakness of this practice that is sometimes overlooked. Jargon is a private language and can amount to an unknown foreign language.

Jargon is usually understood inside a discipline. Computer jargon is known to computer people. Metallurgical jargon is known to metallurgists. However, interdisciplinary understanding is limited.

Clichés are a different ball game.

The cliché "different ball game" is generally understood, as are most clichés. But they are generalizations. They go only part of the way in delivering information.

Use of clichés can be interpreted as an effort to go beyond standard language. It is probably more realistic to regard them as a part of standard language. Overuse could be interpreted to mean that technical writers are short on originality.

Buzzwords are current clichés. About the only thing in their favor is that they are trendy, "in" topics of local, national, and international interest groups.

Example:

"*Productivity* and *high technology* are the key words in welding technology today."

The trouble with the italicized buzzwords and others of their ilk is that they really don't do much for the reader. For the writer, they are great. He feels he has fulfilled his obligation by merely putting the buzzword on paper. There is no need to go farther. To explain. To document. To develop one's premise. Presumably, magical things happen that make the normal writing tasks unnecessary.

There is a fine line between legitimate explanation and a truism. Pointing out the obvious is a truism. It is editorializing when it is not needed or called for and contributes to the verbosity and dullness of the literature.

Sophistry is a practice found largely in learned essays, speeches, and some commercial pieces. Say you use a buzzword to set up a premise at the beginning: "What industry needs to cure what ails it is higher productivity." Henceforth, premise becomes fact. All sorts of productivity panaceas can be promulgated without fear of contradiction. As far as the busy reader is concerned, the manufactured information produced by sophistry is not useful and wastes his time. The information is suspect. It has no more meat than bare generalities and claims.

Jargon should be looked upon as a language that does not always communicate.

Clichés and buzzwords have the same inherent limitations as generalizations and claims.

Truisms are a form of verbosity.

Sophistry is information not supported by fact.

Elimination is the obvious remedy in each instance. However, this advice has an element of sophistry in it. It is not always possible or practical to go all the way. Restraint is recommended.

Practice No. 8—Using Various Forms of Unecessary Repetition

Italicized words in the following example flag a common form of redundancy.
Example:

"Other new automotive applications for *aluminum* are listed by the *Aluminum* Association. These include: forged and cast *aluminum* wheels which are planned for the 1983 Ford Topaz and GM 'T' minitruck; *aluminum* transmission structural supports for light trucks, and the first time use in 1982 of formed *aluminum sheet bumpers.*"

The example illustrates a strong tendency to repeat a key word or key words—aluminum in this instance. In one paragraph, the word is repeated five times. There is no justification for the repetition after the first sentence.

Clarity should not be sacrificed in the name of brevity, but the need for the latter can't be underplayed. A writer should not use one more word than he needs. In the example, he should take advantage of the context set up by the first sentence and drop the three succeeding mentions of "aluminum." When you add up similar savings in words over an entire piece, the result can be worthwhile to the reader.

Brevity is not the sole consideration. The writer who repeats unnecessarily at every opportunity perpetuates the drabness and dullness of technical writing. As in the case of standard information and standard language, there is a sleep-inducing sameness about this literature.

The edited aluminum example reads:

"Other new automotive applications for aluminum are listed by the Aluminum Association. These include: forged and cast wheels planned for the 1983 Ford Topaz and GM 'T' minitruck, transmission structural supports for light trucks, and the first time use in 1982 of formed sheet bumpers."

In comparing the revision with the original, you will note that the words "which are" following "cast aluminum wheels" have been dropped. The redundancy fighter looks for shortcuts. Pronoun phrases like "which are" and "that are" can usually be eliminated without a loss in readability.

Another example:

"The chemical and allied industries continue to place increasing emphasis on the *reliability of process equipment* in an effort to avoid disastrous failures. One approach toward *improved reliability of process equipment* is to *improve the reliability of structural materials* that are used."

In this instance, phrases are repeated unnecessarily.

Dropping "improved reliability of process equipment" at the front of the second sentence is a quick fix. A simple rewrite would read as follows: "One approach is to improve the reliability of structural materials that are used."

Did you spot "that are" at the end of the last sentence? Did you notice "improved" and "improve," standard words discussed previously, in the same sentence?

A simple rewrite:

"The chemical and allied industries continue to place increasing emphasis on the reliability of process equipment in an effort to avoid disastrous failures. One approach is to improve the reliability of structural materials used."

The new version reads more smoothly than the original because the elimination of repetition aids sentence-to-sentence flow.

Once the writer becomes sensitive to the various forms of needless repetition, he finds that the problem usually suggests a solution. Pruning a word or phrase here and there is a typical remedy. However, the writer should often go further. In looking for shortcuts, for example, he may be obliged to rethink and rewrite what he has to say.

Practice No. 9—Using Stilted Language

"Trends toward energy saving products *are exemplified* by ABC Company's new steel."

"The alloy *is characterized* by high toughness."

"In order *to accomplish* the feat of *achieving* a record performance."

The stilted language is italicized. In each instance, the choices illustrate limitations in trying to make a small writing vocabulary cover a kaleidoscope of situations. Further, stilted language like "exemplify, characterize, accomplish, achieve" are more at home on commemorative plaques than in the mundane business of reporting a new steel composition, or describing a material's properties, or suggesting ways to reach a planned goal.

This is standard language. The writer who settles for it only approximates what he has to say. Stilted language also tends to breed verbosity.

Practice No. 10—Overusing Certain Sentence Construction Practices

Consider upside-down construction, piling up modifiers, and passive construction.

Upside-down construction is similar in effect to practices that delay statement of the subject and the start of primary documentation. The message is delivered indirectly. What needs to be said first is said last.

Example:

"*The International Research Organization announced* the development of a new zinc alloy called [trade name] containing 5% aluminum and less than 1% mischmetal."

One alternative:

Start with "The development of a new zinc alloy called . . ." and end with "is announced by the International Research Organization."

The secondary information is essential in this instance and is not dropped. However, the writer is not evaluating his material properly when he starts with what is secondary, particularly when you keep in mind that upside-down construction can be repeated many times in a piece of technical writing, compounding problems relating to organization.

The obvious remedy—turning the sentence right side up—may not always be possible. For example, if what is written by the writer must be cleared by a boss, by a director of public relations, or by lawyers, the writer may be forced into upside-down construction. But what is important to the company—its name, its products, its outstanding record—is of little or no importance to the reader.

Another example:

"*The National Aeronautics and Space Administration (NASA)* has proposed changes in its patent policy to allow industry increased rights to inventions developed under contracts with the agency."

What is important to NASA is up front and italicized. What is important to the reader is at the back.

Rewriting for the reader, you get:

"Expanded patent rights for companies that invent in their work for the National Aeronautics and Space Administration are being proposed by the agency."

Indiscriminate piling up of adjectives, a practice as common as the cold, converts English into a hybrid language that typically slows down the reader and may require him to stop and translate.

Example:

"The largest casting is a *5 lb., 1982 model, low volume* air cleaner."

The writer may claim he is helping the reader via condensation and is packing into one sentence what may require two or more otherwise. The result is overcondensation. Any saving in words is offset by a loss in readability.

A simple rewrite takes 14 words as opposed to 13 in the original.

"The largest casting, a low volume air cleaner for 1982 models, weighs 5 lb."

With overcondensation, the only saving may be the elimination of a word like "of" or "for." Repair is usually simple, as the next example, actually two, illustrates.

"The *Space Shuttle main engine, high pressure fuel turboprop* uses *cast nickel base superalloy blades.*"

The original has 15 words. I add five in a rewrite:

"The high pressure fuel turboprop for the Space Shuttle main engine has blades made of a cast, nickel base superalloy."

Both examples demonstrate the illusory savings obtained via overcondensation. Recommended practice does not run counter to the general rule: "Don't use one more word than you need." In this instance, a few words need to be added to aid readability.

Another example:

"American Management Association Annual Data Processing Conference."

Better:

"Annual Data Processing Conference of the American Management Association."

Passive construction, by comparison, always encourages verbosity and results in another form of indirect expression.

Example:

"Tungsten plating of a metal *can be accomplished* via vapor phase techniques."

The italicized words are the troublemakers. Technically, the problems (unnecessary words and indirect statement) are caused by making the subject of the verb the object of the verb.

Shorter and more direct:

"Tungsten can be plated on metal with vapor phase techniques."
Example:
"A laser beam *is being employed* to vaporize metal."
Better:
"A laser beam vaporizes metal."
Example:
"*Precision cutting* of heavy gage metal *can be accomplished* by this machine."
Better:
"This machine precision cuts heavy gage sheet metal."

Practice No. 11—Overusing Commercialism

From the writing standpoint, commercialism has a variety of shortcomings—overuse of unsupported claims, generalities, laundering of information and data, upside-down construction, and needless repetition. Promotion of a company, its products, and its services usually seem to have a higher priority than the communication of information useful to the reader.

Example:

"ABC Company today announced their new X Series product line of water-based synthetic fluids for all-purpose use in manufacturing operations."

Note the upside-down construction. In the two pages that followed, the company's name was repeated three times, the name of the product seven times.

Readers are routinely subjected to unrestrained selling in the piece.

Example:

"U.S. porcelain enamelers may well be on the verge of the biggest break in operating costs to come along in the last 30 years." The statement is followed by such unsupported generalities as a "unique furnace" and such claims as the "first in the United States." This is an example of sophistry.

Other examples of commercialism in the piece: all advantages are cited and discussed in detail. Not a single disadvantage is mentioned. And gratuitous statements like the following are used indiscriminately:

"The equipment is among the highest powered in the world."

Curiously, readers of technical literature expect commercialism; they are naturally skeptical and seldom taken in; they are practical and realize that someone with something to sell is more willing to talk than someone who is guarding what he feels is a proprietary secret. My purpose is to point out the shortcomings of commercialism from the viewpoint of the writer.

Recap—Writing for Busy Readers

The reader starts a piece by reading its title—a partial statement of the subject. Amplification is needed. But the title may say enough for the reader to decide, "I won't read this piece and stand the chance of wasting my time."

A reader who is still interested after sampling the title to a piece would like the technical writer to be businesslike . . .

. . . to start the statement of the subject at the front end of the text to a piece.

. . . to avoid starting with an introduction or background the reader already knows and sees each time he reads a piece on a given subject.

. . . to avoid intermingling background with what the reader is looking for in the statement of the subject.

Once the writer has fulfilled this part of his obligation, or it appears he has, the reader can make another go—no go decision.

If it's a "go," the reader expects the writer to start his primary documentation immediately, not to delay the start with more background, or intermingle background and primary documentation.

Other practices under discussion are equally unsuitable for a busy reader.

Constant flashbacks and connectors that aren't needed impede reading progress.

Generalities and claims deliver only part of the message. In effect, the writer withholds information—the specifics the reader of technical literature looks for.

Standard information is known to the reader who is looking for an update—for new information.

Standard language is a writing vocabulary that is not equal to the task.

There is always a chance that the writer using jargon is not communicating.

Clichés, in addition to compounding the nonoriginality of this literature, have the same shortcomings as generalities and claims.

Truisms are a form of pompous editorializing.

Sophistry is a word game with an element of deceit.

Pointless repetition adds to the number of nonfunctional words that must be read.

Stilted language is a part of the standard language and is out of place in technical literature.

Upside-down construction slows down delivery of the message.

Piling up adjectives is false word economy because readability is often obstructed.

Passive construction is an indirect method of statement that takes more words than needed to make a given statement.

Commercialism is to be frowned upon, mainly because essential information is withheld from the reader.

Next Chapters Amplify the Busy Reader Style of Writing

I have outlined some problems and suggested countermeasures. In the chapters that follow, practices and remedies will be put under the microscope and examined.

CHAPTER 2

Matters Relating to Organization and Documentation

In this chapter, the topics are: 1. Location of the statement of subject in text. 2. Location of the start of primary documentation. 3. Distinguishing between primary and secondary documentation.

The busy reader style (statement of subject at head of text, followed immediately by primary documentation) is used by some technical writers, but it is not common practice by any stretch of the imagination.

An example of the busy reader style:

Title: "Trends in Protecting Autos Against Corrosion."

Start of text with statement of subject:

"During 1982 and throughout the middle eighties U.S. automakers will be upgrading the level of corrosion protection against surface rust and panel perforation."

Next, start of primary documentation:

"This means that the usage of precoated and galvanized steel will continue to increase at the expense of bare hot and cold rolled steel. One domestic manufacturer already uses one side galvanized steel for all exterior panels, including the roof. By 1984, Chrysler will be using galvanized steel to a similar extent, except for the roof.

"While one side galvanized steel provides excellent protection against perforation, surface or cosmetic corrosion goals are left unbenefited. There will be significant efforts by all automakers to supplement the protection afforded by the prime and color paint films by the use of some type of two side galvanized product. Chrysler will be using a hot dipped G90/A01 product, internally known as $1\frac{1}{2}$ side galvanized steel . . ."

And so on.

The reader's time is not wasted by information secondary to the stated subject. He is given the specifics he is looking for. He may decide after sampling the piece that the information is not new to him or that he is not interested, but putting him in the position to make the decision with some certainty is as important as supplying him with information useful to him in his work.

Another example of the busy reader style, in which data are part of the primary documentation:

Title:

"Sensitization of 2117 Aluminum Rivets and Stress Corrosion Cracking."

Statement of subject in the first sentence of text:

"Studies have shown that 2117-T4 rivets heated to 300 F for 3 hours become highly susceptible to stress corrosion cracking (SCC)."

Immediate start of primary documentation:

"The table below compares the failure rates of unheated stress corrosion test specimens with the failure rates of specimens heated in this temperature range. Specimens were stressed to the level shown and exposed to alternate immersion in 3.5% NaCl for 60 days.

SCC Tests on Heated 2117-T4 Aluminum

Heating Cycle	Applied Stress % Yield Strength	Ksi	Failure Ratio	Days to Failure
None	75	21	0/3	
3 Hr at 300 F	30	7.5	0/3	
3 Hr at 300 F	50	13	3/3	6,6,14
3 Hr at 300 F	75	19	3/3	6,6,14
6 Hr at 300 F	75	19	3/3	6,14,21
3 Hr at 325 F	75	20	3/3	14,20,21
3 Hr at 275 F	75	19	1/3	60

Based on results it will be necessary to limit usage of 2117-T4 rivets (and probably rivets fabricated from similar material) where short time exposure to temperatures in the vicinity of 300 F is expected."

One more example of the style that should be emulated:

Title:

"Environmental Regulations Dictate Materials."

Statement of subject in first sentence:

"Trends toward improved ecology and protection of the environment have spurred moves toward desulfurization of flue gases from powerplants, smelters, and other industrial plants."

Primary documentation next:

"To handle the corrosive condensate often produced in stacks and desulfurization units, high performance stainless and nickel alloys are sometimes specified. These alloys are frequently furnished as solid plate, but because of their high cost, clad plate can be a more desirable alternative. ABC Steel Company can provide many grades such as IN625, Hastelloy G, 317LM, and others in clad form. The same corrosion protection is afforded by the clad, and in addition, stronger backing steels can be specified, resulting in design savings."

So much for the positions of the statement of subject and start of primary documentation for the moment. Selection of information—knowing the difference between primary and secondary documentation—is equally important to the busy reader style of technical writing.

The subject comes up first when you read a title, which the writer uses to define his topic. To put it another way, the writer promises to write about a given subject. This is why introductions or background information should not precede the statement of the subject. As the following example shows, such information is related to the title, but it is in fact secondary information or documentation. Once the statement of the subject is established, it, in combination with the title, establishes what is primary and what is secondary documentation.

In the example, the writer promises to write about a coating called C638.

Title:

"C638, Copper Alloy Coating for Electrical Connectors."

Text starts with an introduction:

"Electrical connectors typically utilize gold on gold contact surfaces. Gold is an ideal connector surface since it is immune to both oxidation and fretting damage, but unfortunately, adds significantly to the cost of these connectors. Historically, less costly materials, such as tin plate, have been used successfully under conditions where heavy contact forces act to minimize contact degradation when fretting damage occurs. With design efforts directed toward connector miniaturization, it is becoming more difficult to maintain the large contact loads required for use of these materials."

Anyone conversant with the subject already knows the secondary information in the introduction. The reader is looking for information on C638. The writer in this instance should have dropped his first paragraph and started with the second:

"In laboratory investigation of lower cost alternatives it has been found that the contact response of a connector pair consisting of bare alloy C638 on gold plate was comparable to a gold on gold connector under long-term fretting conditions."

Primary documentation was next, where it should be. In fact, for what must be regarded as a commercial piece, it has admirable objectivity. The writer starts his primary documentation with a qualification:

"Although additional environmental tests are required, these initial results are promising for applications where fretting is known to be a major failure mode (i.e., edgeboard connectors for printed circuit boards). C638 is a Cu-3A1-2Si-0.4Co alloy, developed by ABC Company."

In the next example, the writer promises to discuss "developments in metallic glasses"; nothing else. However, he waits 132 words—three paragraphs—before he delivers.

Title:

"Developments in Metallic Glasses."

Next, three paragraphs of secondary information:

"Amorphous metal, also known as metallic glass, is a new and exciting comer on the materials scene. It is produced by quenching alloys of carefully controlled, near eutectic compositions, from the molten state at rates of 100 000 to 1 000 000 C per second. This rapid quench enables the solidified alloy to retain the relatively random atomic array of the liquid. Hence the term 'glass.'

"Metallic glasses made by rapid quenching were introduced to the world in 1960 by P. Duwez and coworkers, whose researchers were mainly concerned with noble metal alloys on a small experimental scale.

"The technology began to develop rapidly in 1972, with the introduction by Allied Chemical Corp. of glassy transition metal alloys of modest cost, especially iron based alloys and the development of new continuous rapid quench strip fabrication techniques."

Finally, the statement of the subject:

"Improvements in alloy compositions, alloying methods, strip fabrication, and annealing techniques have led to expanded operations and created diverse new products . . ."

Now it behooves the writer to start documenting "improvements in alloy compositions . . ." and so on.

Once the statement of subject is completed, it is a fairly simple matter to separate primary and secondary information. In the following example, the writer ties down his subject in the last sentence of the first paragraph (italicized). He follows with secondary documentation (546 words' worth).

Title:

Energy Sources for Heat Treating Oil Country Products."

Statement of subject:

"Fuel shortages, rising prices, deregulation of oil, and anticipated deregulation of natural gas have created a tremendous demand for oil country products. Over 70 000 new wells will be drilled this year in the United States. Many of the products required for drilling, production, and maintaining wells require forming and heat treating operations. *Induction heat treating lends itself to most of these operations and provides benefits of efficiency, operating costs, improved working conditions, higher production, less floor space, and more. The following is an overview of the induction heating and heat treating applications in oil country products.*"

Primary documentation should start here. Instead, the writer chooses to present a tour of oil country technology. He writes:

"Let's first familiarize ourselves briefly with a basic knowledge of an oil well. Most of us are familiar with the aboveground portion of a completed well, either in the form of a Christmas tree network of control valves and piping, or a pumping unit commonly called a horsehead. Also the oil and gas companies have recently shown us many drill derricks while advertising their products on television.

"Wells are typically drilled to depths of 5000, 10 000, 20 000, even 30 000 feet. In a given well, the length of casing and tubing products used averages three times the well depth. Looking at the cross sections of a well, we see several types and sizes of casing. The larger, relatively shallow casing sections are called conductor pipe and surface casing. These sections are installed during the drilling of a well and will remain even if the well is determined to be nonproducing. These large diameter pipe and casing reach a few hundred to several thousand feet deep and are used to retain the softer formations, such as sand.

"Production casing is of a higher strength casing and will line the full length of the well. It is only installed if it is decided to produce the well. Usually the well is not produced through the production casing. Instead, small diameter pipe called tubing is placed in the well to serve as the way for oil or gas to flow to the surface. A packer is installed on the outside of the tubing and sealed against the casing to force fluids to enter the tubing to get to the surface. Sometimes an additional tubing string can be run alongside of or inside the first in the same manner to produce another formation. Tubing is supported from the well head so the requirement of strength is great. The above wells are so-called free flowing and will have a Christmas tree network at the top.

"In the case of an oil well that has potential but requires some assistance, a string of rods, called sucker rods, is placed in the well inside the production casing or tubing. The oil is carried up as the rods are moved up and down. The up and down movement is provided by a mechanical pump unit at the surface.

"A rotary drill rig turns the drill pipe sections with a drill bit attached. Additional sections are added as the well is drilled deeper and deeper.

"Drill pipe is made of very heavy wall pipe and requires a great strength. Drill pipe is connected with threaded ends called pins and boxes which have a much heavier wall than does the pipe body.

"By comparison, the thin wall casing and tubing products as well as the sucker rods are generally connected through couplings. Casing and tubing are also of thinner cross sections than drill pipe.

"Now that we have an idea of the products required to drill, produce, and maintain oil or gas wells, let's talk about the manufacture of these products."

Primary documentation was almost started with the last paragraph. However, the writer did not get around to the induction process for another 38 words. He writes:

"Manufacture of pipe for casing and tubing is usually by either electric resistance welding (ERW) or seamless processes. Electric resistance welded pipe is made from strip, formed and welded to a specific diameter called the mother size. *After welding, the weld may be annealed by induction heating to avoid cracking during shipment or subsequent operations. This installation utilizes a 400 kw, 3000 Hertz power supply with a seam weld anneal laminated inductor.*"

Following the dictates of the busy reader style, I would drop everything between the end of the statement of the subject in the first paragraph and the italicized words, which start primary documentation. This is not to criticize this writer or any technical writer who uses a similar slow rambling style of presentation. There are no guidelines for doing otherwise.

The writer also took too much time to state his subject in the first paragraph. In this matter—how long it takes to say something—there are two considerations to balance: the writer should not use one more word than he needs, but the reader does not object to length per se as long as he is being rewarded. The latter is illustrated by the example that follows. It also clears up a point. Depending on the subject, standard information, like history, may qualify as primary documentation.

History is established as primary documentation by the statement of the subject in the following example.

Title:

"Heat Treatment of Tool Steels."

Statement of subject:

"The heat treatment of tool steel is generally overlooked by engineers, purchasing personnel, and designers. Usually a great deal of time is spent on the selection of the proper tool steel grade. Heat treating is specified by a desired hardness range. A number of heat cycles can be used to obtain the same hardness with large variations in properties such as impact strength. Proper material selection should require an equal amount of time given to the heat treatment and the nature of the cycle required. Some heat treatments have hardly changed since the 1600's, yet the nature of these is little understood today. It's not that the information is not available but more a tendency to specify heat treatments by the resultant hardness."

The next two paragraphs, both grounded in history, must be labeled as primary documentation:

"Historically it is interesting to note that some of the earliest principles of tool steel heat treatment were used well into this century. *Natural Magick*, published in 1589 by Gianbattista della Porto, became the bible of heat treating from its first appearance. It became a best seller early and demand required successive editions until 1669. One principle of this work, that tools must not be heated beyond red hot (1400 to 1700 F), persisted well into this century. Some contemporaries disagreed with Porto, but their works were either unpublished or less widely read. For example, Italian foundryman and metallurgist Vannoccio Biringuccio published a book known as *De la Pirotechnia* (1540). In this work he suggested that maximum hardness resulted in some steels when quenched from white heat (1900 to 2100 F). Biringuccio's view, however, would be overlooked until 1898.

"It was this year that Fred Taylor and Maumsel White performed a number of experiments on high speed steel. They individually had been working on the development of chromium-tungsten steels. In an effort to demonstrate their superiority, a number of tools were ordered in this grade. They assembled all the foremen for the demonstration at the Bethlehem Works. To everybody's surprise, the chromium-tungsten tools performed much worse than normal tools.

"After a number of studies, the problem was found to be in the heat treatment. They found that for these steels to develop their maximum strength a much higher austenitizing temperature was required. This was directly opposed to the standard belief (originated in Porto's time) that steel should not be heat treated beyond red hot.

"Clearly, the Taylor-White experiments demonstrated that heat treatment is as important as grade selection. Today the problems are much less because of the amount of scientific data available on heat treating. Still the selection of a proper heat treatment is complex because the engineer is always trading off various properties such as wear resistance, impact strength, toughness, and hardness."

And so on. In the next section, the writer discusses "a strategy for heat treatment selection." He is finished with history. It has served its purpose.

In review, the title gives you an approximate definition of the subject. Statement of subject should be the first element in text, and should not be preceded or interrupted by introductory or background information that is secondary documentation as defined by the title. Primary documentation should start immediately after the statement of subject is completed. It should not be preceded or interrupted by secondary documentation as defined by both the title and statement of the subject. Any secondary documentation should be presented at the back of the text, if at all.

Exercises: Placement of Subject and Primary Documentation Plus Differences Between Primary and Secondary Documentation

Exercise No. 1. I removed the title from a short newspaper article as an experiment. Find the subject is the game. In your sentence-to-sentence, paragraph-to-paragraph hunt, I surmise, you will pick up a first-hand view of the function of the title. To make the point dramatically, the exercise would require you to wander about the entire piece before arriving at "this is it!" However, to drop a hint, the magic moment is located near the halfway mark.

> DETROIT (AP)—For much of the northern half of the United States the first major snowfall of the winter is not much more than a month away.
>
> This means that in many metropolitan areas streets and highways soon will be coated with salt or other highly corrosive chemicals to keep them clear of ice and snow.
>
> While extremely effective in keeping thoroughfares open to traffic, these chemicals greet exposed sections of automobiles and trucks like a rabbit embraces a carrot.
>
> Michigan Bell Telephone Co., which operates a fleet of 4600 passenger cars and light trucks 46 million miles a year, estimates its corrosion maintenance repair costs at $80,000 annually.
>
> Anyone who has ever operated a passenger car during the winter in areas where salt and similar chemicals are used is aware of the problem. Rust seems to develop almost overnight, and whole panels dissolve into red dust.
>
> Gilbert H. Selke, superintendent of motor equipment for Michigan

Bell, has devised a $30 antirust treatment which he believes will cut corrosion maintenance costs sharply and extend the useful life of the vehicles.

The telephone company's aim is to operate passenger cars six years and trucks seven years before replacing them.

After trying all of the usual rustproofing systems the company found that a grease used to prevent corrosion in underground piping would do the job on cars and trucks.

The process amounts to greasing the inside of the car. This is done, Selke said, by drilling an average of 18 holes into a typical two-door sedan and spraying the grease compound into them at high pressure.

"We go inside the car into the boxed-in sections where moisture doesn't readily evaporate," he explained. "We treat the rocker panels, insides of the doors, the corner posts, wheel housings, areas around the headlights and pockets under fenders that can hold mud for any period of time."

The holes are drilled in inconspicious places and are plugged with plastic caps.

The grease forms a pliable film so that it can't be chipped off or washed away, although it has been shown to soften slightly in hot weather.

In two years of testing, Selke reports almost total elimination of rust. He says further testing may show a need for reapplication sometime in the future to reach the aim of six- to seven-year vehicle life.

Michigan Bell has subcontracted the application work to private firms which probably will make it available to private car owners.

My answer—Without the guidance of a title, I read, stop, analyze each sentence. It's a guessing game. Is this piece about the first major snowfall this year in the northern part of the United States? Corrosion of vehicles out on streets and highways maintained with salt? What corrosion costs Michigan Bell in the operation of its fleets of passenger cars and light vehicles? A treatise on the rusting of vehicles in states using road salts during winter months? The invention of an inexpensive antirust treatment by an employee of Ma Bell in Michigan? In traveling past this point, it becomes evident that pay dirt has been reached. Likewise, it's clear that the hunt-and-peck technique of tracking down the subject could be avoided with a title. Something like "At Last, an Inexpensive Antirust Treatment for Vehicles."

Applying the busy reader style, I would begin the piece with the sixth paragraph. "Gilbert H. Selke . . ." etc. This is the statement of the subject. Primary documentation follows, where it should be.

Can anything coming prior to "Gilbert H. Selke" be salvaged?

From the new ground we now occupy, is there any need for the weather forecast in the first sentence? I think not.

Any need to be reminded that salt is scattered on roads in winter and rusts out vehicles in this environment? I think not.

Do we need to know the extent of Michigan Bell's investment in passenger cars, light trucks, etc.? Possibly.

Any need for the general definition of the corrosion process? I think not.

I feel the next to last sentence could be rationalized because it is a specific on the seriousness of the corrosion problem. However, it is secondary documentation and must go to the end of the piece. The antirust treatment is the subject.

The net result is a shorter piece. Due care has been taken to avoid wasting the reader's time with material that does not fit the subject.

Finally, there is no reason why the piece can't be made longer—provided only primary documentation is added. The reader is discriminating. He can exit any time he feels like it.

Exercise No. 2. Now I want to demonstrate that a title is a yardstick for "does this belong?" Find the statement of the subject is the objective, as indicated by the title, "Two New Single Crystal Alloys." The analysis does not require a full understanding of the subject matter.

Title:

"Two New Single Crystal Alloys."

Text:

"A strong trend evident in the aero gas turbine industry is the move to single crystal turbine blade and vane castings from equiaxed or directionally solidified (DS) columnar grian components.

"Single crystal superalloys do not contain the grain boundary strengthening elements such as carbon, boron, and zirconium, thus enabling very high solution heat treatment temperatures to be used without incipient melting problems. High solution temperatures (2350 to 2400 F) produce a homogeneous microstructure, and subsequent fast cooling insures a very fine gamma prime strengthening precipitate size of 0.2 to 0.3 micron. This results, in conjunction with the high volume fraction of gamma prime (65%) in these alloys, in quite dramatic improvements to creep rupture properties-temperature capability gains of up to 65 to 80 F over the DS MAR M 200 hafnium alloy at 1800 to 1900 F testing temperatures.

"These temperature capability gains translate to increased fuel efficiency through corresponding increases in turbine inlet temperatures. Component durability is also significantly improved by perhaps a factor of 2 over DS through improvements to cyclic properties, such as thermal, high and low temperature fatigue. The absence of grain boundaries and carbides in the microstructure also gives significant improvements to the performance of oxidation-corrosion resistant coatings on single crystal parts.

"ABC Company has developed two new single crystal alloys. These alloys offer a unique combination of good foundry performance, a practical solution heat treat window (the gap between the gamma prime solvus temperature and incipient melting point) of 45 to 55 F, microstructural stability, and high strength properties. Extensive evaluation is current with both these alloys by the gas turbine industry."

My answer—If you don't give up because of the jargon and keep your eye on the target (namely, "Two New Single Crystal Alloys"), you will find the first reference to the subject in the last paragraph. The busy reader style is premised on an update, which assumes that the reader is cognizant of everything preceding the last paragraph. Otherwise, why would he bother to read the piece? Also, the background is probably too technical to benefit the uninitiated reader. Finally, the distance from title to statement of subject is more typical in this example than most of the other examples used thus far.

Exercise No. 3. Another instance where the statement of the subject is buried under a flurry of words that don't belong. The example is tougher than the last one. The statement of the subject is scattered over several paragraphs. A rewrite is needed. As usual, secondary information should be eliminated. Once more, judge the material in terms of the title.

Title:

"Evaluation of Heat Resistant Alloys in the Field."

Text:

"Heat resistant alloys find extensive use in the heat treating industry in a variety of applications. These applications typically include surface hardening treatments such as carburizing and carbonitriding, and involve the use of muffles, retorts, fixtures, boxes, trays, and baskets. Such alloys must withstand the effects of various heat treating atmospheres and thermal cycles over extended service lives. Degradation of these materials most often begins with carburization, and usually involves thermal fatigue.

"The choice of the best alloy is seldom based on sound comparative data, but is based on experience and judgment. While good judgment will always be necessary, it can be supplemented by comparative data obtained in field service. Laboratory tests are useful, but of limited value, principally because it is difficult to devise laboratory tests that completely reproduce all of the interacting effects encountered in the field. Laboratory tests, however, remain as useful screening tests.

"One approach in the heat treating industry that is overlooked is the use of controlled test data from the field. To obtain data, materials of interest must be exposed to the same conditions. The conditions that affect alloy performance include not only the obvious ones of atmosphere, temperature, thermal cycling, and service life, but more subtle factors such as loading and mechanical abuse. The problem of trying to isolate extraneous factors from material factors may be simplified by including all materials of interest in one unit. Heat treating bar baskets which contain a number of similar components are possibly best suited for this purpose.

"It was decided to obtain comparable performance data using this approach. Two heat treating shops were selected for the purpose of providing field experience. Both shops perform a variety of heat treatments that are typical of the industry. The heat treatments also involved atmospheres and quenching conditions damaging to heat resistant alloys by a combination of carburization and thermal fatigue."

My answer: The first mention of the subject is in the second sentence of the second paragraph, "While good judgment will always be necessary, it can be supplemented by comparative data obtained in field service." But as it stands alone, we find by continuing to read, that sentence is not a fair statement of the subject.

The first sentence in the next paragraph starting, "One approach in the heat treating industry that is overlooked, . . ." is promising at first glance, but the title indicates that the scope of the subject is broader than the recitation of the pluses of field testing that follows. Once more it seems necessary to move on. We have found some of the pieces, but the puzzle has not been reconstructed.

In my opinion, the second sentence in the last paragraph provides the missing link, "Two heat treating shops were selected for the purpose of providing field experience."

Here's an example of a simple rewrite:

"Controlled field testing is an overlooked technique for the evaluation of heat resistant alloys for heat treating applications. In this instance, two heat treating shops were selected. Both perform a variety of typical heat treatments, including atmospheres and quenching conditions damaging to these alloys via a combination of carburization and thermal fatigue."

I would toss all the other information we looked at.

Exercise No. 4. The same businesslike approach should be adhered to once the subject is nailed down. Secondary documentation should not be tolerated until, if at all, primary documentation has been finished. Analyze the information in terms of both the title and the statement of the subject.

Title:

"Low Density Aluminum Alloys."

Text:

"High strength aluminum alloys containing up to 3% lithium are being developed under a government program with ABC Airplane Co. Various design trade-off studies have shown that density is the material property giving the greatest impact on the weight of aerospace structural components, and lithium additions can be used to produce alloys having a 10 to 20% reduction in density, a 20% increase in modulus, and strength equivalent to that of current high-strength alloys.

"These low density, high modulus alloys are made using a powder metallurgy technique to allow both the incorporation of the high lithium content needed for significant density reduction and the retention of acceptable levels of ductility and toughness in the materials.

"Future programs will examine the possibility of using the rapidly solidified powder approach to increase the lithium content to even higher levels, while also examining the feasibility of producing the 3% lithium alloys using newly developed modifications of conventional ingot technology."

My answer—About a specialized material of interest only to aerospace material engineers and academicians. Assume no one else would bother to read beyond the title. Remember that direct documentation should begin immediately after the statement of subject is completed.

The way I read it, the subject is packaged in the first sentence. The 23 words following it are secondary documentation known to all aerospace material engineers and to academicians who keep up on the subject. Primary documentation begins with ". . . and lithium additions can be used to produce alloys having a 10 to 20% reduction in density," etc. From this point to the end, everything is primary documentation. There is no place for the 23 words of secondary documentation.

Exercise No. 5. The ban on secondary documentation before the statement of the subject, before primary documentation is commenced, and sandwiched between primary and secondary documentation does not stifle the writer's capacity to communicate via the written word. It is a departure from established practice, which means the writer must acquire and develop new habits.

The following is an excerpt from the front end of a piece. Where is the statement of subject? Where is the primary documentation? Where does it start?

Title:

"Thermal Spray Coating Technology."

Text:

"The role of plasma spraying will continue to expand due to the developments in equipment and materials areas. Low pressure spraying of cobalt-chromium-aluminum-ytrium and similar bond coats are becoming state of the art in the aerospace industry. Hardfacing technology will employ plasma spraying to a far greater extent when chamber spraying becomes fully utilized. Increased particle velocity, coupled with increased flame spray temperature and decreased oxidation, will become attractive for the plasma spraying of carbide-bearing systems. The petroleum industry will thus find plasma spraying attractive for a number of drilling applications. Coupled with this is the effectiveness of plasma sprayed stainless steel in improving corrosion-fatigue behavior of drill pipe. This product is in short supply, and plasma spraying will assist in maintaining wells in operational modes."

And so on.

My answer—Subject in first sentence. Start of primary documentation with second sentence. Pure primary documentation for the remainder of the piece. With the counsel of the busy reader standard, one can acquire what is perhaps the greatest virtue of a technical writer: being able to write both briefly and effectively. Some people seem to feel that being brief means shortchanging the reader. An alternative standard is suggested: take as long as you need to, but stop when you are finished. A short piece like the one that follows may be all that is needed to convey the message.

Exercise No. 6. You have heard the remark, "If I had more time, I would write you a shorter letter." The truth in the statement is that a long first draft may be needed to straighten out one's thinking before a shorter second draft can be put together. As a part of this evolutionary process, the writer evaluates his documentation in terms of primary and secondary documentation. By keying exclusively on what belongs, the writer can usually come up with a much shorter piece than he would have otherwise. As the following short example demonstrates, you can come up with concentrated writing that is highly readable.

Once more. Spot the statement of subject and start of primary documentation. Analyze the remaining documentation.

Title:

"High Speed Tool Steel Developed."

100% of text:

"A new type of high speed tool steel has been designed with regard to the metallurgy of several carbide forming elements and its interaction with columbium. Keeping in mind that columbium forms primary carbides of the MC-type, this element does not influence the solubility of multicarbides in austenite. Based on the matrix composition of conventional M2 high speed steels, columbium and carbon have been alloyed in a stoichiometric ratio for a total carbon content of 1%. This new alloy system (1% carbon, 3% tungsten, 3% molybdenum, 1% vanadium, 3% columbium, 4% chromium) showed after hardening at 1200 C and tempering at 560 C its optimal hardness of 68 Rockwell C and a homogeneous microstructure. The results show that despite a reduced total alloying content and the maximal use of less expensive carbide forming element columbium even improved properties can be achieved. This new high speed tool steel helps alloy cost saving of about 30% at current prices."

My answer—Subject in first sentence. Everything after that primary documentation. Nothing wasted. Words or the reader's time. If he opts to read, he gets a decent return for his investment of time. If the subject is not his cup of Sanka, he has three fast chances to get out: title, statement of subject, start of documentation. He can quit reading at any time with some assurance that he is not missing something he should know about.

Exercise No. 7. One more example that shows that "good" and "long" aren't necessarily synonymous. Analyze as usual.

Title:

"Cutting Tool Market Study."

100% of text:

"Gorham International Inc. has announced a multiclient study of the worldwide market for cutting tools in the 1980's. The study will cover computer controlled high speed machine tools, new materials, coatings, shapes, new suppliers and users entering the market, near net shape technologies, and availability of cutting tool materials.

"For information about the ten-month study, contact Dr. Andrew C. Nyce, vice president, Gorham International Inc., Gorham, Maine 04038; 207/892-6761."

My answer—At first glance, I would say that the first sentence, which states the subject, is an example of upside-down construction. To repair, start with "A multiclient study of the worldwide market for cutting tools in the 1980's" and end with "has been announced by Gorham International Inc." On second thought, though, I feel the original construction is proper. For two reasons. First, the piece is from an association magazine distributed among members of the association. Within the fraternity, "who" did what is at least as important as "what" they did. Second, the practice of starting with the source has been established as the style for the section of the magazine in which this news appears. All items are written this way: "GCA/Vacuum Industries has designed . . ."; "Nyby Uddeholm AB expects production. . ."; "Hoeganaes Corporation will spend $4 million . . ."; "American Powdered Metals Company, a wholly owned subsidiary of Aluminum Company of America . . ."; and so on.

In evaluating the documentation for this piece, one must conclude that it is complete, however brief it may be.

CHAPTER 3

Making It
Hang Together

Continuity is the next concern—what I have been calling flow to avoid the jargon of a technical term and still describe the desired result.

The job is to establish and maintain the illusion that the reader is constantly moving forward as he reads. Sentence to sentence. Paragraph to paragraph. From beginning to end.

To get an insight into what is required, it is helpful to recall the roles of the title and statement of subject in determining what is primary and secondary documentation. A similar thread of logic must be maintained sentence to sentence.

To oversimplify, you do this by showing the relationship of what is said in sentence No. 2 to what was said in sentence No. 1. You generally work backward. Sometimes what you are referring to goes back several sentences or several paragraphs. One way to get continuity under these general circumstances is to repeat a word or words from a preceding sentence. Sometimes you can take advantage of context and it isn't necessary to use a continuity device. Sometimes you shift gears. You depart from what you have been talking about or make an abrupt change. Technically, a transition or bridge is needed to make the jump. Such words as "but" and "however" are used.

Here, for review, are some common flow devices:

1. Using tack words like "this" or "therefore."
2. Repeating a word or words.
3. Referring in some way to what was said in a previous statement, such as by paraphrasing or extracting a concept or idea.
4. Answering a question (raised directly or indirectly) in a previous statement.
5. Stating or implying that something follows (related to the device. "Tell them what you are going to tell them, then tell them").
6. Taking advantage of context.
7. Anticipating reader response to something that has been said.

Device No. 1: Using Tack Words

The easiest way to show the connection between what is being written and what has been written is to use a tack word such as "this" or "that."

Example:

"Scientific and technological developments must be anticipated and prepared for in advance. *This* will serve a dual purpose."

The temptation to use words like "this" is great because they are ubiquitous in technical literature. They have shortcomings, and should be used with restraint. To begin with, the reader may be forced to stop and back up to see what is being referred to because a tack word is essentially a say-nothing word. Reference is often obscure. In the example, the writer would have served the reader better if he had tried to be more explicit. The writer might substitute "the approach" for "this" and come up with a sentence that reads, "The approach will serve a dual purpose." The technique, in this instance stating the concept of what is being referred to, is covered in Device No. 3.

In the example, I believe the tack word is misused. Overuse is also a problem. Twenty or so "this's," "thus's," and "therefore's" running through a technical piece can mean 20 or so stops and starts for the reader.

Flow devices, like tack words, are also used where they aren't needed, adding to the tedium of this literature.

For example:

"Manufacturing inventories fell to an all-time low. This pushed supplier inventories to an all-time high."

One alternative is to take advantage of context (Device No. 6), drop "this," and rewrite the second sentence to read: "Supplier inventories rose to an all-time high."

Another possibility is to reduce the second sentence to a phrase such as: "Manufacturing inventories fell to an all-time low, pushing supplier inventories to an all-time high."

In either instance, flow is improved, and the alternatives have more polish than the original.

Device No. 2: Repeating a Word or Words

Note the italicized words in this example:

"ABC Steel Corporation has *licensed Japan Steel Corporation* to produce and market a patented aluminum-zinc, alloy-coated sheet steel which ABC calls [trade name].

"*Japan Steel* ranks as one of the world's largest producers of galvanized sheet steel, accounting for about 2 million tons annually.

"Announcement of the *license agreement* was made here today by Mr. [Vice President Marketing] of ABC Corp.

"Under the *agreement*, the terms of which were not disclosed, *Japan Steel* will be permitted to use ABC Steel's technology, products, and steel (trade name)."

In the second paragraph, the writer referred back to Japan Steel in the first. In the third paragraph, reference is based on "license" in the first paragraph. In the fourth paragraph, "agreement" from the third paragraph is repeated.

In the four paragraph example that follows, note how a man's name, "Mr.

Scarp," and a pronoun, "he," are used. The latter device is a variation on the repetition of a word or words.

Example:

" 'There's probably not a scrap processor in the country who isn't glad that 1981 is over,' *Mr. Scarp* said. 'Two bad years back to back have left the scrap market in a shambles.'

"*Mr. Scarp* said, however, that he expects 'some improvement' early in 1982— 'the earlier, the better.'

" 'We should see a modest improvement in demand levels for ferrous scrap in the first and second quarters of 1982,' *he said*, 'with hopefully a real strengthening in the market during the second half of the year.'

"*Mr. Scarp* said the dismal performance in 1981 could be traced to a unique confluence of economic forces—a general economic recession with high interest rates at a time when the scrap market was already at a low point."

In the following short piece, the words "agreement" and "ABC Corp." are used:

"*ABC Corp.* announced today it has entered into a definitive *agreement* to sell for cash the C Division to XYC Corporation. Terms of the *agreement* were not disclosed.

"*ABC* will retain certain rights to market text-processing products to the U.S. government.

"*ABC* is a high technology company that provides electronic products and professional services to industry and government."

In another example, a single set of words, "Expo/East," is repeated:

"The Manufacturers of America opens its 1982 show season with *Expo/East*, February 28-March 2, at the Sheraton Centre Hotel in New York City.

"This is the largest *Expo/East* ever and the only convention and supplier exposition of its kind in 1982 for the Manufacturers of America on the East Coast.

"*Expo/East* will reflect the 1982 show theme, "Play to Win . . .""

One more short example. Note that "purchase" is repeated:

"The Fife and Drum Corp. here has agreed in principle to *purchase* the Consulting Division of ABC Corporation.

"The *purchase* includes . . ."

Combinations of flow devices are used—in this instance, the tack word "these" and repetition of the word "areas."

Example:

"The companies agreed that the most promising *areas* for cooperation were chemical and metals refining, raw material and energy conservation, microelectronics, robots, and licensing.

"*These areas* will be the main themes of technical symposia . . ."

In commercial pieces, names of companies and products are frequently used as flow devices. In this instance, the latter are featured:

"ABC Chemical Corporation has announced the availability of its new *6000 series products* for the treatment of industrial boiler water systems.

"The *6000 series products* utilize a proprietary chemical to distort and disperse scale crystals and sequester soluble iron . . .

"Several *6000 series products* are blended with Wonderine—ABC's catalyzed hydrazine—to scavenge oxygen and passivate metal surfaces.

"The *6000 series products* provide a flexible line which has a wide variety of applications."

Device No. 3: Paraphrasing a Previous Statement, or Referring to a Concept in the Previous Statement

Overuse of repetition adds to the drabness and sameness of technical literature. Devices like "stating the concept" require more ingenuity on the part of the writer, but they provide a welcome change of pace plus polish.

Example:

"ABC Aluminum Corporation is doubling its capacity in commercial storefront products, which have resisted the sales slump in residential products and outside extrusions.

"The producer of architectural aluminum products also has shifted its furnace mix toward more scrap to take advantage of favorable prices compared with producer ingot.

"The persistence of aluminum storefront sales has defied conventional wisdom, which decrees that a drop-off in residential housing is followed six months later by a decline in commercial product demand.

" 'So far it hasn't happened,' states ABC, adding that no signs of decline are evident."

At the front of the second paragraph, ABC Aluminum could have been repeated. For a change of pace, the writer reflected on what the corporation does and came up with "the producer," an example of paraphrasing.

In the third paragraph, "The persistence of aluminum storefront sales," is borrowed from the concept of a statement in the first paragraph, "storefront products have resisted the sales slump in residential products."

The third word, the pronoun "it," in the fourth paragraph refers back to "decrees that a drop-off in residential housing is followed six months later by a decline in commercial product demand."

In the following example, "at that rate" is a way of referring back to "6 million tons" in the preceding sentence.

Example:

" 'We know purchases of scrap dropped in November and December and should add only about 6 million tons to the year. At that rate,' he said, '1981 will total out at about 44 million tons, 7% (or 3 million tons) ahead of the 41 million tons purchased domestically in 1980.' "

The next example is three short paragraphs long. In the second paragraph, a variation of a word is repeated. In the third, an editorial comment on the gist of the first two paragraphs is used. These techniques are indicated by italics.

Example:

"Area steelmaking is expected to continue at current depressed levels into the new year due to weak order outlooks at district *mills*.

"*Mill* spokesmen throughout the region generally conceded that current steelmaking rates would continue through the next several weeks unless order books improved.

"*Typical of the mill's over-all plight* were ABC Corporation's operations. Layoffs rose to 7100 this week from 6000 two weeks ago."

A similar approach is used in the next example. The first sentence announces completion of a contract. The work done could be called a project, which was the word used in the second sentence.

Example:

"ABC Corporation recently completed a contract for scrapping the machinery and equipment of the former buggy whip factory in Pottstown.

"The project yielded more than 8000 tons of scrap metal for sale . . ."

Device No. 4: Answering a Question

What is said in a preceding sentence may raise a question directly or by implication. Answering the question in the next sentence or sentences establishes continuity, as in the following example. In this instance, a question is raised in the first sentence. The answer is in the next two.

Example:

"The sharp reported rise in London Metal Exchange tin stocks last week reflected an adjustment from a previous reporting mistake rather than an actual increase in metal held in storage."

The next sentence answers, "What was reported?"

"Tin stocks rose last week to 13 325 metric tons, an increase of 4415 metric tons over stocks of 8910 tons at the close of the week ended Dec. 18."

The next sentence explains "the reporting mistake."

"However, the rise in stocks reflected a paper replacement of 3750 tons of tin held in Hull warehouses that had been reported as shipped out in the prior week, but which were actually still held in storage."

Several questions are raised in the next example.

Example:

"The Columbus Division recently started a $160 000 group sponsored research program to analyze the current and projected construction activity in Mexico.

"The program, which is scheduled for completion in 1982, will provide short and long term projections of residential, nonresidential, and nonbuilding construction activity in Mexico. Nonbuilding construction includes such projects as dams, bridges, pipelines, and power plants.

"According to R. M. Thomas and G. B. Rolland of Columbus, who head the study team, Mexico is rapidly and dramatically increasing its construction expenditures as a result of revenues from oil and natural gas.

"In its current state of development, however, the Mexican construction industry does not have the capability nor the capacity to meet these new demands without assistance from foreign manufacturers of building materials, heavy equipment, and other construction related products."

The first sentence raises the question: "Why does Mexico want an analysis of current and projected construction activity?" Answers are in the third and fourth paragraphs: "Mexico is rapidly and dramatically increasing its construction expenditures . . ." and "Mexican construction industry does not have the capability nor the capacity . . ."

Delaying the answers for one paragraph did not cause a problem. It's safe to assume the reader can remember that far back. Of course, the questions could be answered immediately. In this instance, other matters were taken care of first.

Note that the word "program" was repeated to get continuity from the second to first sentence.

You can anticipate another question in the second paragraph. What is "nonbuilding construction?" In this instance, the answer is in the next sentence— "dams, bridges," etc.

A short example:

The technical committee on cold finished bars is updating the cold finished bar chapter of the Institute's Product Manual. Changes include a new cut-to-length tolerance table, a definition of tempering, and revisions in definitions for thermal treatment."

In the first sentence, "updating" raises the question, "What's being done?" Examples are given in the next sentence, starting "Changes include . . ."

The statistical report for a week in December in the next example suggests, "What does the whole year look like from here?" "How does it compare with the year before?" The answers are given in the second paragraph.

Example:

"Freight traffic on major U.S. railroads during the week ended December 19 totaled an estimated 16.5 billion ton-miles, 11.7% below the corresponding week a year ago.

"Cumulative volume for the first 51 weeks of 1981 was an estimated 903.6 billion ton-miles, 0.3% below the comparable period of last year."

Device No. 5: Stating or Implying That Something Follows

The technique is to say you are going to do something, then do it. If you use a list or summary to set this up, it can be used as a pattern for the organization of what follows.

An example illustrates the technique:

"A task force of the Institute has identified four areas in manufacturing for potential research with the Association. These are: improvement of power delivery in the early stages of melting; induction reheating of semifinished steels; plasma slab conditioning by superficial melting; and electric conditioning of surface in preparation for coating."

If the writer continues beyond this point with a discussion of the four areas, he can use the listing to set up an organizational pattern. He has, in effect, promised the reader that he will write about these four things. He can start with the first item and work forward, or he can start with the last item and work backward.

In the next example, the statement of policy in the second sentence is set up by the statement in the first sentence.

Example:

"The president restated the company's public relations credo: to work with diverse publics in the best manner possible; to cooperate to the best of its ability to serve their needs and the needs of the company, its sister companies and its parent organization; and to be as honest and open with those organizations as is humanly possible."

Here is an example that sets up the next two sentences:

"The scrap metal industry suffered its second consecutive bad year in 1981, but expected an improved market in 1982.

"Exports of carbon steel and iron scrap in 1981 are expected to be off more than 40% from the 10.9 million net tons exported in 1980.

"A pickup in exports is expected to arrive in early 1982."

Device No. 6: Taking Advantage of Context

Read the following two paragraphs. It is apparent that it is not necessary to show a connection between the second paragraph and the first. Because of context, the second amounts to a continuation of the first.

Example:

"The nearly nonexistent demand for aluminum in refrigeration, automotive, construction, and aerospace industries caused shipment of aluminum by distributors to plunge in November to the lowest level recorded in 1980 and 1981.

" 'Business conditions are flat,' commented an east coast distributor. 'We were excited without our sales pickup in September and October, but experienced a nosedive in November. The only bright spot is the electronics industry which is doing well because of the increasing use of computers."

Another example:

"The conference and workshops will provide a forum in which participants can interact to develop an action-oriented framework for examining and solving issues and problems.

"A wide diversity of issues will be explored, including assurance of adequate supplies of strategic materials and standby conservation measures . . ."

It isn't necessary to add "at the conference and workshops" after the word "explored." The relationship is clear from context.

Tack words can sometimes be eliminated by taking advantage of context.

A simple example from a brochure for an adult education program:

"Also, we are interested in what you may recommend as an area of interest, so there is a place for that on the card below."

Put a period after interest. Drop "so." Capitalize "there."

Here is a more complicated example:

"Customers bought more welding supplies from distributors in 1981 than in 1980, according to the Association.

"According to the Association's Index, the demand for welding supplies from distributors in November rose 12.8% above the figure reported in 1980. Year to date sales now stand 9.9% above the 1980 level."

Drop the first six words in the second paragraph and start with "demand." I would rewrite because of the unusual amount of repetition: "according to," and "welding supplies from distributors."

I suggest: "Demand in November rose 12.8% above the figure recorded in the Association's Index in 1980." The next sentence should stand as is because it takes advantage of context.

Device No. 7: Anticipating Reader Response

Example:

"Tool & Die Works Inc., one of the largest producers of high quality custom tools and dies in the upper midwest, has expanded its manufacturing capabilities with the recent acquisition of Short Run Metal Stamping Inc. "The reason for the acquisition, say company officials, is to expand into a market it now serves as a supplier of tools and dies.

"Short Run Metal Stamping will continue to operate under its own name."

The writer can anticipate that the reader will want to know why the company is going outside its field (which is taken care of in the second paragraph) and whether the acquired company will lose its identity (which is responded to in the next paragraph).

The device is similar to Device No. 4 (answering a question) in some respects; both provide an opportunity to get continuity by suggesting what to say next. However, they should be thought of as separate techniques. One may help where another doesn't, as in the following example.

Example:

"The outlook for 1982 is that it will be a mediocre year, states [Mr. Executive]. 'I don't anticipate an increase in demand until at least June. The year 1983, however, looks super and will probably be a big year. The aircraft industry looks like it will be particularly strong.' "

Following the prediction for 1982 with the one for 1983 is the result of anticipating reader response: "If 1982 is bad, what about 1983?" I doubt if 1983 would have come up if the speaker had been thinking strictly in terms of the question-answer technique.

Or take this example, which presumes knowledge that the U.S. steel industry claims that imports are to blame for many of its problems.

Example:

"November steel imports of 1 921 000 net tons increased 2.7% from October levels, according to Imported Steel Company.

"The president of the company remarked that the increase reflects a strong business climate that commenced in early summer when steel was ordered from mills operated at 80% of capacity. The figures also reflect increased demand for tubular goods and semifinished steel from domestic mills.

"According to the company president, 'imports are not injuring the domestic industry. Imports are the result of American industry's problems. They fill the gap in U.S. demand not met by domestic industry.' "

Obviously, "imports are not injuring the domestic industry" is in anticipation of a negative response to a report from a company that imports steel into the United States.

The writer takes advantage of anticipated reader response four times in a row in the following example.

He starts:

"ABC Aluminum Corporation said today that it will take the fourth and last producing potline out of production at its West Virginia works."

A discerning reader would want to know "when?" The writer obliges in the next sentence:

"Preparatory work for closure of the line, with an annual rated capacity of 40 750 tons, will get underway immediately . . . with complete shutdown scheduled for January 10, 1982."

The following sentence anticipates the question, "Why is the facility being shut down?"

It is stated:

"The action is the result of an inability to reach agreement with hourly employees on changes necessary to make the plant competitive, a company spokesman said."

What did the company ask for?

"The company had offered to keep the line in operation through 1982 in return for proposed changes in work practices and seniority provisions."

Weren't wages and benefits involved?

"Wages and benefits were not an issue."

One more example. The writer starts with:

"A federal administrative law judge has granted a request from ABC Safety Company for a ruling that the case against its Respirator be dismissed only with prejudice."

It's safe to assume a reader would want an explanation of "only with prejudice." The next sentence adds:

"Dismissal with prejudice means the case cannot be reopened with the same evidence."

Wanting to know who brought the case is another probable reader response. The answer is provided next:

"The National Institute for Occupational Safety and Health had begun proceedings to revoke approval of the Respirator."

What did the judge rule?

"Administration Law Judge H. S. Smith, writing in a 20-page opinion, upheld ABC's contention that the Respirator equipment had been unfairly accused of substandard performance."

In review, when a piece of technical literature hangs together from beginning to end, it has flow or continuity. Think of starting with a stack of assorted, often unrelated, sentences. The writer's challenge is to assemble the piece of literature sentence by sentence, in some logical manner. Several techniques are available. In some instances, something extra—like using a tack word or repeating a a word—is needed to make sentence "B" fit sentence "A," or sentence "C" fit sentence "B." This is a kind of mechanical way of building continuity into a piece. You can also create continuity, so to speak, with techniques like anticipating reader response, answering questions, and saying you will say something, then saying it. In some instances, "B" and "A" fit together naturally. Nothing extra needs to be done. Finally, in some instances, sentence "B" is different enough from sentence "A" to call for a word like "but" or "however" to indicate the break to the reader.

Exercises

Exercise No. 1. Identify the three continuity devices used.
Example:
"The Seventh Biennial Conference will be held July 27-29 at Harper's Ferry, W. Va. The theme will be: 'Strategies for Coping with Critical Issues Related to Engineering Materials and Minerals.

"General chairman of the Conference is Dr. Agee of ABC Institute.

"The conference will define the key elements common to critical issues related to materials and minerals; review what has been accomplished thus far by legislation and other methods, and determine what remains to be done in coping with these critical issues.

"The conference will provide a forum in which participants can interact to develop an action oriented framework for examining and solving these issues and problems.

"A wide diversity of issues will be explored, including assurance of adequate supplies of critical and strategic materials, development of substitution preparedness, and stand-by conservation measures . . ."

My answer—Note the colon after the fourth word in the second sentence. The device is related to the "tell them what you are going to tell them, then tell them" technique. In this instance, the theme is introduced. In the next three paragraphs, "conference" is picked up from the first sentence and repeated. A continuity device is not needed for the last paragraph. The writer takes advantage of context.

Exercise No. 2. What's the technique used here?
Example:
"Two new home learning programs are now available through the Institute.
"*Applications and Properties of Castings* is intended for designers, specifiers,

and users of castings. It takes the practical approach to the metallurgical, mechanical, and physical characteristics of cast ferrous metals and their relationships.

"The course covers the effect of welding on cast irons and cast steels when joining castings together or to wrought products to simplify casting design, or to overcome foundry or material handling limitations.

"Flaws that can occur in casting, their detection by NDT methods, and repair by welding are described. Protective, decorative, and sealant type coatings are also included.

"Through *Tool Materials: Metallurgy and Applications*, participants will gain a practical understanding of tool and die materials, their compositions, and guidelines for selecting and processing these materials of optimum performance."

My answer—In the first sentence, you raise the question, "What are the two new home learning courses?" The writer obliges with the first course immediately, but is a little slow in getting around to the second one. A suggestion: Name both courses immediately after the first sentence. This sets up your organization, and it won't bother the reader if you are slow in getting around to a discussion of course No. 2. He already knows its title.

Exercise No. 3. Find the four techniques used in the following excerpt from a statement to stockholders.

Example:

"Last week ABC announced an agreement to acquire Fastships Inc., a producer of large aluminum marine vessels. These ships are used primarily in offshore oil and gas support service and military applications. Fastships also produces steel vessels and provides vessel repair services.

"Because this acquisition is a major step in ABC's transition into a more broadly based company, I want to take this opportunity to review with you ABC's directions and progress. As one measure of progress, we expect that almost half of ABC's revenues in calendar 1983 will come from activities that were not part of ABC a year ago. However, many investors still consider ABC to be primarily a uranium company. Because of this perception, we believe the stock market has overreacted to the weak uranium market and has not reflected ABC's real value.

"Uranium sales will provide less than 20% of ABC's revenues next year. These sales are expected to be profitable. In fact, we expect each of ABC's major business groups to be profitable next year.

"Our diversification program is aimed at making use of the technical and management skills of the Company . . ."

My answer—The writer picks up on "vessels" and the name of the company, "Fastships," in the first sentence and uses "these ships" in the second sentence, which could be counted as a paraphrase or repetition. "Fastships" is repeated next. In the first sentence of the next paragraph, the word "acquisition" refers to "acquire" in the first sentence. A continuity device isn't needed for the next sentence because it is within the context of the preceding one. But the word "prog-ress" at the end of the sentence is picked up and repeated in the next one. "How-ever," at the front of the next sentence flags a change in direction from acquisition and progress to the image of the company in the eyes of investors. The writer uses the repetition device in the next sentence by paraphrasing an aspect of the preceding sentence, "because of this perception . . ." This sentence in turn suggests that curious reader will want to know "What is the extent of ABC's participation in the uranium market?" The answer is in the next sentence, as well as the word "sales," which is picked up in the next sentence. The word "profit-able" at the end of this sentence is picked up in the next one. Finally, the first three words in the next paragraph, "Our diversification program," refer all the way back to the idea behind the first sentence in the second paragraph, starting "Because this acquisition is a major step in ABC's transition into a more broadly based company . . ."

Exercise No. 4. Only two devices are used in this excerpt from a rather long news release issued by the National Aeronautics and Space Administration.

Example:

"A long range research program getting underway at NASA Lewis Research Center here is designed to help reduce the nation's dependence on foreign countries for its supply of essential, high-performance metals.

"Focused sharply on the aerospace industry—which is the prime consumer of a number of strategic metals—the ongoing research program is aimed at providing alternative materials and concepts which will substantially reduce that industry's consumption of cobalt, columbium, chromium, and tantalum.

"Named COSAM for 'Conservation of Strategic Aerospace Materials,' the program is managed by the Lewis Center's Strategic Materials Section. Heart of the program: basic research on the effects of strategic metals on the performance of aerospace alloys and identification of potential substitute materials.

"The pre-eminent aerospace position these metals have achieved within the metallurgical hierarchy . . ."

My answer—"Program," the fifth word in the first sentence, is picked up and repeated once in the second paragraph and twice in the third. In the last paragraph cited, another form of repetition is obtained by referring to the idea behind the program: the importance of strategic metals in the aerospace industry. The paraphrase reads, "The pre-eminent aerospace position these metals have achieved . . ."

CHAPTER 4

Delivering the Message

In this chapter, conventional technical writing language is the subject. Common practices that sanction and/or promote inefficient or improper use of words by the technical writer. We will concentrate on generalities and claims, standard language, clichés and buzzwords, jargon, stilted language, verbosity, truisms, and sophisms.

Generalities and Claims

The generality and the claim are identical twins, and there probably isn't an overwhelming reason to give them separate identity, but I prefer to call the first example a claim and the second a generality. A claim usually has a commercial tone, low key or blatant.

Example No. 1:

"The alloy is characterized by excellent cold formability, dimensional stability after aging, and high strength."

Example No. 2:

"It was a pretty fascinating phenomenon. The view out of the window was unbelievable."

In both instances, the writer stops short of delivering his message to the reader. "Excellent cold formability" is not documented. "A pretty fascinating phenomenon" is not explained.

In other words, be specific.

Suppose I write this claim:

"ABC Corp. has a carburizing process said to be gaining favor among heat treating shops that work less than 24 hours a day or seven days a week."

Why?

Unless I follow with specifics, the only contributions to the reader are a hint and an unanswered question.

In this instance, the writer continued with:

"Operators can keep furnaces filled with parts while a normal cycle is suspended. Parts are maintained heated under a nitrogen-hydrocarbon atmosphere for periods ranging from several hours to many days without altering case profiles or depths. Productive furnace time is increased by eliminating the need to cycle empty trays."

To repeat, a claim-generality typically raises a question, direct or implied, that is left unanswered. The practice is universal in technical writing.

Example:

"The advantages and near-term potential in the titanium and superalloy areas are very questionable."

Example:

"While most of the equipment is not revolutionary in the area of near net shape technology, it does represent significant incremental steps in efficiency of material utilization."

Example:

"The alloy shows great promise for this application as well as exceptional corrosion resistance."

The writers did not go any further in each instance. The reader is left hanging in midair.

Why are "the advantages and near-term potential in the titanium and superalloy areas very questionable"?

Explain "significant incremental steps in efficiency of material utilization."

Why does "the alloy show great promise for this application"?

In each instance, as always, the message lies someplace in the dark and mysterious silence beyond the generality-claim. What the reader paid his money for he didn't get.

I am not suggesting that generalities-claims should be banned. What I recommend is: if you use them, document. After writing each sentence, ask: "Did I raise a question that should be answered?" "Should I explain?" "Did I stop too soon and leave out specifics which I have?"

Shortcomings of Standard Language

Standard language is usually the first word or phrase to come to mind because practically all technical writers use the same word or phrase in a given context practically every time. What are the beefs? Three at least, as illustrated by the following.

Example:

"This has been *achieved* by quenching and tempering. The result of these *improvements* is a heavier gage steel with *improved* resistance to temper embrittlement and *improved* low-temperature toughness."

The standard words are italicized. One complaint, as indicated, is that there is an annoying sameness about this language. Word selection is another obvious shortcoming, as illustrated by the first sentence. "Achieved" is hardly the word for what was done by "quenching and tempering." The writer should have given it at least one more try. Even "this was done by quenching and tempering" is an improvement. This brings us to the next example. Standard language, like generalities-claims, prevent the writer from doing his job. Questions are raised and left unanswered. What are "these improvements?" Document "improved resistance to temper embrittlement." Document "improved low temperature toughness."

The challenges of word selection and documenting should not be discounted. In my opinion, however, recognition, the necessary first step in doing something constructive about standard language, is the sleeper as far as degree of difficulty is concerned. Standard words are all about us. They are accepted. They look right. They sound right.

Here, for example, is a random list of standard verbs:

increase	engage	served	occur
conduct	result	accomplish	rise,rose,risen
obtain	attain	cut	effect
achieve	eye	acquire	possess
continue	employ	affect	grow
use	perform	provide	manifest
utilize	indicate	decrease	found

Add standard phrases like the following:

is characterized by	increasing use of
will continue to grow	can be achieved
has been developed	achieve various beneficial effects
studies were performed	studies were conducted
as a result of	according to
is actively engaged in	in case of
has successfully passed	has successfully withstood
was successfully completed	in the form of
on the other hand	in order to
on the basis of	the big picture
in conjunction with	in the area of

Anytime you run into a word or phrase with the limitations of a generality-claim, you probably have a standard. Check out this short example:

"ABC's highly basic submerged arc flux allows the welding engineer the flexibility of increasing production and improving mechanical properties."

The culprits are "increasing" and "improving." It's up to the writer to flesh out what is meant by "increasing" and by "improving."

Anytime you run into a word or phrase that makes you think, "Hey, this word or phrase is always used in this manner or context," you probably have a standard. Check out this example with the word "achieve" in mind:

"Weld metal Charpy V-notch properties of 100 foot pounds are achievable at the service temperatures required for Arctic gas service. These high-impact properties are achieved with welding fluxes and wires which meet the stringent performance criteria required for line pipe. These properties and excellent performance are achieved with either two or three wire welding systems. ABC's wire and flux are a matched combination providing metallurgical properties to achieve these outstanding results."

I count one "achieve" per sentence.

In another example running 1¼ pages, typewritten and double spaced, I found 14 standards: revised, utilize, permitted, in order to, improved, offer, increases, found, used, provide, called, offered, reduced, and available.

Greater variety is desirable in this instance. However, the writer should go beyond playing around with substitutes. Stop and think of what you have to say or want to say. Give the entire sentence another try. Stop. Sit back. Evaluate it. Take advantage of your reading and speaking vocabularies.

What to do About Clichés and Buzzwords

Here is a cliché:

"Zinc batteries are still *in the spotlight*."

Here is a buzzword:

"Additional savings will be realized from the decreased amount of *energy* required to produce the new coin."

In terms of time, clichés can go all the way back to Shakespeare and the Bible. Buzzwords are like clichés but are of current origin. They would have required explanation in Shakespeare's time. They may not be remembered by the next generation.

Both clichés and buzzwords have limitations similar to those of standard language. What is meant by "in the spotlight"? Unless the phrase provides some light, the reader is left in the dark. Buzzwords also require more explanation than they normally get.

In the case of both buzzwords and clichés, the writer relies on them to do more than they can. There is the feeling that I have fulfilled my obligation as a writer if I merely put "in the spotlight" or "energy" on paper.

At this time in the early 1980's, everyone's attention is focused on the energy crisis. The term "energy crisis" is becoming a cliché. Writers seem to think that all they have to do is refer to the "energy crisis" in some manner and all sorts of things will pop out of the woodwork. In the example, the "additional savings . . . realized from the decreased amount of energy required" are probably insignificant. The writer makes them sound like a big deal. We can't be sure because no documentation is provided. We merely have the bare, unsupported statement. In this sense, buzzwords are like clichés.

Here is a partial list of clichés:

skating on thin ice	see eye to eye
sour grapes	acid test
square peg in a round hole	all things to all men
step in the right direction	at loose ends
straight and narrow	blessing in disguise
straight from the shoulder	blaze a trail

swan song	in the same boat
take the bull by the horns	bull in a china shop
tighten one's belt	bury the hatchet
vicious circle	buy a pig in a poke
weather the storm	call a spade a spade
in the last analysis	clean bill of health
in the same boat	conspicuous by its absence
irons in the fire	cool as a cucumber
know the ropes	drug on the market
last but not least	exception that proves the rule
leaps and bounds	face the music
leave no stone unturned	few and far between
lion's share	ax to grind
man in the street	fly off the handle
mark time	from A to Z
new broom sweeps clean	handwriting on the wall
put all one's eggs in one basket	draw in one's horns
read between the lines	hit the nail on the head
rob Peter to pay Paul	sell like hotcakes

Buzzwords of the day center on subjects like inflation, high technology, high interest rates, innovation, quality, imports. The word "crisis" is often twinned with a buzzword, as in materials crisis, nuclear crisis, profitability crisis, cost crisis, ecology crisis, energy crisis, and environmental crisis.

As in the case of clichés, if the writer would document the buzzword in a reasonable fashionable, no harm would be caused. However, this is not always the case. Too many writers put down the buzzword, then stop, or at least stop too soon.

The Two Sides of Jargon

When jargon used by the writer is not understood by the reader, jargon becomes a private language.

Readers of business news probably do not have trouble translating the following message:

"The market for coated steel product now may be even softer than the steel market as a whole, industry sources believe, due to increasing imports and the depressed state of all major markets for the product."

The term "softer" is jargon. The writer means: of all steel products, sales of coated steel are the lowest. Note that in the case of standard language, jargon goes only part way. "Softer" is a generality. The reader looks for specifics.

If you are acquainted with computer jargon, you should not have any trouble with this:

"The less expensive RAM is used as a scratch pad to make temporary calculations before storing data in EAROM."

To the initiated, "RAM" stands for random access memory. "EAROM" is defined as an electrically alterable read-only memory. For the uninitiated, still further translation is necessary. The point, in this instance, is that the writer should use language that reaches the reader. Further, the writer should care about reaching the reader. Some prefer to show off.

Stilted Language, Verbosity, and Truisms

In this instance, one thing seems to lead to another. Where you have stilted language you can probably find verbosity and truisms.

For example:

"Sophistication in the design of new facilities is manifested in increasingly detailed considerations of a material's properties."

The word "manifested" is obviously out of its element. But before one can come up with a replacement, one must try to re-create what the writer had in mind: a sure sign of verbosity. The verb "shown" is a possible quick fix, but I would recommend going back to the drawing board with the sentence. From context, it is apparent that the writer is referring to the "sophistication in the design of new facilities" for making alloys. But until the facts are checked out, the verbose statement is not clear.

The writer goes to the opposite extreme with truisms by overexplaining the dreadfully obvious.

In this example, the truism is in the form of a prediction:

"Capacity will double between 1965 and 1970. So as capacity is expanded during that period you can expect the deficit in capacity to ease up."

In this example, the truism takes the form of an interpretation:

"Sales jumped from $700,000 in March to $800,000 in April, meaning they rose $100,000 during the period."

The cure for truisms is to eliminate them because they contribute nothing to the reader, add unnecessary wordage, and enlarge the dullness of this literature.

Stilted language, like standard language, calls for extensive policing because you can't escape it. We use "possess" instead of a more ordinary word like "have." We "accomplish the painting of a machine" instead of "painting a machine." We always "conduct a survey" instead of "we survey." We "perform three operations" instead of "doing three operations."

It's a Dickensian language that's out of place when we are writing about technology.

I question the suitability of the following language:

". . . a flux which has most of the excellent characteristics of both types of products."

". . . has achieved outstanding impact strength."

". . . we are investigating some advanced methods to achieve this end."

". . . a major improvement in corrosion resistance, as measured by the salt spray test, has been accomplished."

". . . the alloy is expected to experience increasing demand during the 1980s."

Simplicity in language is the recommended alternative. More likely than not, this can't be done by merely juggling words. Sentences must be rethought and recast; as in the case of verbosity.

In fact, verbosity is often a telltale sign of first-draft thinking. The writer is fumbling. He does not know exactly what he wants to say or how to say it.

I would take another crack at the following:

"This operating strategy is a bit different from the ABC tradition of acquisition, which could stand as testament that the entrepreneurial spirit is alive and well in U.S. business."

Or:

"Because turbine blade processing cost and primary growth angle misorientation are interrelated, parametric studies oriented to better define the primary growth angle processing influences are underway at the space flight center."

Or:

"The dearth of actual new technology planned for 1982 introductions may, in fact, be the result of past reductions in R&D expenditures, as evidenced by a now emptying pipeline of material and fabrication innovations."

Or:

"A method has indeed been discovered which, though particularly inexpensive, allows very high strength of the bond to be achieved."

In the trial-and-error process of writing, the writer may have to take several cracks at a sentence to discover and express what he has to say. The following example, titled "Keep It Simple," is appropriate:

> Some years ago, a friend of mine who is a plumber made a discovery. He found that in cleaning out clogged drains, hydrochloric acid did a mighty good job. He wrote the good news to the National Bureau of Standards in Washington, told them what he had discovered.
>
> The bureau wrote back: "The efficacy of hydrochloric acid is indisputable, but the corrosive residue is incompatible with metallic permanence."
>
> My plumber friend wrote the bureau that he was glad they agreed with him and he immediately received this reply: "We cannot assume responsibility for the production of toxic and noxic residue with hydrochloric acid and we suggest you use an alternate procedure."
>
> My plumber friend was happy to receive the second compliment, as he put it, and to know that the bureau again agreed with his idea. He wrote them his thanks.
>
> The next letter he received from the bureau was as follows:
>
> "Don't use hydrochloric acid, it eats Hell out of the pipes."

Sophisms: Using Language To Sell

As stated previously, generalities and claims are used to sell. My complaint is that the writer stops too soon. He fails to document. Sophism is a variation of the generality-claim practice. The writer again has something to sell, overtly or covertly. He starts with a generality or claim, such as the familiar "better, faster, cheaper" theme, which is subsequently treated as fact by the writer who uses it, in a sense, to fabricate documentation. In some contexts, the same technique is used to predict or speculate. Readers of technical literature are not conned by such practices; they are skeptical by nature. From the viewpoint of the technical writer, I have two reservations: if your mission is to sell, you may fail because the technical reader is quick to disenchant. Of equal importance, the value or quality of the information (documentation) is often suspect.

Here is an example of a generality-premise:

"The high cost of even low-alloy steel such as SAE 8620 is going to encourage the use of more weldments."

What one must keep in mind is that the writer (or his employer) in this instance is selling welding.

In the next example, the premise is: greater usage of microprocessors in heat treating will result in greatly improved reproducibility and quality. Documentation is built on the premise.

Example:

"Use of microprocessors for in-process control will continue to catch on rapidly. As a result, reproducibility and quality in heat treating will be greatly improved.

"This expanded use of microprocessors will permit production use of new and innovative metallurgical processes, or those that previously could be handled only in a laboratory, using skilled personnel."

Another common form of sophistry is used to extol the virtues of a product. Here are several generality-premise examples.

Example:

"The improved material cleanliness will enable all mechanical property and quality requirements, including ultrasonic examination, to be met."

Example:

"ABC Corporation has developed a flux which has most of the excellent characteristics of both types of products."

Example:

"We can eliminate most of the problems that have led welding engineers away from sub-arc as a stainless steel joining process."

Example:

"It is the leading electrode in the fabrication of North Sea offshore drilling platform components."

Example:

"The seamless flux cored wires offered by ABC Corporation deliver the lowest all weld hydrogen content in the industry."

Example:

"The advantage of relatively thin, and formable, coatings which give good production makes the (trade name) product desirable in many applications."

Example:

"There is no other alloy available which meets the design requirements for this key component."

Example:

"Product X is superior to conventional galvanizing alloys in corrosion resistance, paintability, ductility, weldability, formability, and provides better edge and scratch protection than Product Y."

Example:

"The new materials offer superior engineering properties and are radically different from conventional zinc die casting alloys."

Example:

"Many foundries are specifying the easy-to-cast alloy to replace cast iron and malleable iron to save on secondary machining and to eliminate distortion and finishing problems."

Sophism is common in speeches. In the following example, the speaker sets up "scientific management" as a miracle drug and proceeds to perform a number of minor miracles. Here is a sampling.

Example:

"Scientific management is the touchstone that can bring great benefits to the underdeveloped nations in the world . . .

"It is evident that without scientific management neither technological progress nor investment of capital can cause the potentially attainable maximum increase in output . . . thus greater benefits both of a strictly economic nature and of a general social character that could be yielded by such a maximum fail to materialize.

"If it is true that scientific management is able to increase the productivity of both labor and capital by making possible the best combination of those two factors, and if it is true that the increase of profit thus obtained cooperates to reinforce the economic position . . . it is also true then that the logical consequences of the thus determined processes create the bases for a further expansion of employment for the increase of earnings. . . ."

And so on.

A Summing up: From Generality to Sophism

What we are talking about is the use of language to deliver a message. In a variety of ways, the common practices discussed limit the capacity of the writer to communicate effectively.

Generalities and claims deliver only part of the message and typically raise questions that are left unanswered or call for explanation that is not given.

The writer who relies on standard language has a 1-foot vocabulary for a 10-foot job.

It is assumed that clichés and buzzwords say more than they actually do. Like generalities-claims, they lack the specifics sought by readers of technical literature.

Stilted language is as inappropriate as an umbrella in a blizzard. It is typically accompanied by verbosity and truisms. Verbosity clogs up the communication pipeline. Truisms have no value to the reader.

Using sophisms is a form of word con game.

Exercises: Some Do's and Some Don'ts

Exercise No. 1. Spotting generalities-claims. Look for statements that raise questions or require explanation.

Example:

"A unique bond between art and industry was recently established in Milwaukee.

"A Milwaukee Company and a local sculptor teamed to produce a bronze sculpture for the new headquarters of ABC Mutual Life Insurance Co. in downtown Milwaukee."

My answer—The first sentence, a generality, begs for explanation. What is the "unique bond"? The answer is given in the next sentence.

Exercise No. 2. More on hunting generalities. The example is a verbatim comment from an early astronaut. Underline generalities wherever you feel the speaker raises questions but leaves them unanswered or does not explain where explanation is in order.

Example:

"The view out of the window was unbelievable. You can't take your eyes away from that window for the first few seconds of weightless flying. It's incredible. There aren't enough words in the English language to describe the beauty. I felt this during the last part of powered flight. I was supposed to monitor the inertial guidance system's performance, but it's really a chore to get your . . . as soon as the spacecraft pitches over and you ride on the horizon. It's just a tremendous effort to get your head back in the cockpit and look at those instruments. I think it's the sort of thing that one is really fortunate to get to be able to do. I was impressed."

My answers—A beautiful example of the limitations of expressing oneself with generalizations. The writer is frustrated. The reader is shortchanged.

Start with "unbelievable." Documentation is limited to "can't take your eyes away from that window for the first few seconds of weightless flying." The best the speaker can do at the moment is another generality, "incredible." Then he cops out with "there aren't enough words in the English language to describe the beauty." He follows with three tries at describing his experience, and winds up with another copout, "I think it's the sort of thing that one is really fortunate to get to be able to do," and concludes with another generality, "impressed." Both generalities and standard language are dead-end streets. In fact, use of generalities and standard language is a sign to the experienced writer that he is in trouble. He does not have command of his subject. He is groping for what he has to say and wants to say. Time permitting, one device that often works is to leave the writing for something else for a day or so. When you come back fresh you may find that your vision has cleared. The words, the specifics, come easily. Another technique is to ask yourself, "What do I mean by 'incredible', by 'unbelievable'?" Once you get down to specifics, it may take several sentences, several paragraphs, even several pages to spell out and fully document "incredible" or "unbelievable" —the sort of thing the reader is looking for. He does not object to length as long as you are delivering. At the least, if he is not interested, he can stop reading with the assurance that he has a pretty good idea of what you are writing about and he isn't missing anything of value to him.

Exercise No. 3. Identify the general statement that is a question, then the answer given.

Example:

"Introduction hardening was deemed not suitable for tank track parts (called end connectors and center guides) subject to abrasive wear. The shapes have complex geometries which, for heat treating, are complicated by a slotted hole."

My answer—The second sentence answers why "induction hardening was deemed not suitable."

Exercise No. 4. How far one goes with documentation varies with circumstances. Length may be determined by how many specifics are available, a writer's judgment as to how far to go, or, as often happens, documentation may raise new questions, as in the example below. Identify the general statement and the documentation.

Example:

"ABC Corporation also foreseees increasing importance of surface treatments, citing boronizing and ion nitriding because they efficiently use energy and materials."

My answer—The first question, "What are the surface treatments?", is answered by the two examples "boronizing and ion nitriding." What commends them? "Because they efficiently use energy and materials."

In this instance, the writer continues to document "efficient use of energy in his next paragraph:

"One ion nitriding system uses an insulated hot wall vacuum chamber instead of cold wall design; it also features forced gas quenching, pumpdown in less than 15 minutes, and a dedicated microprocessor based controller. The strip chart recorder, for example, provides a record of temperature, pressure, power, process mode, and alarms."

Exercise No. 5. Advocates, including those with something to sell, often do not bother the reader with documentation of claims. Identify the claims in the example.

Example:

"ABC Company has developed a welding flux which has most of the excellent characteristics of fluxes that give excellent performance or those of fluxes suitable for multipass welding. The product is relatively inactive and will not add excessive amounts of manganese to the weld deposit under the full practical range of welding parameters."

My answer—Several unanswered questions, starting with the first sentence. Explain "most." What are the "excellent characteristics" referred to? What are the comparative products "that give excellent performance"? Name those "suitable for multipass welding." What is meant by "relatively inactive"? What is the significance of not "adding excessive amounts of manganese to the weld deposit"? What are the "welding parameters"?

Exercise No. 6. Users of claims typically say all the nice things they can about a product, process, or service. Why is this example different?

Example:

"Bearing research conducted by the International Organization shows that Alloy A provides a lower coefficient of friction, superior load-bearing capabilities, and greater wear resistance than Alloy B. However, Alloy A must be employed where bearing temperatures will not exceed 250 F."

My answer—What's unusual is that the author makes a comparison with another alloy in the first sentence instead of stopping with "lower coefficient of friction, superior load-bearing capabilities, and greater wear resistance." In addition, a disadvantage is cited in the second sentence.

Exercise No. 7. One more set of claims. Underline the words.

Example:

"The method has the advantage of being able to coat any surface after suitable activation, and enables any thickness of coating to be obtained simply by prolonging the process."

My answer—"Coat any surface"? "After suitable activation"? "Any thickness"? "By prolonging the process"? Explain.

Exercise No. 8. Identify the standard language in the following example.

Example:

"Today many steel plate specifications utilize higher strength carbon steels. These material changes have permitted weight savings in several applications.

"Emphasis on greater strength has been accompanied by greater concern for notch toughness. In order to meet the demands of both higher strength and improved resistance to impact failure, steel plate producers will offer heat treated carbon steels in most cases. This heat treatment, of course, increases the steel plate price."

My answer—"Utilize," "in order to," "improved," "offer," and "increases" are standard language. Aside from a lack of originality, it is a first-draft language that limits the writer's ability to express himself. Finding alternatives is not a solution. The writer should rethink what he is trying to say and take another crack at expressing himself—as often as necessary to become reasonably satisfied.

Example No. 8. A cliché may be interpreted as an attempt by the writer to dress up his act. The motive is worthwhile, but the method is questionable. Spot the cliché in this example.

Example:

"While many new products are being offered, it is worth noting that many of the workhorses of previous years are being continually improved."

My answer—The trouble with words like "workhorses" is that they are generalities. Unless the writer continues by putting some sort of cart behind the horses, he does little for the reader.

In the example cited, the writer did follow up with specifics. He wrote, in part:

"A case in point is A387-22, a chromium-molybdenum steel for high-temperature service. To provide better resistance to failure under a variety of conditions involving both low and high temperatures, A387-22 has been upgraded by reducing the residual content and lowering the sulfur level."

If you have the specifics, you may find that the cliché is not necessary. Of course, there is no law against clichés. I suggest restraint and warn that clichés can make the writer look pretty ridiculous, as in the following example.

The writer is talking about a company being pursued by a predator looking for acquisitions. He writes, "Being branded a target for takeover connotes to many a falling star in need of a facelift." The dual cliché adds up to a mixed metaphor, instead of being a clever turn of phrase.

CHAPTER 5

Economical Use of Language

In this chapter we talk about ways of locating excessive wordiness and needless repetition along with ways of repairing the damage.

Both functions are built into these guidelines:

Guideline No. 1—Look for shortcuts.

Say you feel too many words are used to say something, as in this standard phrase:

"In spite of the fact that . . ."

Is there a shorter way of saying the same thing?

Think about what is being said.

In this instance, one word, "although," can take the place of six.

To do this sort of thing, one must weigh each word, each phrase, each sentence, and study the relationships of words, phrases, sentences.

Guideline No. 2—Remember what you have said.

What to look for here is a form of repetition: saying the same thing two different ways and perhaps not being aware of it, as in:

"The machine will inspect small and minute parts."

Repetition of this sort can be in the same phrase, the same sentence, or succeeding sentences. Upon inspection, it should be obvious that "small" and "minute" say the same thing. Better still, substitute a specific, such as "parts up to 100 kg or 100 mm long."

Guideline No. 3—Avoid unnecessary explanation, such as using words to define a well-known word, as in:

"Combination washer-dryer."

Needless repetition again. "Combination" should be exorcised.

Unnecessary explanation is one of the weed patches of technical writing. There is, as we shall see, a bumper crop of hearty species.

Guideline No. 4—Avoid indirect expression that promotes unnecessary length, as in:

"Tin acts as a barrier."

This is a form of wordiness. If you think about the function of the standard verb phrase "acts as" you see that "is," for example, says the same thing.

Guideline 5—Don't use words that can be eliminated without loss of meaning, as in:

"All bolts within the range of the machine can be held to close tolerances."

Examine "within the range of the machine." There is no reason to say this. No function is served. "All bolts can be held to close tolerances on this machine" is enough. It's possible, depending on context, that "on this machine" can also be erased.

Guideline 6—Now is the time, unless otherwise stated, as in:

"Today it is estimated."

A form of verbosity. Unless otherwise indicated, we write in the present tense. Now. This is indicated by verbs like "is" and "are." Words like "now" and "today" aren't needed. In some instances, "nowness" or "isness" is implied, as in "universal practice at present." In this instance, "at present" won't be missed.

Guideline No. 7—Take advantage of context, as in:

"We have completed our annual marketing analysis, the results of which are encouraging.

"According to the analysis, steel shipment levels for 1982 should be 20% higher than the rate in the last half of 1981."

A form of repetition. Check out "according to the analysis," starting the second paragraph. The phrase could be dropped. In fact, the transition would be smoother than it is when "the analysis" is repeated from the first paragraph. At this point, it is established that the subject of the piece is "our annual marketing analysis."

Let's back up for a longer look at each guideline.

Guideline No. 1: Look for Shortcuts

Verbosity is found within phrases, within sentences, within succeeding sentences.

More phrases. Note they are usually a part of the rubber-stamp language mentioned before.

Example:

"In excess of."

Think about it. "Over" says the same thing.

Example:

"For the reason that."

Try a word like "because."

Example:

"Within the realm of possibility."

How about "possible"?

Example:

"Is in a position to supply."

If you want to eliminate the first five words, you can substitute "can."

Example:

"An actual fact."

If there are no "nonactual facts," one could write "a fact" without loss of meaning.

Example:

"Is definitely probably."

The second word is suspect. The survivors say it.

Here are some ways the same thing happens within a sentence.

Example:

"Kaiser bid for and successfully built the plant."

The second, third, fourth, and fifth words aren't needed. "Bid for" is a detail that is understood. If Kaiser had not won the bidding competition, it would not have built the plant. "Successful" is a parasitic standard word often attached to another word when it isn't needed. Same reasoning as "bid for." A plant built is successfully built. If not, give the particulars. My recommendation: try "Kaiser built the plant."

Example:

"It was used to process the parts discussed in this article."

Study each word and the whole. The idea is to find a shorter way of doing the same thing.

Think about "was used to process."

How about this slight rewrite: "It processed the parts discussed in this article."

Example:

"The machine is fast, operating at a rate of 60 strokes per second."

A combination of a generality, "fast," and a wordy phrase, "operating at a rate of 60 strokes per second." Try a little rewrite: "The machine runs 60 strokes per second." Note that in this instance we have a generality and specifics. Two ways of saying the same thing. Always go for the specifics.

Even more hay can be made if one looks at two or more succeeding sentences. The challenge is to consolidate all the information into a single sentence.

Example:

"The company will buy equipment for its Windsor plant. The new machinery will be used to produce component parts."

If you add up the two sentences, you find the writer is trying to say something like: "The company will buy new parts-making equipment for its Windsor plant." One sentence replaces two. Eleven words instead of 19. The wordiness indicates first-draft thinking. The combination "component parts" says the same thing twice. "Component" and "part" are one and the same. It's like saying "part part."

Example:

"The company will build a new plant in June. The new plant will cost $200 million."

As the two sentences now stand, the poor old busy reader is forced to contend with 16 words to get a tad of information. A writer on the prowl for shortcuts will come up with a replacement of this ilk: "The company will start building a $200 million plant in June." Eleven words. Redundancies like "build a new plant" are common. "Build a plant" is enough. But in this instance, "new plant" is repeated in the second sentence. You can't build an "old plant." You could refurbish or restore one.

Example:

"Highway mesh has been moving slowly in New England this year. The same is true of building mesh."

To the writer tuned to economical use of language, there is no reason for two sentences to convey this message. You can do it with, "Highway and building mesh have been moving slowly in New England this year." Further, it is probably clear from context that the time is now, "this year." The phrase, "has been moving slowly," takes care of the time element.

Example:

"Professionals who sold wheat short earlier in the year must enter the market now to cover. The price of wheat is a minor consideration to them as they must get sufficient quantities to cover commitments."

The second sentence repeats the first and does not add much. Why not extract what new information the second conveys and consolidate it with the first sentence? We get:

"Price is no consideration to professionals who sold wheat earlier in the year and must enter the market to cover."

The two sentences contain 35 words. The same meaning is conveyed with 21.

Example:

"A new grade of resulfurized free machining steel was introduced to the marketplace in 1980. Extensive field trials verify that the new steel is equivalent to premium free machining steels such as those containing lead and tellurium. The product contains bismuth, which enhances machinability in the same way as lead. Melting during machining provides lubrication and encourages embrittlement and void formation, thus leading to easier chip formation."

A realistic example containing 69 words. Look for wordiness and unnecessary repetition, word for word, sentence for sentence. Potentials for saving are indicated by the following rough rewrite. Start with the second sentence:

"Extensive field trials of a resulfurized free machining steel containing bismuth introduced in 1980 verify it is equivalent to premium grades containing lead and tellurium. Bismuth enhances machinability in the same way as lead. Melting during machining provides lubrication and encourages embrittlement and void formation, leading to easier chip formation."

Fifty words instead of 69. The first sentence contains information taken from the first three. It isn't necessary to say "new" if you also have "introduced to the marketplace in 1980." "Thus" is a standard connector that isn't needed.

One more example:

"The parts are required to meet stringent mechanical and metallurgical properties to insure structural integrity. Heat treatment plays a significant and vital role in achieving the required properties. It is essential that all thermal operations carried out be performed under precisely controlled conditions to avoid property deficiencies. Rigid process controls are normally instituted to insure conformity with specified requirements. Temperature and furnace atmosphere are commonly monitored on a continuous and automatic basis."

An example that is both wordy and repetitious. In this instance, I cannot compact the five sentences into one. One procedure is to look at the whole first. You may be able to do more than you would with the one-sentence-at-a-time approach. For one thing, it is evident from key words like "stringent properties," "structural integrity," "heat treatment," and "close control" that the first four sentences have a lot in common. They are related to the same concept, suggesting that the meat of the four sentences can be condensed into one. For example:

"To get the mechanical and metallurgical properties needed to insure the structural integrity of these parts, they must be heat treated under precisely controlled conditions."

The original, to this point, contains 59 words. The rewrite says the same thing in 25. It's also easier to read. Less tedious. Less boring.

The last sentence, starting with "Temperature and . . ." can be shortened. Anytime you see the standard word "basis," you can be pretty sure of finding an indirect expression. One way to handle this sentence:

"Temperature and furnace atmosphere are commonly monitored continuously and automatically."

It could be argued that "continuously" and "automatically" are tantamount to the same thing. In this case, I would opt for "automatically." My argument: "automatic" implies "continuously." However, "continuously" could be handled manually.

I should point out that in consolidating the first four sentences, I passed up other opportunities to prune. If you took the sentence-by-sentence tack, for example, you could redo sentences two and three in this manner:

"Heat treatment plays a vital role in meeting these requirements. All thermal operations must be carried out under precisely controlled conditions."

Both "significant and vital" are not needed in the first sentence. "Meeting these requirements" avoids unnecessary repetition of "properties" and the standard "achieve." Taking advantage of what was said in the second sentence, it is possible to eliminate "It is essential that" from the start of the third sentence. "Must be carried out under precisely controlled conditions" replaces the standard and redundant "be performed." Finally, "avoid property deficiencies" isn't necessary if we take advantage of context.

Guideline No. 2—Remember What You Have Said

Needless repetition is the big game we are hunting. The species in question can be flushed from word combinations sentences, and succeeding sentences.

Example of a word combination:

"Repatriated back to his native land."

If you know what "repatriated" means, you don't need the words that follow.

Example:

"Distorted out of shape."

Are the last three words necessary?
Example:
"Camaraderie and team spirit."
Either one will do.
Example:
"Bellwether sign."
See how Webster defines "bellwether."
Example:
"General, across-the-board price increase."
I question the need for the first word.
Example:
"Some slight evidence."
Either "some" or "slight" but not both.
Example:
"Invisible to the naked eye."
No kidding.
Example:
"Smorgasbord mix."
Try "smorgasbord."
Example:
"Narrow minded, insular attitude."
Either "narrow minded" or "insular." One is enough.
Example:
"Foreign imports from overseas."

Unless reference is to imports from this hemisphere, you can get by with "imports" alone. "Foreign" is normally coupled to "imports" as a standard combination. It is never necessary.

You have to pay closer attention to catch needless repetition of this type within a sentence.

Example:

"To protect the surface of finished and semifinished steel parts from decarburization, carburization, and intergranular oxidation during heat treatment, protective environments are used when heat treating steel parts above 1400 F."

In this instance, the author had two lapses of memory. He said "steel parts" and "heat treatment" at both the front and back of the sentence. Starting with "when," seventh word from the end of the sentence, I would substitute, "when temperatures are above 1400 F."

One more example:

"The problem was in a previous operation at a prior station."

To me, "previous operation" and "prior station" come close to being one and the same. I would drop "in a prior operation." To me, "prior station" is the more definitive of the two.

The big prizes are won when succeeding sentences are put under surveillance.

Example:

"Surface hardness of gears made from rare earth boron steel is comparable to that obtained with the present standard material (chromium steel) for the part. After carburization, quenching, and annealing, the hardness number is around Rockwell C 59-62 in both instances. We therefore can meet the desired requirement."

The last sentence is not needed. The author has already explained.that the hardness values ("the desired requirement") of the two steels are about the same.

You often encounter this type of repetition in commercial material, as in:

"ABC Corporation has announced the introduction of a new high-impact technology for its entire line of TBP material. The new TBP high-impact technology—TCV—adds superior toughness to the other proven properties of ABC resin."

Note "new high-impact technology" in both sentences. One possibility:

"ABC Corporation announces TCV, a new high-impact technology for its entire line of TBP material, which adds superior toughness to the other proven properties of ABC resin."

I don't care for "has announced the introduction of." I feel that "announces" does it. I left in some commercialism, "superior" and "proven properties."

Guideline No. 3—Avoid Unnecessary Explanation

The term MIL indicates military specification. If I write "military specification MIL-H6875" I am in effect writing, "MIL-H6875 MIL-H6875." Obviously, there is no need to do this for clarity. Yet unnecessary explanation of this type is epidemic. Sometimes it appears that writers are overly concerned about making their readers understand. Sometimes it appears that writers are seeking originality. Whatever the motive, the end result is an infraction of the busy reader style of writing. The writer uses more words, at least 100% more, than he needs to say something. "Retrospection" becomes "historical retrospection," or we write "cool to a temperature of 700 F" when "cool to 700 F" is sufficient.

From my viewpoint, the biggest challenge is in demonstrating to you that the practice is ubiquitous. Once you are put on the right wavelength, spotting the bloopers becomes fairly automatic. The remedy is implicit in the concept: say it only once.

Example:

"Converting swords into plowshares for peaceful purposes."

Think about the meaning of "swords into plowshares." There is no reason why the cliché should be explained with "for peaceful purposes." To be snotty about it, the writer assumes that the reader isn't any smarter than he is. In any event, too much of this doesn't make a good impression on the reader; and remember, the problem is widespread in literature he must read as part of the job.

Example:

"Doomed to failure."

Once more, what does "doomed" mean? "To failure" is a gratuitous explanation.

Example:

"Guinea pig will act as a proving ground."

No need to spell out the "guinea pig" cliché.

Example:

"It is factually correct."

If you are "correct," it has to be "factually."

Example:

"ABC Company has eliminated its pricing premiums which had prevailed."

Unless you have a guideline like "say it only once," it probably would not occur to you that "which had prevailed" is not needed.

Example:

"The jet was going in a downward direction when it passed us."

With a little mental detective work, it is apparent that "downward" is a specific word and "direction" is a general word. In this instance, a slight rewrite is called for, as: "The jet was going downward when it passed us."

Example:

"A special patented device."

Writing "special patented" is like writing "patented patented."

Example:

"Used as a fulcrum for leverage."

If you know what "fulcrum" means, "leverage" is redundant.

Example:

"The wheel is made of hundreds of pieces of cloth-coated abrasives factory formed into a wheel."

Repetition of "wheel" is the tipoff. You could drop "factory formed into a wheel." If upon investigation it is found that "factory formed" should be retained, you could do something like this: "Hundreds of pieces of cloth-coated abrasives are factory formed into a wheel." As far as this example is concerned, the saving in words isn't dramatic. But remember we are talking about a discipline. If you take your finger out of the hole in the dike, the result is a flood of useless words.

Example:

"Frustrating dilemma."

A "dilemma" is "frustrating."

Example:

"In the month of June."

Eliminate the first four words.

Example:

"Current status."

The second word makes the first unnecessary.

Example:

"Successful antidote."

If it is not "successful," it is not an "antidote."
Example:
"Currently pending."
Try "pending" alone.
Example:
"Past history."
The culprit is "past." "Past record" and "past experience" are also used in this manner.
Example:
"Still remains."
If you write "remains," the word "still" isn't needed. You also see phrases like "still pending."
Example:
"Additions to the existing plant."
A tricky one involving a word, "existing," seen in many places where it does not belong. We write "existing technology" or "existing capital goods" or "existing knowledge." Because this is standard language, it is not often questioned. In the example, think about the concept behind "additions." It would not be possible to add on to something that is not "existing."
Example:
"We are in a business recession."
Is "business" necessary?
Example:
"Brief résumé."
You see this often. Perhaps the first word is added with the best of intentions.
Example:
"This allows freedom for future expansion."
The gist of the sentence puts "expansion" into the future, making the word "future" unnecessary—the "future" seems to encourage redundancy. You see, for example, "we anticipate greater success in the future." I don't think that "in the future" is needed because "anticipate" takes care of this. In "labor's goals for the future," I feel that the last three words can be dropped. In "It may evolve into a separate industry in the future," I would not use "in the future." The phrase "may evolve" indicates "future," which is any time beyond now, including the second after the one we are in.
One more example of doubling up:
"Light gray in color."
Drop the last two words. "Light gray" is a color.

Guideline No. 4—Avoid Indirect Expression That Promotes Unnecessary Length

You frequently see standard phrases like "in the form of," "acts as a," "in case of," and "in addition to." They are signs of indirectness of expression, meaning that excess language is present.

We write a phrase like "in the form of plates." This is a long and indirect way of saying "plates."

Ditto "in case of emergency." Let "emergency" do it alone.

You see sentences like "In addition to sheets, most specials are also sold." Drop "in addition to," and write something like "Sheets and most specials are sold." Note that I also dropped "also." It is not needed. The rewrite is a more direct statement than the original.

Another example: "Medium-sized plants, in many cases, have backlogs." I would do something like this, "Many medium-sized plants have backlogs."

In the next chapter we will talk about sentence construction practices that encourage indirect expression and use of words that are not needed.

Guideline No. 5—Don't Use Words That Can Be Eliminated Without Loss of Meaning

What I have in mind here are practices typically involving qualifications, comparisons, and descriptions.

Example involving a qualification:

"The total cost of the 11 wage demands would be 48 cents per hour, if granted."

I can't justify "if granted" because the subjunctive "would be" takes care of this qualification.

Example of a comparison:

"That was an increase of 9.2% over 1962 shipments."

I feel that "over" makes "increase" unnecessary. I suggest, "That was 9.2% over 1962 shipments."

Example of a description:

"These telescopes are designed for the inspection of inaccessible surfaces."

I question the need for "designed." Drop it and leave the rest of the sentence as is.

One more example of qualification:

"The defect does not influence strength in any way."

I don't see how "in any way" makes the sentence more precise than "does not influence strength."

Guideline No. 6—Now Is the Time Unless Otherwise Stated

All too often, in my opinion, technical writers waste the reader's time by reminding him what the time is, as in:

"It is now estimated."

The verb "is" indicates present tense. Adding "now" is the same as writing "is is."

You see this sort of thing in many places.

Example:

"Is in general use today."

I doubt the need for "today." Remember what "is" does for us.

Example:

"We are now about to buy a new lathe."

Get rid of "now."

Example:

"It is universal practice at present."

With "is" you do not need "at present."

Example:

"In 1 or 2 seconds time."

A variation on the theme. "Time" is superfluous. In this instance, we have specifics, "1 or 2 seconds," and a general term, "time." The latter can always be dropped without loss of meaning.

Example:

"May and June of this year."

The last three words are not needed if you mean "this year."

Guideline No. 7—Take Advantage of Context

Each subject has certain key words. Once they are established in text, it is possible for the writer to take advantage of context to avoid unnecessary repetition—a practice which adds unnecessary length and promotes a tedious, humdrum style of writing.

In the piece I am looking at, "laboratory" and "test" are key words. The following example is from the middle of page 4:

"The *laboratory* has eight hot wall *test* units, 12 Cortest units, 12 wick *test* units, two slow strain rate *test* units, and two corrosion fatigue *testing* units."

To this point in the text, the word "laboratory" has been used seven times— several times without purpose—and "test" 41 times.

In the example, the key words are italicized. I would substitute a word like "facility" for "laboratory" to cut down on the number of times the word is used.

I would let only the first "test" stand. If you take advantage of the context of this sentence, the others aren't necessary. Note that in one instance, "Cortest," use of a trade name with the word "test" in it removes the need for "test" here, but I have seen examples like "Cortest test."

My rewrite:

"The facility has eight hot wall test units, 12 Cortest, 12 wick, two slow strain rate, and two corrosion fatigue."

Another example from the same source, starting at the bottom of page 4 and continuing onto the next page:

"The *corrosion testing* program at the *laboratory* focuses on four basic forms of *corrosion*:

"Uniform *corrosion.*

"Intergranular *corrosion.*

"Pitting and crevice *corrosion.*

"Galvanic *corrosion.*"

Note that in addition to "testing" and "laboratory" I have italicized "corrosion."

In evaluating this example, we should keep in mind that all of the work done by
this laboratory concerns corrosion testing. It is a corrosion testing laboratory.

Now for the first sentence. Is it possible to eliminate at least one "corrosion"?
What about "testing" and "laboratory"?

By now, at the end of the fourth page, it has been well established that what is
being written about is a corrosion testing laboratory. For this reason, I would not
hesitate to amend the first sentence as follows:

"Program focuses on four basic forms of corrosion."

From context, it is clear that we are talking about "corrosion testing at the
laboratory." The "corrosion" at the end of the sentence is necessary. Working off
it, however, removes the need for subsequent mentions of "corrosion."

I suggest:

"Program focuses on four basic forms of corrosion: uniform, intergranular,
pitting and crevice, and galvanic."

How much of a problem is this form of repetition? In the piece in question,
which covers 12¾ pages, typewritten and double spaced, there are 128 mentions
of "laboratory" and "test." The high is 16 on one page; the low is six. I feel that at
least 50% of the duplication can be eliminated.

Unnecessary repetition of key words is particularly blatant in commercial
pieces. In a 1½ page news release, the key words are "ABC Control Valve" and
"valve." The former is repeated three times, the latter 16 times.

In a short paragraph near the bottom of the second page, the company stated:

"ABC Company is a worldwide manufacturer of process control *valves*. The
new *ABC Control Valve is ABC's* latest addition to its line of *valve* products,
which includes its flowing control *valve* and cage type offerings."

There is special motivation to repeat company and product names where selling is involved. I use the example to point this out and to indicate where repetition can be avoided if there is a desire to do so.

My rewrite:

"ABC Company is a worldwide manufacturer of process control valves. The
new product is the latest addition to its line, which includes flowing control and
cage types."

An example in which the writer avoids unnecessary repetition:

"Nickel-base foils, another product of rapid solidification, are replacing gold-
nickel brazing foils. High temperature, mechanical, and metallurgical properties
are said to be comparable to those obtained with gold-nickel."

The writer let context work for him and did not repeat "nickel-base foils" in the
second sentence. Yet there could be no doubt that he was comparing "nickel-
base" and "gold-base."

In summary, here are the points to keep in mind.:

Guideline No. 1: If you run into wordiness and unnecessary repetition, look for
shortcuts, particularly for ways to consolidate two or more sentences into one.

Guideline No. 2: If you are saying the same thing twice in difference ways, eliminate one. If one word or phrase is general and the other specific, always go for the latter.

Guideline No. 3: Avoid explaining what is well known.

Guideline No. 4: Some phrases promote indirect expression. Think about what is being said, and find a more direct route to your message.

Guideline No. 5: If a word or phrase doesn't help you or the reader in any conceivable way, eliminate.

Guideline No. 6: Don't give the "time" if you don't have to.

Guideline No. 7: Once you have established your subject and its key words, take it easy on repeating the key words. Take advantage of context where you can and eliminate.

Exercises

Exercise No. 1. In this example, we see combinations of general and specific terms along with some repetition. See if you can identify the problems and find a shortcut.

Example:

"Growth in net income from continuing operations was reduced by the severe effects of currency fluctuations in 1981. For the year, the impact reduced growth from 43% to the reported increase of 8%. The negative effect on growth is equivalent to approximately $90 million."

My answer—Analyzing word for word, I note "severe effects" first. A general term and often a sign of first-draft thinking. At the end of this sentence, we have "1981." This means "for the year" isn't needed at the front of the next sentence. We can take advantage of context. The rest of the sentence gives the specifics behind "severe effects." The phrase "negative effect" in the next sentence is in the same ball park as "severe effects." Specifics at the end of the sentence spell out "severe effects."

One way to consolidate the information:

"Currency fluctuations reduced growth in net income from continuing operations from 43% in 1980 to 8% in 1981. The loss is equivalent to about $90 million."

Exercise No. 2. A similar example in which a similar patching technique is used. How would you condense the following?

Example:

"The first tape controlled transfer line for engine blocks has been installed. The line contains 16 machines and is designed to handle five different types of blocks."

My answer:

"The first tape controlled transfer line to be installed handles five different types of engine blocks and contains 16 machines."

Looking at the two sentences together, the first one is general and the second contains specifics—"16 machines" and "five different types of blocks." I merely tried to see if I could transfer the specifics to the first sentence. Little rewriting was needed.

Exercise No. 3. In this instance, the writer repeated what he had just said in a different way. Identify the unnecessary repetition.
Example:
"It was announced that the meeting was voluntary, not mandatory."
My answer—"not mandatory" repeats "voluntary."

Exercise No. 4. Two more examples similar to that in No. 3. What are the redundancies?
Example:
"He was extremely candid in his convictions."
"We need cheaper Space Age materials at less cost."
My answer—In the first example, "in his convictions" repeats "candid." In the second example, "cheaper" makes "at less cost" unnecessary.

Exercise No. 5. Explaining or defining a word when it can be assumed it is known is another form of redundancy. See if you can find the culprits here.
Examples:
"Forewarns of trouble."
"Regaining lost confidence."
"Due to a lack of business in sufficient volume."
My answer—No. 1, "of trouble" should be dropped. No. 2, "lost" is superfluous. No. 3, "lack of business" need not be explained with "in sufficient volume."

Exercise No. 6. Time is usually indicated or implied by verb tense. Which word isn't needed in the following example?
Example:
Is presently being used."
My answer—"Presently." Time is also indicated unnecessarily in the following sense: "In a few month's time." The word "time" is not required here.

Exercise No. 7. Indirect expression promotes verbosity. Take a crack at underlining the wordy parts of this statement.
Example:
"On the basis of rumors spread by speculators, it now seems pretty certain that the price of metal will increase in response to the stimulus in the near term future."
My answer—These phrases are signs of verbosity: "on the basis," "in response to the stimulus" and "in the near term future." Some rewriting is needed. Think about what is being said.
I suggest something like:
"Speculator rumors will probably push up the price of metal soon."

Exercise No. 8. In this example, see if you can shorten the second sentence by taking advantage of context—what was said in the first sentence.

Example:

"Freight traffic on major U.S. railroads during the week ending January 16 was 20.2% below that the corresponding week a year ago. This sharp decline was due to severe weather conditions in many sectors of the country which affected production of goods that move primarily by rail."

My answer—In the first sentence, we say shipments are down 20.2%, so "this sharp decline was due to" is repetition. The word "conditions" is not needed after "severe weather." The word "sectors" is a stilted way of saying "parts," and the phrase, "affected production," is a sign of wordiness.

I suggest:

"Freight traffic on major U.S. railroads during the week ending January 16 was 20.2% below that the corresponding week a year ago. Severe weather in many parts of the country reduced production of goods that move primarily by rail."

Exercise No. 9. As you go through this short paragraph, underline repetition and signs of redundancy. Then think about the message being delivered and try your hand at a rewrite.

Example:

"A businessman engaged in reading is either generating gains or losses for his firm. He generates gains when his reading either satisfies his personal needs, keeps him in touch with current affairs of all types, or gives him the procedural tools to improve the economic situation of the firm. He generates losses when his reading is not successful in these areas. He is not successful either through his own fault, the fault of the writer-publisher team, or the fault of the reading and external communication program of his firm."

My answer—In the first sentence, "engaged in reading" is a long way of saying reading. In the second sentence, "generates" is repeated from the first sentence, as is "reading." The rest of this sentence is wordy. In the third sentence, "generates" is repeated, as is "reading." In the fourth, "successful" is repeated from the third and "fault" at the end of the phrase is repeated twice in the next two phrases.

I feel that more than patchword is needed. At least study what is being said and experiment with ways of condensing and being more direct.

For example:

"A businessman spends his reading time profitably when it satisfies his personal needs, keeps him in touch with current affairs, or gives him procedural tools to improve his firm's profits.

"His reading may not be profitable through his own fault, that of the writer-publisher team, or that of his firm's reading program."

Exercise No. 10. Expression is often indirect in sentences starting with "there." How would you fix this example?

Example:

"There is a great demand for skilled workers."

My answer—"Demand for skilled workers is great."

CHAPTER 6

Sentence Construction Practices Clog Communications

Topics in this chapter are piling up modifiers, upside-down sentence construction, and indirect or passive construction.

An example of piling up modifiers:

"Temperature-actuated Model 25981 fire extinguishers . . ."

The problem in this instance, which could be called overcondensation, can be solved by adding a couple of words. For example, start with "Model 25981 fire extinguishers," then add something like "which are temperature actuated." This is not an exception to the rule: don't use one more word than you need. You need more here for readability. In fact, a reader may be forced to stop, find the noun, then translate.

An example of upside-downism:

"ABC Power Cooperative, which serves 150 000 customers in central and southern Iowa, will build a $500 million, coal fired electrical generating plant in Fish County."

The news is stuck at the back of the sentence, while the source and other secondary information are placed first, in the honored spot. You could reconstruct the sentence this way:

"A $500 million, coal fired electrical generating plant will be built in Fish County by ABC Power Cooperative, which serves 150 000 customers in central and southern Iowa." Perhaps something should be done with the three compound modifiers preceding "plant" in the front of the sentence. One way: move $500 million to this position," A coal fired, electrical generating plant costing $500 million will be built . . ."

An example of indirect construction:

"A laser beam is being used to vaporize metal."

The problem is caused by the passive voice, which results in indirect statement. "A laser beam vaporizes metal" is the desired alternative. Technically, the object of the verb in the active voice becomes the subject in the passive voice. Active: practices promote indirect statement. Passive: indirect thinking is promoted by practices.

A Closer Look at Piling up Modifiers

"I am condensing to save some words" is one defense when the practice of piling up modifiers is challenged. The saving is illusory. A typical tradeoff is a 50% or more loss in readability to avoid a word like "of." We write "U.S. firms' business proposals" instead of "Business proposals of U.S. firms." The result is the clumsy genitive case.

I am concerned about two piling up techniques:

1.   A number of modifiers, of the twinned variety, are placed in front of a noun. You often see this in a description, as in: "The double acting, steam powered, double clutched, computer controlled machine."

2.   The writer goes to the genitive case, a possessive case, and piles up modifiers, supposedly to save a word or so, as in: "Ways and means committees' meeting."

You get a similar result by preceding a name with a long title and other modifiers, as in "ABC Corporation's Deputy Assistant General Counsel for Labor Relations Samuel F. B. Morse."

One solution is to put the modifiers behind the noun, in this case the man's name. For some reason, length is no longer a consideration.

For example:

"Samuel F. B. Morse, Deputy Assistant General Counsel for Labor Relations at ABC Corporation . . ."

Another fix, as mentioned before, is to move only part of the modifiers behind the noun.

For example, this can be done by changing "aluminum producers' annual reports" to "annual reports of aluminum producers."

Or convert "an International Organization of Standards Committee" to "a committee of the International Organization of Standards." Once more, length is not a problem where modifiers are in apposition to the noun.

"For" is another word that is "saved." For example, "color television shadow mats" instead of "shadow mats for color television." Or "experimental motor laminations" instead of "laminations for experimental motors."

Other examples where genitive construction causes trouble: we write "mill stored stocks" instead of "stocks stored at mills" or "the part's 168 pound weight" instead of "the part weighs 168 pounds."

The bottom line of this widespread practice is unnatural language with a readability disability.

More examples:

In place of "a National Association of Purchasing Agents Committee," try "a committee of the National Assocation of Purchasing Agents."

In place of "ABC Corporation's employee stock purchase plan," try "ABC Corp.'s stock purchase plan for employees."

In place of "oil country goods producers" try "producers of oil country goods."

In place of "the U.S. Metal Forming Association Illinois District will present," try "the Illinois District of the U.S. Metal Forming Association will present . . ."

Instead of "Model SF2600 braided ceramic fiber sleeving," try "braided ceramic fiber sleeving of Model SF2600."

The other form of piling up calls for a different solution.

For example:

"This is a constant current, midfrequency, solid-state converter. It costs about $2000."

In this instance, as in the example at the beginning of this chapter, you have to add some words or, in some cases, split one sentence into two.

You want to move one or more of the compounds away from the front of the sentence.

Here is one solution:

"This is a midfrequency, solid-state converter. The constant current unit costs about $2000."

In the following example, you could make two sentences out of one for the sake of readability.

Example:

"San Francisco's 75-mile high-speed computer controlled transit system was completed recently."

A possible fix:

"San Francisco's 75-mile transit system was recently completed. The high-speed line is computer controlled."

Example:

"A specially formulated, molybdate based inhibitor for closed cooling systems."

You could say this is a borderline case and forget about the double-double modifiers. But notice the gain in readability when one double modifier and the noun are moved to the front and one double modifier is put into apposition.

For example:

"A molybdate based inhibitor specially formulated for closed cooling systems."

Example:

"A high performance, yet cost efficient, real-time spectrum analyzer."

Eight modifying words, two doubles, one triple. The same technique used in the preceding example is suggested: move the last four words to the front and rewrite accordingly.

For example:

"A real-time spectrum analyzer combining high performance and cost efficiency."

Always remember, you are not stuck with the original words or the original sentence construction. If you merely play around with the words, you may not make much of a contribution.

A Closer Look at Upside-down Sentence Construction

Of the three practices discussed in this chapter, this one is without question the most common. Unfortunately because thinking and expression are indirect, message delivery is impeded, and the practice can add several points to the critical reader's blood pressure.

Once more the news in a sentence is buried by incidentals and other miscellany that belong in the sentence but should be riding in back instead of driving up front.

Example:

"ABC Metals Company announced today that International Metals Company Southeast Asia, a majority owned subsidiary of ABC Ltd., will reduce production at its Hong Kong plant, starting February 1."

Eighteen words are consumed to reach the last ten, which hold the news. In this case, as in others though, you usually can't blame the writer. He works for a company or client who likes or insists on having his name up front. However, I would recommend something like the following for consideration:

"Production will be reduced starting February 1 at the Hong Kong plant of International Metals Company Southeast Asia, a majority owned subsidiary of ABC Ltd., the parent company ABC Metals announced today."

For a change of pace, here is an example of right-side-up construction.

Example:

"The latest technical and marketing data to aid both businesses and municipalities in implementing recycling programs are contained in the new edition of the publication from the ABC Association, *Alumimun Recycling Handbook*."

Typically, the piece would start with something like this, "ABC Association's newest edition of its publication, the *Aluminum Recycling Handbook*, contains the latest . . ."

Here is a doubleheader from succeeding paragraphs in an announcement from a university.

Example:

"The U.S. Institute of Alchemists has honored Dr. Isaac Newton, professor emeritus at ABC University, by naming him a 1982 Alchemical Fellow."

Later, in the next paragraph:

"According to ABC University's Alchemy Department Chairman, Dr. I. N. Organic, this is the tenth award Professor Newton has received in the past eight months."

In the first instance, I suggest starting with something like: "Dr. Issac Newton has been named a 1982 Alchemical Fellow by the U.S. Institute of Alchemists, the tenth award he has received in the past eight months . . ."

Sometimes the news is pushed out of the spotlight by the names of a top officer and the company, as in: "Edward G. Robinson, chairman and chief operating officer of ABC Corporation, announced that fiscal 1981 sales, net earnings, and earnings per share were the best in the corporation's history."

Stockholders would prefer a construction like this, "Fiscal 1981 sales, net earnings, and earnings per share were the best in the history of ABC Corporation, announced Edward G. Robinson, chairman and chief operating officer.

Companies tend to follow the same practice in talking about products or services, as in: "A new line of ABC washers offers consumers two added benefits—much quieter operation and a choice of front panel inserts."

Given a free hand, the writer should concentrate on what the reader, not the company, is interested in.

For example, "Much quieter operation and a choice of front panel inserts are consumer benefits which have been added to the new line of ABC washers."

Upside-down construction is everyplace, and you can bet you will always find it where there is a report of a meeting.

A typical treatment:

"At its meeting held in Boston, ABC Board of Directors voted in a series of adjustments in the 1980 budget, originally proposed at a Finance Committee meeting in Chicago June 13–15, to provide funds for potential tax liability for 1979 and to come closer to a break-even budget for the year."

If you remember to start reading from the back of the sentence, you shouldn't have too much trouble with these babies.

For example:

"Providing funds for potential tax liability in 1979 and for a closer to break-even budget for the year was among adjustments in the 1980 budget voted in by the ABC Board of Directors at its meeting in Boston. The adjustments were originally proposed at a meeting of the Finance Committee in Chicago June 13–15."

A similar example:

"The Pacific Area Standards Group at its fourth plenary meeting (CSAP IV) unanimously endorsed a plan for the reorganization of the governing Council of the International Organization."

Start the rewrite when you reach the 15th word, as:

"A plan for the reorganization of the governing Council of the International Organization was unanimously endorsed by the Pacific Area Standards Group at its fourth plenary meeting (CSAP IV)."

You would think people wouldn't do this when they have something to promote, but they do.

For example:

"The ABC Agents National Association has scheduled its one-day regional workshop seminars for 1982 entitled, 'How to Work With Manufacturers' Agents for Profit.' "

Why not start with the title, as in:

'How to Work With Manufacturers' Agents for Profit' is the title of the one-day regional workshop seminars for 1982 scheduled by the ABC Agents International Association."

Another one of the same:

"ABC Heat Treatment Center's three-day course, 'Understanding Heat Treatment Today,' will be repeated for the umpteenth time in May."

Once more, title first, as in:

'Understanding Heat Treatment Today,' a three-day course, will be repeated for the umpteenth time in May by ABC Heat Treatment Center."

Finally, names of people in the news are also buried, as in:

"ABC Corporation has announced the appointment of Michael G. Baltimore as manager of manufacturing for its plant in Maryland."

Start with the person's name, as in:

"Michael G. Baltimore has been named manager of manufacturing for ABC Corporation's plant in Maryland, the corporation announces.

Upside-downism, it should be noted, is typically found at the front of the opening paragraph of a piece when the subject is being stated.

However, a variation should be acknowledged. In the example that follows, the portion of the sentence that normally leads off is italicized:

"Metallurgists were besought today, *in a paper at the 92nd annual meeting of the World Association*, to meet the oil industry's need for drill string steels that have better fatigue properties and higher corrosion resistance."

To get the direct style of expression I recommend, I would transplant the italicized phrase to the back of the sentence, as in:

"Metallurgists were urged to meet the oil industry's need for drill string steels with better fatigue properties and higher corrosion resistance in a paper given today at the 92nd annual meeting of the World Association."

I would go further with the rewrite if I were preparing the sentence for publication, but the mere fact that I was forced to take another look to rewrite prompted a couple of constructive changes. I substituted "urged" for "besought" because I think the latter is stilted word that is out of place. I had second thoughts about using "today." Where it was, it should be eliminated because context indicates "now." Instead, I moved "today" behind "in a paper given" because it is functional there. The piece is an excerpt from a newspaper account of the association meeting. In this instance, it seems that "today" adds urgency or nowness to the message. Lastly, I dropped "that have" in front of "better corrosion properties" in favor of "with," a shorter way of saying the same thing.

Awareness of practices promoting indirect expression tends to focus attention on directness. The resulting discipline encourages the writer to be ever on the alert for shortcuts.

Another practice related to upside-downism should be mentioned in passing. I call it a sentence parted down the middle. To make sense of what he is reading, the reader must connect what is said at the front of the sentence with what is at the back of the sentence, buried under a pile of intervening words. In the example, the key words are italicized:

"*An elastic orifice*, which automatically regulates gas flow to gas lubricated bearings, appears to be more stable and provides greater bearing surface (especially at higher loads), lower power consumption, and greater load carrying capacity per unit of power consumption *than a fixed orifice*."

I recommend something like:

"In comparison with a fixed orifice, an elastic type which automatically regulates gas flow to gas lubricated bearings appears to be more stable and provides greater bearing surfaces (especially at higher loads), lower power consumption, and greater load carrying capacity per unit of power consumption."

No sense in making the reader wait for 36 words before he is told that an elastic orifice is being compared with a fixed orifice.

A Closer Look at Indirect or Passive Construction

Because the practice is so widespread and rarely questioned, it is beneficial to get a feeling for how indirect or passive construction is being used.

Example:

"Quench severity tests were conducted on 2 and 2.50 inch diameter bars of 4340 steel to determine the effects of austenitizing temperature on midpoint hardness."

You can't repair the damage by snipping or lopping here and there. You have to get the gist of what is being said and try other more direct routes.

For example:

"Bars of 4340 steel 2 and 2.50 inches in diameter were tested for quench severity to determine the effects of austenitizing temperature on midpoint hardness."

I know of no formulas for these repairs. You take what you think is a better, more direct way. In effect, you are editing what you have written, and you run into a new problem: the original thinking, the original words, the original sentence construction exert a powerful influence on the "editor," in this case the writer, trying to fight an established practice. You must break yourself away from these influences. Be able to do the sentence over in your own thinking, own words, own sentence construction.

Example:

"They can perform such tasks as monitoring internal wall surface smoothness and gaging wall thickness and uniformity."

In examining the sentence, I see that I want to get rid of "perform," and I want to come up with a construction that permits me to replace "monitoring" with "monitor" and "gaging" with "gage." I don't want to change the meaning of the sentence, which means I must find a way to retain the thought expressed by "can perform such tasks." I used "for example."

Here is a possibility:

"They can, for example, monitor internal wall surface smoothness and gage wall thickness and uniformity."

An easier repair job:

"The five-channel tester performs accurate gaging of every inch of the high-pressure steel tubing examined."

What we want to do is drop "performs" and substitute "gages" for "gaging." That's easy enough: "The five-channel tester accurately gages every inch of the high-pressure steel tubing examined." However, the sentence has other problems. The last part of the sentence is verbose and needs to be unscrambled. It seems to me that "accurate gaging of every inch of the high-pressure steel tubing examined" is a long way of saying "any part of the tubing examined is gaged accurately," a fairly ridiculous thing to say. On second thought, I recommend this revision:

"The five-channel tester accurately gages steel tubing."

Here is another example where you have to rethink what was said in the original. Once more, the double verbs are italicized for a reminder.

Example:

"Fictional control in sheet metal forming *is achieved by using* a lubricant having the correct combination of film shear strength and film thickness."

One possibility:

"A lubricant with the correct combination of film shear strength and film thickness controls friction in sheet metal forming."

Have you noticed how often standard verbs like "achieve" and "perform" are involved in the examples? Of course, other word combinations are used.

Example:

"A visual inspection of housings and shafts is made."

Try something like this instead:

"Housings and shafts are inspected visually."

Example:

"The vinyl is resistant to mechanical abrasion."

More direct:

"The vinyl resists mechanical abrasion."

Example:

"One of our prime objectives will continue to be to effect needed price restorations."

Another instance where wordiness is also involved.

Why not something like this:

"We need higher prices."

Besides constructions with dual verb forms like "is achieved by using," you encounter closely related constructions like "imported cars are showing a sales increase." These are indirect and tend to be wordy.

Example:

"Hafnium in bar, rod, sheet, and plate form is being produced on a commercial basis."

Alternative:

"Hafnium bars, rods, sheets, and plates are being produced commercially."

Notice the indirect standard phrase "in bar . . . form." The word "basis" has the same evil influence.

Example:

"Customer stocks are at a low level."

Try:

"Customer stocks are low."

Example:

"It is not unusual for a downtrend to occur at this time of the year."

Here is a shortcut:

"A downtrend at this time is not unusual."

Example:

"Purchasing agents are not achieving their full potential in the earning positions of their companies."

If you rethink that one, you get something like:

"Purchasing agents can do more to help their companies boost earnings."

One more:

"ABC Corporation and ABC Company have entered into an agreement to develop and market a new coal slurry technology."

Shorter and more direct:

"ABC Corporation and ABC Company have agreed to develop and market a new coal slurry technology."

One more:

"The certificated seminar session has been designed exclusively for the manufacturer who uses or is evaluating the use of manufacturers' agents in his marketing program."

Phrases like "have been designed exclusively for" are invariably a long way of saying something that can be handled by a simple word like "for," as in this example:

"The certificated seminar session is for the manufacturer . . ."

To recap, piling up of modifiers produces a hybrid language that reduces readability. Two forms of piling were discussed. In one, multiple modifiers, often in twins, are placed in front of the noun—as in a description of a piece of equipment. In the other form, the genitive case is used, as in a long title preceding a name. A word like "of" is saved. Readability is lost.

In upside-down construction, what is important is placed behind what is less important in a given sentence. A slow, indirect way of delivering a message.

Indirect construction is also a slow, passive mode of communication. In a typical application of this construction, you have two sets of verbs, as in "were able to attain."

Exercises: Piling up, Upside-downism, and Indirect or Passive Construction

Exercise No. 1. An example of piling up modifiers. How would you fix this one?

Example:

"Second Senior Vice President of the Huber Clay Division Jack Rogers."

My answer—Put the man's name first and the title in apposition:

"Jack Rogers, second senior vice president of the Huber Clay Division." When the title is in apposition, its length is immaterial.

Exercise No. 2. A slight variation of the previous example. Use of the apostrophe after "s" in "Resources" indicates the genitive case is used. Fix this one.

Example:

"Department of Natural Resources' Georgia Environmental Protection Division."

My answer—The example illustrates that this is unnatural language. In this instance, I would switch front and back and add "of the." Result: "Georgia Environmental Protection Division of the Department of Natural Resources."

Exercise No. 3. Two sentences illustrating the second form of piling up modifiers. Identify each and rewrite the sentences.

Example:

"The ingot forging operation will be centered around a new fully automated horizontal forging press. The new press is a 1380 ton rated, computer controlled, double-acting ABC forging machine."

My answer—In the first sentence, you have two sets of twins, "fully automated" and "horizontal forging." In the second you have four sets of modifiers preceding "machine."

The first sentence isn't too objectionable as is, but I think you can improve readability by moving one set of modifiers, as in: "The ingot forging operation will be centered around a horizontal press that is fully automated." I don't think the word "forging" needs to be repeated. It was established at the front of the sentence.

In the second sentence, I suggest something like this:

"The ABC machine is rated at 1380 tons, computer controlled, and double acting."

I didn't repeat "forging" because I am taking advantage of context—what was said in the first sentence. I also avoided repeating "the new press."

Exercise No. 4. One more example of piling up. Rewrite the sentence.
Example:

"ABC's eastern steel division carbon steel flat rolled operations lost $14 million last year."

My answer—"Flat rolled carbon steel operations of ABC's eastern steel division lost $14 million last year."

Exercise No. 5. An upside-down construction. How would you redo it?
Example:

"ABC Steel Company, a leading producer of cold finished and special purpose steel bar product, recently has introduced another comprehensive revision of its popular *Value Analysis Manual.*"

My answer—Start at the back, where the message is located.

For example:

"Another comprehensive revision of its popular *Value Analysis Manual* has been announced by ABC Steel Company, a leading producer of cold finished and special purpose steel bar products."

Exercise No. 6. Upside-downism again. How would you fix it?
Example:

"The president of the National Association of Associations said today that the business community will give its broad support to the economic program outlined last night by President Reagan in his State of the Union Message."

My answer—You read 29 words, up through "President Reagan," before you get an idea of what is being talked about. I suggest something on this order.

"The business community supports the economic program outlined in President Reagan's State of the Union Message last night, states the president of the National Association of Associations."

The president of the United States rather than the president of the association should get top billing in this instance.

Exercise No. 7. An upside-down construction that buries the news. Locate the news. Rewrite.

Example:

"Ben Hur, chairman of the board of ABC Diversified Industries Inc., today announced a substantial loss for fiscal 1981."

My answer—The news: "a substantial loss for fiscal 1981." I would start there, as in:

"A substantial loss for fiscal 1981 is announced by Ben Hur, chairman of the board of ABC Diversified Industries Inc."

You must recognize, though, that some top officers will insist on getting first mention in such announcements. Companies play the same game. In some instances, the construction I complain about can be rationalized. Say ABC Company is important to a depressed industry and the following statement is released:

Example:

"ABC Company will invest approximately $244 million during the next 24 months in the construction of a 110 000 square foot titanium melting facility and the installation of an ingot forging operation."

In this instance, ABC Company is making the news and what would normally be upside-down construction is justified.

However, a company can do itself a disservice by putting its name first in an upside-down construction, as the following example illustrates.

Example:

"ABC Inc., Measurement Systems Division, a leading manufacturer of temperature and pressure transmitters, will feature its electronic temperature transmitters at the Sixth Symposium and Exhibit in Washington."

I feel that the company will get more mileage out of this rewrite:

"ABC Inc. will feature its electronic temperature transmitters at the Sixth Symposium and Exhibit in Washington, announces the Measurement Systems Division."

Exercise No. 8. An indirect construction. Try your hand at a rewrite.

Example:

"Titanium ingot to billet conversion, as well as other typical cogging operations, will be performed on the new ABC machine."

My answer—I wanted to eliminate "will be performed" and concentrate on another way of stating the front end of the sentence. I came up with:

"The new ABC cogging machine converts titanium ingots to billets, for example."

Exercise No. 9. An easy one. Fix it.
Example:
"A banquet was given by them."
My answer—Turn it around: "They gave a banquet."

Exercise No. 10. One more simple example. Rewrite it.
Example:
"The point of vaporization occurs in most metals at 10^5-10^6 watts/cm^2.
My answer—"The vaporization point of most metals is 10^5-10^6 watts/cm^2."

CHAPTER 7

Cons and Pros of Commercialism

In this chapter we will review certain common practices in commercially motivated literature that I have criticized, then shift gears to the positive and discuss the unique role of commercial literature in technology transfer, and, finally, offer several suggestions for adding reader values to this brand of literature.

We'll start with generalities-claims, upside-down construction, unnecessary repetition, and sophism.

In what I call commercial literature—it may range from a press release published in a trade magazine to a technical talk presented at a conference—the "sell" may run from hard to soft and may take a variety of forms.

Example:

"More and more welding applications will switch to semiautomatic, automatic, and robotized welding processing."

The observation, which as the appearance of a prediction or interpretation of a trend, was actually a competitive claim. The comparison is with manual welding.

Typically, though, competitive comparisons are straightforward.

Examples:

". . . flux cored electrodes now compete favorably with low hydrogen electrodes."

". . . when small diameter, flux cored electrodes replace coated electrodes, considerable savings are realized."

More, from the reader's point of view, classic examples of unsupported claims. Left unexplained are why flux cored electrodes now compete favorably with low hydrogen electrodes, and the considerable savings realized when flux cored electrodes replace coated electrodes. That's the main beef. The reader is baited with scraps of information, but in the end he is left unrewarded, or partly rewarded at best. If he "buys" whatever is being sold, he must go elsewhere to get the additional information he needs. More than likely, he will have multiple questions, as the following example, excerpts from an eight-line news release, indicates:

"effective in steel and copper . . . under a wide range of operating conditions . . . can provide results comparable to those obtained with Brand X without associated environmental hazard . . . remains stable at high temperatures . . ."

The last line of the release bears out a point I made above. If more information is wanted, the reader will have to go elsewhere. It is not supplied within the piece. In this instance, the source of the news release advises "additional information may be obtained by getting in touch with . . ." The practice is common.

Another example, this time from a part of one paragraph, to give you a feel for the size of the problem we are dealing with. In this instance, generalities-claims are italicized.

Example:

"Brand Y, ABC Company's trade name for a steel coating, *is a promising development for new container materials*. Potential applications include welded side seam containers, draw-redraw cans, and drawn and iron containers. *Significant cost reductions are possible* in production because coating thicknesses *are substantially below* other levels, and the coating material *is considerably cheaper*. In addition, production rates in recent trials *were at least as high as those of other coatings*. These factors add up to competitive performance at reduced cost."

The piece stops at that point. The reader is left dangling. He is alerted to something he may want to follow up, but he will have to go elsewhere to satisfy his curiosity. How significant are the cost reductions? What are the coating thicknesses? How much cheaper? What are representative production rates? Answers are missing. In one instance, the piece went part of the way with documentation. The listing of potential applications at the front end helps to document "a promising development for new container materials." The remainder of the documentation is in the form of generalities-claims.

Note the logic of the last sentence. The piece is saying, "If you buy the above claims, they add up to competitive performance at reduced cost."

So far we have been looking at excerpts or short pieces. I will review a piece that is two typewritten pages long to present a realistic, overall view of what the busy reader must contend with: generalities-claims, upside-down construction, and unnecessary repetition.

In the first sentence, as expected, we have an upside-down construction: company name, followed by the first mention of a new product line it is introducing.

Second sentence, a claim for the product line: " . . . can be used as a coolant-lubricant for the full range of operations."

Third sentence, new paragraph, an upside-down construction, starting with the name of the company's president, who repeats the trade name of the product line and piles on more claims.

New paragraph: more mentions of the name of the product line, more claims, such as "fully biodegradable, nontoxic, and nonhazardous . . . meets all current and pending EPA and OSHA requirements . . . handled by conventional waste disposal methods . . . due to the long life of the product, waste disposal can be virtually eliminated."

New paragraph, more claims, trade name repeated twice.

New paragraph, more claims, like "as smoke and oil misting are eliminated. . . . Machines and floors stay cleaner. Rancidity control is excellent."

New paragraph, trade name repeated, more claims, "excellent lubricity . . . improved tool life and productivity."

Last paragraph, trade name repeated and more claims. The ending: "The controlled system will thus last indefinitely, requiring minimal replenishment."

Why does the reader bother to read such pieces?

As we shall see shortly, he may not have any other way of finding out about new technology or developments in technology useful to him in what he is doing. But first a brief look at sophism, another common practice that is seldom talked about.

You start by setting up a premise built with generalities-claims but stamped with the fiat of fact, such as:

"Rapidly rising capital equipment expenditures, shortages of competent research staff, and dynamic technologies have placed growing emphasis on the need for efficient and innovative R&D executives.

From this point, you can go anyplace with immunity from fact. What you have, in the case of commercial literature, is a neat little sales pitch.

In this instance:

"With R&D spending in this country running $70 billion yearly, managers must be better prepared to manage their resources, including forecasting their needs, then selecting projects and staffing to meet those requirements."

The source of the information in this instance is selling the know-how needed to cope with that challenge. Of course, there is a question about credibility. But the reader is not taken in by generalities-claims and sophisms. He knows enough about his subject to separate wheat from chaff. He is upset when a piece of technical literature does not give him enough documentation to make up his mind whether a given new product or service is worth pursuing. He wants the evidence needed for him to weigh it and come to his own conclusions. To him, it does not matter whether he agrees or disagrees with what is said.

What I have labeled commercial technical literature has a unique role in technology transfer. The day has passed when what the technical person needed to know was supplied largely by simon-pure journals not carrying paid advertising and erudite technical talks with the stamp of preapproval by a jury of peers. The commercially motivated piece has become a major segment of technical literature. The "sell" ranges all the way from hard-hard, which you find in some paid advertising, to soft-soft, which is close to a journal piece.

The reader is willing to put up with whatever he has to in reading commercial literature because he may have no other way of learning about happenings he should know about in technology. It could be a new process that makes obsolete what his company and competitors are using. Or it could be an application for a material that is new to him and appears to have some advantage in his work. This sometimes sensational, sometimes mundane technological news is not as available as it once was for a variety of reasons. "It's proprietary information" has become a common turndown for a refusal to share technological information. You also hear, "Our public relations people feel that divulging this information would be against company policy." "The company lawyer feels that publishing this information could possibly get us into some trouble with product liability." "My boss is worried about what his bosses will say if I publish this." "Our staff has been cut to the bone, and we just don't have time to write anymore."

In other words, users of technology are talking and writing less than they did in the past, giving sellers of technology, who are always willing to talk and write, a golden opportunity to step in and fill the widening gap in technical communications. The commercial source of the information does not seem to make any difference to the reader. In fact, he does not make a distinction between advertising and editorial matter in the magazines and journals he reads. He looks on both as sources of information, leads, ideas.

In my opinion, sellers of technology are not taking full advantage of their opportunity to communicate with the technical community. The exceptions I know of take a posture similar to that of the author of a classical technical paper. The sales talk they forgo is more than offset by a gain in good will.

These noncommercial commercial pieces have two things in common: the company maintains a low profile and the reader gets a reasonable amount of unbiased documentation.

In low-profile pieces, company names and trade names are used with restraint. Names of company executives and quotes from them are used only where there is a purpose that serves the reader, not the company alone. Also missing is the PR love story. It's obvious that every effort wasn't made to say all the nice things possible about a new product or service.

Eliminating the negative tends to turn off readers. They are experienced and knowledgeable. They know there are no panaceas. What suppliers of commercial pieces don't seeem to realize is that they can make points with their readers if they talk about disadvantages as well as advantages.

Example:

"Electroslag remelting (ESR) provides no mechanism for the removal of atomic hydrogen from the melt. However, a carefully operated ESR process does not add hydrogen either, any hydrogen detected being almost totally derived from the electrodes.

"Therefore, precautions taken to prevent hydrogen pickup during any primary melt or casting operation, such as vacuum melting, vacuum degassing, solid state thermal treatments, or even vacuum arc remelting plus ESR duplex remelting are effective in eliminating excess hydrogen."

Example:

"ABC Steel Company has been using polymer quenchants as a replacement for oil in heat treating since January 1980. Advantages over oil realized include lower cost, more uniform properties, and ability to change cooling rates by modifying quench concentrations to produce structures ranging from quenched and tempered martensite to ferrite-pearlite.

"Disadvantages include the need for close control of concentration and temperature for proper heat treating response; and special precautions must be taken to reduce cracking of higher carbon alloy grades."

The seller who would launder or omit disadvantages is fooling himself that he would be hurting his cause in the eyes of the reader if he gave him the straight skinny. Regardless of the source of information, the reader is naturally skeptical, critical, and slow to accept. Present the facts. Let him be the judge. If you leave any gaps and he has enough interest, he will find out from others on his own.

The message is: you can use claims and generalities, but document them. Do a thorough job. Lastly, get through to those in the clearance procedure.

CHAPTER 8

Summing up the Busy Reader Style

The busy reader style of technical writing is anchored on four points:

1. How a busy person—technical or nontechnical—reads technical literature.
2. Why he reads.
3. How he would like to read.
4. How widely accepted practices in technical writing often prevent the writer from giving the busy reader what he expects and needs.

1. *How would this person like to read?*

He expects and needs the most possible new and useful information and ideas, in the shortest possible reading time, and in a readable manner.

2. *Why does this person read?*

Reading is a part of the job or profession. The technical person and his boss depend on reading as a way of keeping up with key developments and trends in their fields.

I assume this person is experienced and knowledgeable in his field, and is looking for new information and ideas. In other words, this person knows the fundamentals of a given subject or technology and wants an update. He does not want to waste his time reading reference book or textbook information, or current information in his field that is known to him. We have a different ball game in selecting information for the reader who is new to a field.

3. *How does this person read?*

Title first.

He wants to know: Is the subject of interest to me? What does the writer promise to say about the subject?

At this point, the busy reader may decide he does not want to read the article for a number of reasons. The subject may be outside his field, or it does not seem that he will learn anything new.

There is a message here for writing titles. Be informative. Putting the busy reader in a position to conclude "No, I don't want to read this" at this point is as important as giving him an article that is helpful.

If the decision is "Yes, I want to read this piece," the reader goes to the beginning of text.

The title, at best, is a shorthand statement of the subject or theme. It is seldom possible for the writer to say everything he would like to in ten or so words in a title.

The reader looks for amplification in text—immediately. He is annoyed, for example, if the full statement of the subject or theme is delayed by an introduction or background material that is old hat to him.

Ideally, this part of the writer's business should be taken care of in the first one or two paragraphs of text.

Once more, the busy reader is in a position to exercise his "read or don't read" option.

If the decision is a "go," the reader now expects the writer to get right down to business of documenting the subject or theme. Once more, background or introductory material should not be mixed with the documentation the reader expects and needs.

After sampling the writer's documentation, the busy reader will often exercise his third "read or don't read" option. Once more, this is important. The reader does not want to waste his time—or, to put it another way, he does not want to be forced to waste his time.

4. *Examples of widely accepted practices in technical writing that prevent the writer from giving the busy reader what he expects and needs.*

For example, the reader likes to feel he is making progress, getting somewhere as he reads. Overuse of connectives—relating a sentence to a preceding sentence or paragraph where it isn't necessary—has a cumulative effect and downgrades readability; a piece doesn't read smoothly. Too many stops, starts, too many backups. There is no feeling of flow or progress.

For example, the reader expects specifics. Generalities and unsupported claims shortchange him. The writer stops before he tells the reader as much as he could or should.

For example, the reader prefers a direct, simple writing style. Such practices as upside-down sentence construction—putting the unimportant first and the important last—and indirect sentence construction—backing into the message—give the reader the opposite of what he wants.

For example, the reader wants a decent return on his investment in reading time. Redundancy in its many forms—saying something more than once when only once is necessary—lowers the return on his reading time.

In brief, to develop the busy reader style:

Make the title informative. State the subject or theme as closely as you can in a few words.

Confirm and amplify the statement of the subject or theme immediately in text—in the first sentence if possible; certainly within the first two paragraphs.

Do not delay the statement of the subject or theme with an introduction or other secondary information. Do not mix primary and secondary documentation in the statement of the subject or theme.

Next, start to develop the subject or theme immediately. Develop each point reasonably well. Leave no questions unanswered. Make sure the piece is self-sufficient.

Do not delay the start of documentation with secondary information.

In developing documentation, do not intermingle primary and secondary documentation.

Do not consider using secondary documentation until after all primary documentation is finished. At this time, you will usually find it is not necessary.

Assume the reader knows standard information of the type found in text and reference books, and that he is looking for an update.

Assume the reader is well aware of current topics; this information is known to him. He tends to encounter it each time he reads a piece on a given subject or theme. He is looking for new information.

Continuity is closely allied to readability. Too much technical literature is boring and difficult to read. Continuity must be maintained from the first to the last sentence in a piece. Maintaining a thread of logic is important, but avoid overdoing it.

A variety of common writing practices in technical literature limits the ability of the technical writer to communicate effectively.

Practices that must be dealt with include generalities-claims, standard language, clichés and buzzwords, jargon, stilted language, verbosity, truisms, and sophism.

Other practices that must be dealt with sanction wordiness and unnecessary repetition. To serve the busy reader, the technical writer should not use one more word than he needs to deliver his message.

Other practices promote indirect expression or impede readability. A simple, straightforward style is recommended.

In each instance, there is an appropriate countermeasure for the common practices.

The busy reader style is an example of writing for an audience. You determine the amount of and kind of interest, if any, a specified audience has in a given topic. The intelligence provides guidelines for what to say (documentation) and the order in which it is said (organization). Maintaining continuity, managing language like a writer, and other parts of the busy reader style are tools of the trade and are recommended for all audiences.

Finally, the busy reader style is not intended to be competitive with any of the 13 formats in Part II. Regard it as a 14th format, for audiences wanting the most possible new and useful information and ideas in the shortest possible reading time and in a readable manner. The format is related to that of a summary, but it is closer to that of a technical article than of a technical paper.

Part II

Customs, Practices and Standards for Technical Writing

CHAPTER 9

Introduction to Part II

My preparation for Part II of this handbook was started with a five-page questionnaire to technical people who write and read technical literature regularly as part of their job. I asked that these people—they are employed in manufacturing, service, metals, aerospace, and research industries; in government; in the armed forces; and in universities in the United States, Canada, and England—to also send along documents that illustrate company (or organization) preferences and practices for any of the forms of technical literature (13 in all) which were covered in the survey.

The 13 forms covered are technical reports, abstracts and summaries, technical papers, technical talks, technical articles, technical news releases, committee reports, technical books, written technical directions, standards and specifications, patent applications, proposals and idea submissions, and theses.

My hope was that by going directly to the sources—rather than relying on published literature on technical writing seasoned with my opinions as editor of a technical magazine—I could put together a different kind of report on the customs, practices, and standards of technical writing in the English language.

My objective was to build a supermarket of guidelines where one could shop for and pick up the components needed to handle a specific assignment, be it a technical report or a specification. Information center is also a valid analogy— where one may ask "How do others do this particular thing?"

I have not tried to present a consensus, nor do I recommend. In fact, I did not go beyond reporting except where I did not succeed in locating the required raw material, and I was left to my own resources.

For example, in the questionnaire I did not ask for information on writing memos and letters. The chapter on the topic (No. 20) represents my views. Likewise, I considered my raw materials for Chapter 22 (Guidelines for Writing Theses) to be too few in number and supplemented them with my views. In one instance, I was obliged to supplement my materials (Chapter 18, Writing Technical Directions) with published information.

My goal from the start has been to be complete without being exhaustive. This explains why I have chapters on "What the Technical Writer Should Know

About Libel Law," "What the Technical Writer Should Know About Patent and Trademark Law," and "What the Technical Writer Should Know About Copyright Law." (Special thanks to Timothy L. Gall, attorney, for assisting me with searches for the chapter on libel.)

In addition to the survey of readers-writers and other sources just cited, I received valuable help on customs, practices, and standards from technical organizations like the Society of Automotive Engineers, American Society for Testing and Materials, and the American National Standards Institute—all are listed at the end of this chapter. A separate survey was sent to a selected list of technical magazines to get a sampling of experience and opinions on writing and editing. These publications are identified at the end of this chapter.

The reader-writer survey was sent to about 400 selected individuals—members of the American Society for Metals and the Aerospace Metals Engineering Committee of the Society of Automotive Engineers. There is no attribution to these people. In many instances, however, their companies are listed at the end of the chapter.

In the reader-writer survey, I asked for two things: common faults found in reading the 13 forms of technical literature previously cited and opinions on what commends those pieces of technical writing that please the reader.

The publishers and editors were asked: What are the usual reasons for rejecting a manuscript? Is content a typical problem? Is content ever too shallow? Or lacking in specifics? Or underdeveloped? Or unclear and confusing? Or lacking in originality? Or too commerical? Or too long? Or cluttered with unnecessary detail? Is organization of text a typical problem? Because of poor organization do articles ever tend to be confusing and contradictory? Are rewrites often required because of poor organization? Are these rewrites typically shorter than the original manuscript? Does language typically present problems? Is needless repetition comon? Overuse of language that does not inform? Language that is ambiguous or otherwise confusing? Do you encounter excessive verbosity and jargon?

The 16 chapters that follow, with the exceptions noted, are composites of the views expressed in the two surveys and the materials supplied by respondents and their employers, plus about 100 organizations, technical societies, trade associations, universities, and government agencies. These sources are listed at the back of the chapter.

To round out this body of information, I have tacked on a fairly broad-gage bibliography for the technical writer. Many of the references were supplied to me by contributors. References are not given at the end of chapters as is the custom because of problems in classification. However, attribution is used within chapters where it seems appropriate.

I want to acknowledge the gratifying cooperation and top-quality assistance of the hundreds of individuals and organizations who have made this book possible. As mentioned before, there is no attribution to individuals. I did not ask for permission, in the hope of getting off-the-record viewpoints and experiences. In this I feel I have succeeded. I attribute sparingly where individuals and organizations supplied materials with the knowledge that portions would be used in this book. I do this with some trepidation. A reference can unleash an avalanche of requests from around the world that a source may not always be in a position to honor. I apologize in advance for any embarrassment I may cause and also hope for due restraint on the part of seekers of more information.

In the chapters on writing technical directions, libel, patents and trademarks, and copyrights, sources are cited. Printed information is readily available.

In closing, I believe this book-within-a-book is more complete than any I have seen on the subject. I also believe the do-it-yourself approach taken is both practical and useful. I must confess, however, I was not so confident early on in my preparations. I sent a questionnaire to a peer whose opinions I respect. He replied with a long letter instead of completing the questionnaire. The gist of his message was that my plan to attack the subject format by format was defective in concept. His argument: technical writing is technical writing, whether it is a report or specification. His remarks, of course, are well taken. I agree in principle, but I have also learned in my researches that each format tends to be a subdiscipline of technical writing. Over the years, customs, practices, and standards for each format have evolved and are in common everyday usage. It is not my purpose to pass judgment on these formats. I feel a useful purpose is served to report them without undue personal comment or bias. Any criticism comes directly from readers-writers and from the publishers and editors surveyed.

Finally, in writing—particularly technical literature—the most predictable happening is a question: "I wonder how others do this?" This book, I trust, supplies some answers.

Special thanks to the following for supplying materials for this book.

These private organizations, technical societies, trade associations, branches of the U.S. government, and universities:

> Aerospace Metals Engineering Committee, Society of Automotive Engineers
> American Chemical Society
> American Foundrymen's Society
> American Institute of Chemical Engineers
> American Iron and Steel Engineers
> American Iron and Steel Institute
> American Management Association
> American National Metric Council
> American National Standards Institute
> American Nuclear Society
> American Society of Mechanical Engineers
> American Society for Metals
> American Society for Testing and Materials
> American Welding Society
> Battelle Memorial Institute

Copyright Office, U.S. Library of Congress
Department of Defense, U.S.
Department of the Navy, U.S.
Electric Power Research Institute
IEEE Professional Communication Society
McGill University (Montreal)
Metal Powder Industry Federation
Metallurgical Society of AIME
Michigan Technological University
National Academy of Sciences
National Bureau of Standards
Oak Ridge National Laboratory
Ohio State University
Society of Automotive Engineers Inc.
Society of Manufacturing Engineers
Society for Technical Communication
U.S. Patent and Trademark Office
Wire Association
Wright-Patterson Air Force Base

Thanks to the editors and publishers of these technical magazines:

American Scientist
Chemical Engineering
IEEE Transactions on Professional Communication
International Journal of Powder Metallurgy and Powder Technology
Iron and Steel Engineer
Machine Design
Manufacturing Engineer
Metal Progress
Modern Casting
Oil and Gas Journal
Scientific American
Technical Communication
33 Metal Producing
Wire Association International

Thanks to following companies for supplying materials:

Ajax Electric Co.
Aluminum Co. of America
Babcock & Wilcox
Bendix Corp.
Bethlehem Steel Corp.
Boeing Commercial Airplane Co.

British Aerospace Aircraft Group
Can-Eng (Niagara Falls, Ontario)
Canadian General Electric
Cannon-Muskegon Corp.
Carpenter Technology Corp.
Charles River Associates
Chevrolet Motor Div., General Motors Corp.
Climax Molybdenum Co. of Michigan
Crucible Inc.
De Havilland Aircraft Ltd.
Fairfield Manufacturing Co.
Fansteel
General Dynamics
General Motors Corp. Research Laboratories
Hinderliter Heat Treating Inc.
Honeywell
Imperial Clevite
Inland Steel Co.
Kolene Corp.
Ladish Co.
Leybold-Heraeus Vacuum Systems Inc.
Lockheed-California Co.
Lockheed-Georgia Co.
Lockheed Missiles and Space Co. Inc.
Martin Marietta Aerospace
Metalta Technology Ltd.
Nooter Corp.
Northrop Corp.
Perfect Circle Div., Dana Corp.
Rockwell International
Steel Co. of Canada Ltd.
Teledyne Allvac
Teledyne Vasco
TRW
United Technologies Research Center

Bibliography for the Technical Writer
Group A

Adelstein, Michael, *Contemporary Business Writing*, Random House, New York, 1971.
Anderson, J. A., "The Preparation of Illustrations and Tables," *Transactions American Association of Cereal Chemists*, 3 (2), 74 (1945). (Reprints available from the American Association of Cereal Chemists, St. Paul, Minn.)

Andrews, Deborah C., and Blickle, Margaret D., *Technical Writing*, Macmillan Publishing Co. Inc., New York, 1978.

Bander, Robert G., *American English Rhetoric* (2nd ed.), Holt, Rinehart and Winston Inc. New York, 1978.

Barzun, Jacques, *Simple and Direct: a Rhetoric for Writers*, Harper & Row, New York, 1975.

Bernstein, Theodore M., *Miss Thistlebottom's Hobgoblins: The Careful Writer's Guide to the Taboos, Bugbears and Outmoded Rules of English Usage*, F.S.&G., New York, 1973.

Beyer, W. H., editor, *Handbook of Tables for Probability and Statistics*, The Chemical Rubber Co., Cleveland, 1966.

Brown, Leland, *Effective Business Report Writing*, Prentice-Hall, Inc., Englewood, Cliffs, N.J., 1973.

Brusaw, Charles T., et al, *Handbook of Technical Writing*, St. Martin's Press, New York, 1976.

Chandler, Harry E., *The "How To Write What" Book*, American Society for Metals, Metals Park, Ohio, 1978.

Dawe, Jessamon, and Lord, William Jackson, Jr., *Functional Business Communications* (Parts II and III), Prentice-Hall Inc., Englewood Cliffs, N.J., 1973.

Day, Robert A., *How to Write and Publish a Scientific Paper*, ISI Press, Philadelphia, 1979.

DeGise, Robert F., "Writing: Don't Let the Mechanics Obscure the Message," *Supervisory Management*, April, 1977.

Evans, B., and Evans, C., *A Dictionary of Contemporary American Usage*, Random House, New York, 1957.

Ewing, David W., *Writing for Results*, (2nd ed.), John Wiley and Sons, New York, 1979.

Fieser, L. F., and Fieser, M., *Style Guide for Chemists*, Reinhold Publishing Corp., New York, 1960.

Flesch, R., *ABC of Style*, Harper and Row Publishers Inc., New York, 1965.

Flesch, R., *The Art of Readable Writing*, Harper and Row Publisher's Inc., New York, 1974.

Foster, J., Jr., *Science Writer's Guide*, Columbia University Press, New York, 1962.

Fowler, H.W., *A Dictionary of Modern English Usage* (2nd ed.), revised by E. A. Gowers, Oxford University Press, New York, 1965.

Gallagher, William J., *Report Writing*, CBI Publications, 1980.

Gensler, W. J., and Gensler, K. D., *Writing Guide for Chemists*, McGraw-Hill Book Co., New York, 1961.

Goeller, Carl, *Writing to Communicate*, Doubleday and Co., 1974.

Gowers, E. A., *Plain Words, Their ABC's*, Alfred A. Knopf Inc., New York, 1957.

Gunning, R., *The Technique of Clear Writing*, McGraw-Hill Book Co., New York, 1952.

Habeck, Charles, "The Key to Readability," *Best's Review* (Life/Health), September, 1978.

Halverson, John, and Cooley, Mason, *Principles of Writing*, Macmillan Co., New York, 1965.

Hanft, P., and Roe, M., "How to Write Well in Business," *CA Magazine* (Canadian), July, 1978.

Hayakawa, Samuel I., *Language in Thought and Action*, Harcourt Brace Jovanovich Inc., New York, 1972.

Hicks, Tyler Gregory, *Successful Technical Writing: Technical Articles, Papers, Reports, Instruction and Training Manuals, and Books*, McGraw-Hill Book Co. Inc., New York, 1969.

Hodges, J. C., and Whitten, M.E., *Harbrace College Handbook* (8th ed.), Harcourt Brace Jovanovich Publications, New York, 1977.

Houlehen, Robert, "Writing Reports That Communicate," *Industry Week*, September 22, 1975.

Hupp, A. H., *The Mechanics of the Sentence*, Van Nostrand Reinhold Co., New York, 1968.

Jacobi, Ernst, *Writing at Work: Do's, Don't's, and How To's*, Hayden Book Co., Inc., Rochelle Park, N.J., 1976.

Janis, J. Harold, *Writing and Communicating in Business*, Macmillan Co., New York, 1978.

John, Richard, "Improve Your Technical Writing," *Management Accounting*, September 1976.

Kapp, Reginald O., *The Presentation of Technical Information*, Macmillan Co., New York, 1957.

Katzoff, S., *Clarity in Technical Reporting*, Langley Research Center, Scientific and Technical Information Div., National Aeronautics and Space Administration, Washington, D.C., 1964.

King, Lester S., *Why Not Say It Clearly—A Guide to Scientific Writing*, Little, Brown and Co., Boston, 1979.

Kriloff, Lou, "Most Business Letters Could Be Improved," *The Office*, April, 1973.

Lannon, John M., *Technical Writing*, Little, Brown & Co., Boston, 1979.

Leggett, Glenn, et al., *Prentice-Hall Handbook for Writers*, Prentice-Hall Inc., Englewood Cliffs, N.J., 1970.

Lesikar, Raymond V., *Business Communication: Theory and Applications*, Richard D. Irwin Inc., 1980.

Lesikar, Raymond V., *Report Writing for Business*, Richard D. Irwin Inc., 1977.

Maizell, Robert E., et al., *Abstracting Scientific and Technical Literature*, Wiley International, New York, 1971.

Mathes, J. G., and Stevenson, Dwight W., *Designing Technical Reports*, Bobbs-Merrill, Indianapolis, 1976.

Mathies, Leslie, and Mathies, Ellen, "A Word to Word Originators," *The Office*, February, 1975.

Menning, J. H., and Wilkinson, C. W., *Communicating Through Letters and Reports*, Prentice-Hall Inc., Englewood Cliffs, N.J., 1976.

Mills, Gordon, and Walter, John, *Technical Writing* (4th ed.), Holt, Rinehart and Winston Inc., New York, 1978.

Mitchell, John H., *Writing for Professional and Technical Journals*, Wiley, New York, 1968.

Natrella, M.G., "Experimental Statistics," *National Bureau of Standards Handbook 91*, Washington, D.C. 1962.

Nicholson, M., *Dictionary of American-English Usage*, Oxford University Press, New York, 1957.

O'Connor, Maeve, *The Scientist as Editor*, John Wiley and Sons, New York, 1979.

Parr, William M., *Executive's Guide to Effective Letters and Reports*, Prentice-Hall Inc., Englewood Cliffs, N.J., 1976.

Partridge, Eric, *You Have a Point There (A Guide to Punctuation and Its Allies)*, Routledge & Kegan Paul Ltd., London, 1977.

Phillips, Bonnie D., *Effective Business Communications*, Van Nostrand Reinhold, New York, 1977.

Ray, Charles M., "Management Reports: How Many Takes?" *Management World*, January, 1975.

Rhodes, F. H., *Technical Report Writing*, McGraw-Hill Book Co., New York, 1961.

Riebel, John P., *How to Write Successful Business Letters*, Arco Publishing Co. Inc., 1971.

Ryan, Charles, W., *Writing: A Practical Guide for Business and Industry*, John Wiley and Sons, New York, 1974.

Schmid, C. F., *Handbook of Graphic Presentation*, The Ronald Press Co., New York, 1954.

Sigband, N. B., "Basic Principles of Effective Writing," *Risk Management*, December, 1976.

Skillin, M. E., and Gay, R. M., *Words Into Type*, Appleton-Century-Crofts Inc., New York, 1964.

Stafford, A. R., and Culpepper, B. J., *The Science-Engineering Secretary*, Prentice-Hall Inc., Englewood Cliffs, N.J., 1963.

Strunk, William S., Jr., and White, E. B., *Elements of Style*, Macmillan Publishing Co., New York, 1959.

Style Manual, Publications Board, American Institute of Physics, New York, 1973.

Style Manual, rev. ed., U.S. Government Printing Office, Washington, D.C., 1962.

Tichy, H. J., *Effective Writing for Engineers, Managers, and Scientists*, John Wiley and Sons Inc., New York, 1966.

Tracy, R. C., and Jennings, H. L., *Handbook for Technical Writers*, American Technical Society, Chicago, 1961.

Uris, Auren, *Memos for Managers*, Thomas Y. Crowell, 1975.

Van Hagar, C. E., *Report Writers' Handbook*, Dover Publications Inc., 1970.

Young, A. E., "Mean Business in Your Writing," *Administrative Management*, April, 1979.

Group B

Biology Editors Style Manual, CBE, 1978, Council of Biology Editors Inc., American Institute of Biological Sciences, 1401 Wilson Blvd., Arlington, Va., 22209.

Effective Lecture Slides, No. S-22, Kodak Motion Pictures and Education Markets Div., Rochester, N.Y. 14650.

A Guide for Preparing Manuscripts, National Academy of Sciences, Washington, D.C., 1975.

Handbook for Authors, ACS, 1978, American Chemical Society, 1155 Sixteenth St. N.W., Washington, D.C. 20036.

A Manual for Authors of Mathematical Papers, American Mathematical Society, P.O. Box 1571, Annex Station, Providence, R.I. 02901 (Reprint of 1970 edition).

A Manual of Style, 12th ed., University of Chicago Press, Chicago, 1973. (Also, 13th ed., 1982.)

A Manual on the Teaching of Scientific Writing (Scientific Writing for Graduate Students), CBE, 1976, American Institute of Biological Sciences, 1401 Wilson Blvd., Arlington, Va. 22209 (Reprint of 1968 edition).

Planning and Producing Visual Aids, No. S-13, Kodak Motion Pictures and Education Markets Div., Rochester, N.Y. 14650.

Publication Manual, 2nd ed., APA, 1974, American Psychological Association, 1200 Seventeenth St. N.W., Washington, D.C. 20036.

Group C (Proposals)

The Anatomy of a Win, Beveridge, J. M., J. M. Beveridge and Associates Inc., Playa Del Ray, Calif., 1969.

Boiler Room Proposal Writing, Honeywell DSD Communications Department, Minneapolis, 1972.

How To Create a Winning Proposal, Ammon-Wexler, J., and Catherine àp Carmel, Mercury Communications Inc., Santa Cruz, Calif., 1976.

How To Start on STOP, Honeywell DSD Communications Department, Minneapolis, 1970.

Proposal Writers Handbook, Honeywell DSD Communications Department, Minneapolis, 1973.

STOP—A Way of Managing Proposals, Honeywell DSD Communications Dept., Minneapolis, 1972.

Group D (Standards)

All standards cited below are available from the American
National Standards Institute, 1430 Broadway, New York, N.Y. 10018

American National Standard Guidelines for Format and Production of Scientific and Technical Reports, ANSI Z39.18-1974.

American National Standard for Writing Abstracts, ANSI Z39.14-1979.

American National Standard Illustrations for Publication and Projection, ANSI Y15-101-1979.

American National Standard Abbreviations for Use on Drawings and in Text, ANSI Y1.1-1972.

American National Standard Glossary of Terms Concerning Letter Symbols, ANSI Y10.1-1972.

American National Standard Letter Symbols for Quantities Used in Electrical Science and Electrical Engineering, ANSI Y10.5-1968.

American National Standard Letter Symbols for Chemical Engineering, ANSI Y10.12-1955 (R1973).

American National Standard Letter Symbols for SI Units Used in Science and Technology, and Certain Other Units of Measurement, ANSI-IEE, 260-1978.

American National Standard Mathematical Signs and Symbols for Use in Physical Sciences and Technology, ANSI Y10.20-1975, including Supplement ANSI Y10.20a-1975.

American National Standard Reference Designations for Electrical and Electronics Parts and Equipment, ANSI/IEEE Std 200-1975.

American National Standard Instrumentation Symbols and Identification, ANSI/ISA S5.1-1973.

American National Standard Metric Practice, ANSI/IEEE Std 268-1982.

American National Standard of Bibliographic References, ANSI Z39.29-1977.

American National Standard for the Abbreviation of Titles of Periodicals, ANSI Z39.5-1969 (R1974).

American National Standard Graphic Symbols for Electrical and Electronics Diagrams (Including Reference Designation Class Designation Letters), ANSI/IEEE Std 315-1975.

American National Standard Graphic Symbols for Fluid Power Diagrams, ANSI Y32.10-1967 (R1979).

American National Graphic Symbols for Process Flow Diagrams in the Petroleum and Chemical Industries, ANSI Y32.11-1961.

American National Standard Graphic Symbols for Logic Diagrams (Two-State Devices), ANSI/IEEE Std 91-1973.

CHAPTER 10

Guidelines for Writing Technical Reports

What are the common faults of technical reports? What are the basic requirements for acceptable reports? What are the common practices in industry and government?

Answers to those questions were obtained by surveying a cross section of industry, government, and academia in the United States, Canada, and Great Britain. They are reported in this chapter. The over-all objective is to give technical people some specific guidelines for writing and evaluating technical reports and/or to enable companies to compare their technical reporting practices and policies with the practices and policies of others.

Some Views on Technical Writing, Including Reports

A sampling of common faults as reported in the survey:

"A lack of organization . . . ambiguity . . . a stiff, stilted style of writing . . . poor grammatical construction . . . no point made . . . too many undigested tables . . . figures detailed but not summarized . . . not written for specific readers . . . verbosity . . . jargon . . . poor spelling . . . unnecessary length . . . writing to peer audiences in jargonese . . . intended to impress, rather than inform . . . too much supporting data subtract from text rather than support it . . . conclusions not adequately supported by documentation or reasoning . . . too many assumptions, not enough facts . . . trying to cover too many parameters . . . dull impersonal writing style . . . overuse of acronyms and unfamiliar abbreviations . . . poor presentation of data . . . points are obscured because of excessive length and rambling discussion . . . inaccuracy . . . poor logic and flow . . . lacking in basic writing skills . . . a lack of clear thinking . . . too technical for use by sales or production . . . bad planning . . . excessive editorializing . . . convoluted sentence structure . . . many technical writers get bogged down in dreary test data and specific detail . . . practical value to nontechnical reader ignored . . . esoteric language . . . performing rather than informing . . . lack of substance . . . failure to distinguish between the significant and insignificant . . . lack of genuine desire to communicate."

Ten common writing problems stand out, in this order:

1. A lack of clarity. Clarity includes use of language, sentence construction, and grammar.
2. Poor organization.
3. Verbosity and unnecessary length.
4. Excessive use of jargon, acronyms, and unfamiliar abbreviations.
5. Problems with spelling and punctuation.
6. Dull, impersonal writing styles.
7. Lack of preparation, which includes thinking, planning, and outlining.
8. Excessive editorializing.
9. Lack of accuracy.
10. Lack of substance.

Twelve ingredients of acceptable technical writing are given. Two are tied for first. Here are the rankings:

1. Clarity.
1. Conciseness.
2. Organization, including format, which is covered separately in this chapter.
3. Language skills.
4. Accuracy.
5. Preparation for writing.
6. Pertinent documentation.
7. Writing for both peer and nonpeer readers.
8. Grammar.
9. Lack of bias.
10. A definite desire to communicate.
11. Selective use of information.

A sampling of verbatim answers to "What are the main requirements for a well-written technical report?"

> "Good grammatical construction . . . readability . . . a piece that is complete but short . . . writing that has continuity, a logical flow to it . . . where necessary background is provided . . . the writer's evaluation leads summarily to a conclusion . . . a piece that is written so someone outside the peer group can get something from it . . . both the scope and limitations of the report are defined . . . language is clear, simple, direct . . . tables are clear and understandable . . . good quality photography for figures . . .

reader is led logically to the point of the report, concisely and completely . . . clarity, clarity, clarity . . . clear interpretation of data."

Writing for an audience or audiences in an understandable manner is a matter of paramount concern.

Comment:

"They don't write to any audience, attempting only to make themselves look knowledgeable."

Comment:

"Many technical reports are not written with the semitechnical or nontechnical reader in mind."

Comment:

"Writing is for a small peer group."

Comment:

"Much writing is in jargonese for a peer group and is intended to impress rather than inform the larger audience."

Comment:

"Reports are too technical for sales or production."

Technical reports are read to pick up information and ideas that can be used by the reader on the job. There seems to be a general feeling that technical reports often do not meet this requirement. Some seem to be written, for one reason or another, when there is nothing worthwhile to report. One reader put it this way, "The important should be obvious. If it is not, then maybe there is nothing important."

Technical Report Formats and Views on Writing Them

Formats are usually skeleton outlines and as such provide some guidelines for organization by establishing a rough order in which certain specified things are said. Formats also dictate or influence what is said and how much is said on a given topic.

There is no single format for a technical report to which a majority points and says, "This is the model we follow." Basically, though, there seems to be a common intent, and differences tend to be over amount of detail.

For example, a simple format like the following highlights the purpose of technical reports and touches all bases:

- Report what we did.
- Report why we did it.
- Report results of what we did.
- Explain what results mean to us.

Another short format stated in the jargon of report formats:

- State the subject of the report.
- State conclusions and recommendations.
- Report details, supporting conclusions, and recommendations.

This brief format adds directions to the writer:

- Clearly define the objectives of the work.
- Establish parameters meticulously.
- Document carefully.
- Provide an evaluation that leads summarily to conclusions.

Now we begin to get into more detail.
Example:

- State objective of work.
- Describe procedures used.
- Give findings.
- Report conclusions based on findings.
- Cite prior work on the same subject.

Note that the last requirement, which relates to a literature review and/or prior work at the company doing this work, adds a new element, as does the next short format:

- Define object (or purpose) of work.
- Provide assessment of outcome of work.
- Give conclusions.
- Provide recommendations related to conclusions concerning the practical implications of the work.

The last statement, translating the technical report for a nontechnical audience, comes up frequently. In fact, the next example starts with this requirement:

- Provide a background for the nontechnical reader.
- Define the scope and limitations of the report.
- Describe experimental procedures.
- Document data.
- Giving findings and conclusions.
- State assumptions used to arrive at conclusions.

The last requirement reflects another common concern. The reader relies solely on what the writer puts on paper to get the message. Writers sometime tend to forget this and omit something vital to understanding by the reader. Not stating "assumptions used to arrive at conclusions" is an example.

Still more specifics are called for by this format:

- Write an abstract covering—
  - Objectives of work.
  - Summary of results.
  - Conclusions.
- Text of report covering—
  - Objectives of work.
  - Procedures used.
  - Results.

- Findings.
- Conclusions in detail.
- References.

Abstracts, summaries, and executive summaries are sometimes regarded as riders on technical reports to reach the audience beyond the peer group. The subject is treated in detail in the next chapter. At this time, my purpose is to give you an idea of the spectrum of formats in use. One point is obvious: there are many acceptable ways of writing technical reports, and there are any number of formats available to suit your needs. The following example, for instance, includes the requirement for an executive summary.

Example:

- Write a clear statement of the purpose of the work.
- Write a concise statement of what was done.
- Give a clear description of conclusions and recommendations.
- Provide a brief executive summary.

Some formats, including the following, show a concern for objectivity.
Example:

- Define the objective.
- Clearly describe experiments performed.
- Report results and compare with those obtained by others in the literature.
- Come to modest conclusions.

You also see a similar degree of caution in this format:

- State problem clearly.
- State current state of knowledge (from literature) clearly.
- Describe work undertaken and why it should aid in the solution of the problem.
- Report results clearly.
- Discuss results in reference to the problem.
- State recommendations clearly.

If there is a classic format for technical reports, the next one cited is from that mold. There are some new items, but those we have examined differ largely in terms of amount of detail. Many things the writer should do are spelled out explicitly. As we shall see in the section on company practices, formats prescribed by large corporations, the federal government, and the armed services set standards for a number of items within technical reports. For example, some say abstracts should not be more than one paragraph long. Summaries should not exceed one page. In the example, items are numbered consecutively to indicate order.

Example of a classic format:

1. Title page.
2. Abstract.

3. Table of contents.
4. List of tables.
5. List of illustrations.
6. Introduction.
7. Experimental procedure.
8. Results.
9. Discussion of results.
10. Conclusions.
11. Recommendations.
12. Bibliography.

There does not seem to be a strong sentiment for any of the formats we have looked at. One may be used for one occasion, another for another occasion. There is also the possibility of building your own from the different format components in use. But this latitude does not extend to permissiveness. Given a choice of a format or no format, the former is favored unanimously.

Among the common faults cited, the handling of abstracts, summaries, and introductions gets top billing. These items are either omitted or they are executed in a slipshod, noninformative manner.

Comments on common faults in the text of reports include:

". . . Data not sufficiently documented . . . there is no clear statement of the problem or its resolution . . . documentation and references to supporting literature are too sparse . . . accuracy of references is questionable . . . internal reports are too detailed . . . too much data . . . conclusions are not supported by the data presented . . . conclusions and recommendations are often buried in reports . . . objective of the work is not clearly stated . . . reports are padded and too long . . . purpose of the work, description of the work, and conclusions are not always clear . . . statement of why work was done is lacking . . . poor introductions, with the writer assuming the reader knows as much about the subject as he does . . ."

Common problems with graphics include:

". . . tables and figures are not edited . . . data are poorly presented; they are not thought through or refined, leading one to conclude that the writer is not aware of his obligation to communicate with the reader. . . ." A test is suggested: tables and figures should be up to publication quality.

Here is an overall criticism:

"Recommendations in summaries are not specific. Review of the literature is not up to standard because it is not properly focused on the subject at hand. Discussion is not confined to the discussion section. Appendixes are not used when detail in text is unnecessarily lengthy."

There is also some concern about a general decline in the quality of technical reports. In part, the problem is attributed to a reluctance to consult with peers and to submit reports for review. Among those who buy research from universities and research organizations, there is a feeling that too many reports are being written. They explain: pressures to publish may be academic or commercial. In both instances, too many interim reports of limited value see the light of day.

Report-Writing Practices in Industry, Government, Armed Services

The preface to a document on laboratory reports issued by a supplier to the steel industry is a suitable introduction to this section.

"The primary function of the laboratory," it is stated, "is to develop technical information that will be useful to our customers, our colleagues, and our management. The communication of our results and ideas is an important part of the technical information transfer mechanism. Thus the results of our efforts are transmitted to readers outside the laboratory via letters, memoranda, and reports. A report is by far the most important method for data transfer. A report can provide an in-depth analysis of a process development, a new analytical technique, or work performed under a research assignment. Because we are judged by our results, we should make every effort to produce reports that are readable, factual, and provide highlights and insight to the problem tackled and the results achieved."

Four different types of reports are used by the company. Full technical reports, short form reports, monthly reports, and reports of call are its vehicles for conveying information. A format for the full technical report is described here in some detail, following the brief descriptions of the other formats below.

Short Form Reports: "Some minor assignments do not require the standard laboratory report. A simple presentation of the experimental results without technical interpretation may be sufficient for some intracompany service jobs. . . ."

Another supplier's form for its short form report has the features shown in the box on the next page.

Monthly Reports: One monthly report of the research manager to the director gives a synopsis of the more important projects and highlights significant research developments. The report is attached to a list of laboratory reports issued during the month, organized by market category, and a chronological listing of industry contacts.

A second monthly report, from the appropriate research manager to the appropriate development manager, gives the status of various service jobs in the laboratory. The report gives a brief statement of progress to date and an updated estimate of the completion date.

Reports of Call: These are summaries of technical discussions with visitors at the laboratory.

Full Technical Reports: The company has standardized what it terms an outline for its full technical reports, explaining "It is desirable to present our reports in a consistent manner so the reader is guided rather routinely to the important features of the report. Items are: title, abstract, introduction, body of report,

Short Form Reports

ABC Steel Corportation
Research Laboratories

Distribution: Report No.
                                                           Date of Report:
                                                           Prepared by:
                                                           References: (notebook and
                                                                        page or other
                                                                        references)

TITLE

Summary and Conclusions:

Action To Be Taken: (use only where appropriate)

Background:

Details: (If this report requires more than one page, show report
           number, date, and writer in upper left-hand corner of each
           page; page number in upper right-hand corner.)

____

Signatures as follows at end of text:

                                          ____
                                          (Name of writer)

Approved:

____
(Name and title of supervisor, manager,
or associate manager)

____
(Name and title of director)

laboratory procedures and results, discussion, summary or conclusions, and recommendations.

Title

In most cases, the same as the subject of the assignment. "In some cases, a revised title may be more descriptive or more concise."

Abstract

Appears on first page, with title, author, date, company for whom work was done (for service jobs), and title of the laboratory investigation or research project as a footnote.

Introduction

The investigator is instructed to write this item "before embarking on the research work. In this way, the investigator will emphasize the objective of the research and method of approach." The background information written at this time may later serve as the introduction to the technical report. In this case, it is supplemented by a brief statement of the experimental approach.

Body of Report

Some individuality is desirable. "In some reports it will be appropriate to have the procedures section precede the results and discussion section. In other reports, it may be more appropriate to combine procedures and results and use a separate discussion section."

Laboratory Procedures and Results

Should include "relatively detailed descriptions of procedures and results obtained. The latter is presented in tabular form or as illustrations in the form of photographs, charts, or photomicrographs. Where possible, place tables and figures at end of report, not intermingled with text. Tables should appear before figures."

Discussion

Essentially a restatement of the objectives and a discussion of whether they have been reached in the experiment reported. A critical interpretation of test results should appear in this section, along with a comparison with results in "previous work at this laboratory or elsewhere."

Summary or Conclusions

Conclusions are appropriate "whether the objectives have been met or they can't be met. If results are tentative or inconclusive, a summary rather than conclusions is appropriate. Results are tabulated and progress is reported. Recommendations for a revised approach to the objective may be given in a separate paragraph."

Recommendations

"Recommendations for additional research toward the objective, or corollary objectives made evident by the research, may be included for future reference. In general, the commercial application of the results or recommendations regarding markets or marketing methods should not appear in the technical report but can be carried in the letter of transmittal."

Format Prescribed by Department of Defense

MIL-STD-847A establishes format requirements for scientific and technical reports prepared by or for the departments and agencies of the Department of Defense.

Under "Detailed Format Requirements," items are summarized with the qualification, "All reports do not necessarily contain all of the following elements."

Front matter includes:

Front cover
Report documentation page
Summary
Preface
Table of contents
List of illustrations
List of tables

Body of the report includes:

Main text
Conclusions
Recommendations

Reference material includes:

References
Bibliography
Appendixes
Glossary of terms
List of abbreviations, acronyms, and symbols
Index
Distribution list
Back cover

Front Cover

Reports are numbered in upper left-hand corner using an alphanumeric designation, as: FML-RD-68-100.

Title should be prominent and "clearly and briefly" indicate subject of report.
Name of author.
Name and address of performing organization.
Date of report.
Type of report and period covered. Indicate whether interim or final.

Inside Front Cover

Such items as reproduction limitations, legal information, and safety precautions.

Introductory Section

Report documentation page is a printed form that duplicates some of the information on the cover and new items such as distribution of report, key words, and an abstract.

Summaries are optional. They may be used, for example, where it is felt that the abstract should be amplified.

A preface, also optional, "may show the relation of the work reported on to associated efforts, give credit for the use of copyrighted materials, or acknowledge significant assistance received."

Tables of contents are "seldom used in a report of eight pages or less. Headings should be in order they appear in report, and page numbers for each should be given."

A list of illustrations is used "only if considered essential." Give figure number, legend, and page number for each illustration. Abbreviate lengthy legends.

List of tables is also optional. Include table number, head, and page number for each table. Abbreviate lengthy headings.

Body of Report

Start first section on new page. Subject matter is usually background information and work objectives.

Succeeding sections may describe work procedure, apparatus involved, tests performed, results obtained, and related matters "as appropriate." Terminal sections usually present conclusions and recommendations.

Headings for sections should stand out from text.

Headings and paragraphs may be numbered "for clarity."

Reference Material

References are "completely identified" at the bottom of the page where they are cited "to aid in reading from microform." Where references are numerous, they may be repeated in a separate listing headed "References" at the end of the report. Bibliographic references are handled the same way. The style for presenting references and bibliographies should be uniform. Items include author, title, source, identifying number, publication dates, and applicable security classifications.

Appendixes are titled, as "Appendix A," "Appendix B." The first one should be on a new page. Figures, tables, and equations should have the same letter designation of the appendix in which they fall.

Glossaries are needed when many unusual terms are used in text. List alphabetically and define in a glossary. Also, unusual terms should be defined in text the first time they are used—either in text or in a footnote.

Abbreviations, acronyms, and symbols are defined the first time they are used in text. List and define at end of text in a separate section if they are numerous.

An index is optional.

A distribution list may be within a report. If so, it should be at the end.

Guidelines for Illustrations

"Treat illustrations consistently throughout a report. Prepare them so details and callouts are clearly legible after final reduction. When practical, crop or mask photographs to eliminate insignificant detail. Do not add a border or frame or use background tones in line drawings unless they contribute substantially to clarity."

As far as practical, callouts should be placed horizontally, unboxed, and near the item called out.

Color should be used only as a means of presenting data clearly. Screens, crosshatching, reverses, dots, and similar techniques are often effective substitutes for color.

Tables should be as simple as possible so the reader can easily grasp the meaning of the data. Avoid vertical and horizontal lines where spacing can be used effectively.

Technical Report Format Prescribed by an Automotive Company

This company has a set of guidelines for reviewing technical reports. The order of elements in a standard format is followed:

1. Is the title descriptive yet brief?
2. Is the purpose of the study clear and concise?
3. Is the significance of the work apparent?
4. Check conclusions and recommendations:
    a. Are they clear and concise? Is there any chance for misunderstanding?
    b. In logical order?
    c. Nontrivial?
    d. In correspondence with purpose and title?
    e. Clearly supported by results in text?
    f. The only conclusions and recommendations suggested by results in text?
5. Does the body of the report read easily?
6. Is organization logical? Is it consistent throughout report?
7. Is the introduction informative and adequate?
8. Is the significance of the work apparent?
9. Has the direction of future work been indicated?
10. Do references reveal appropriate knowledge of technical literature?

Other items that "impede easy reading and lower report quality" are put into a checklist:

- Incorrect grammar and spelling.
- Unclear statements.
- Repetition.
- Long paragraphs.
- Involved sentences.
- Mixed structures and tenses.

- Excessive detail.
- Technical inaccuracy.
- Unclear tables.
- Busy figures.
- Excessive data in figures not mentioned in text.

The company regards the research report, research memorandum, and research publication as "the main arteries in its report retrieval system."

The research report is the "preferred vehicle for describing a sizable project or a series of related projects—especially if broad interest in the subject exists within the company." The research memorandum is for short reports, particularly for minor projects. The research publication is a paper for a journal or for oral presentation and subsequent publication in proceedings, for example.

Required elements for a research report are a cover, title page, statement of purpose, summary or conclusion, body of report, and distribution list. With the exception of the cover, the items also apply to a research memorandum. Its first page is both title page and first page of text.

Optional items for both: abstract, recommendations, acknowledgments, reference list, illustrations, and appendix(es).

Practical guidelines highlight discussions of individual items.

Titles

Should be brief yet definitive. The basic function is "to inform the reader, to the extent practical, of the report's subject field and of what distinguishes it from other reports on the same general topic." Many of the company's writers, it is added, "succeed in expressing the subject of their reports in 10 to 15 well-chosen words. Some manage very well with even fewer."

Another point is made: "Libraries commonly file documents under key words in the title. Be sure your title accurately reflects the subject area being covered."

Some methods of obtaining brevity are discouraged. Abbreviations, symbols, acronyms, and jargon are often unknown to readers. Stacking modifiers in front of subjects to avoid prepositional phrases can impede readability.

A tip: write the title last, rather than first. You'll do a better job in fewer words.

Writing Purpose of Research

Length of this item is geared to the length of the report and to the nature of the reading audience.

If reports are ten pages or less and written for readers with a high degree of expertise in the area, purpose can be expressed in a sentence or two.

When reports are longer than ten pages and the reading audience is made up of peers and others such as managers, modified approach is recommended. If there is no abstract, the "why" aspect of the project should be dealt with in some detail.

The last plateau is made up of reports running 30 pages or more for distribution to several audiences. If there is no abstract, "why" should be amplified as above and the writer should supply background information on previous work in the same area and state the potential significance of the new work.

Length of this item will range from one sentence to about a quarter page.

Follow-up reports, it is advised, should include a complete statement of purpose.

Summary or Conclusions

What you call this section, it is advised, depends upon where the emphasis falls. "Summary" and "conclusions" are not interchangeable terms. A summary is an abridgment of important thoughts from the body of the report, such as the most significant findings or the main ideas gleaned from a state-of-the-art review.

A conclusion, it is stated, "is a reasoned judgment or inference made on the basis of results or findings."

A report may have only a summary, or only a conclusion section, or a combination, depending upon content. Whatever the section is called, its content should flow logically from and respond to the Purpose section. At the minimum, the Summary or Conclusions should tell whether the specific goal cited in Purpose has been reached. "More typically," it is stated, "it means delivering the important results or significant conclusions which the reader should expect after reading Purpose."

Body of the Report

Wide latitude is allowed. Typical items or topics include:
- Introduction or background
- Scope of work
- Theories, principles
- Glossary of terms
- Method of approach
- Equipment, materials, apparatus
- Presentation and discussion of results, including interpretation and qualification if necessary
- Detailed conclusions

Optional Elements

Optional elements include abstracts, table of contents, recommendations, glossary, illustrations, acknowledgments, references, and bibliographies.

Abstracts are supposed to capture the essence of a report in a few words.

A table of contents is advisable if the report has many divisions and subdivisions. The same guideline applies to illustrations and tables.

Recommendations "can be a logical, effective follow-up to a Conclusions or Summary section. That's because these two sections frequently suggest that some kind of continuing effort or decisive action is in order. If this is the case, the Recommendation section pinpoints what should be tried or done."

The preferred location of a glossary is just before the start of text in the body of the report. If the section is several pages long, it should follow the last section.

Researchers are advised: "Think in terms of graphs, diagrams, maps, and photographs when you plan your project. By looking ahead, you should be able to arrange for and complete some of the illustrations as you work on the project. If you follow through this way, you won't have to start from scratch when you begin writing."

If you have a large number of figures, consider reducing the sizes of illustrations and combine two, three, or more on one page.

It is common practice to acknowledge all persons contributing substantially to the completion of a project.

Format Prescribed by a Maker of Specialty Steel

Reports, the company declares, must be concisely and clearly written and should take a minimum of time and effort to prepare and to read. The accent is on simplicity, logical presentation of data, and sound, accurate data analysis.

Reports are prepared to the following format.

Abstract

Elements of the first page are: title, abstract, author's name, project number, and date.

Introduction and Summary

There are five items:
- Brief background on origin of job.
- Review of literature of state of the art.
- Purpose of project.
- Brief description of experimental procedure.
- Summary of results, conclusions, and proposed work and/or recommendations, if any.

Procedure

"Be brief yet sufficiently descriptive for someone else to exactly duplicate your procedure." The alternative, if a procedure is long and involved, is to put this information into an appendix, rather than the body of the report.

Results and Discussion

The directive to the writer is to go only far enough to explain his interpretation of results.

Conclusions

State briefly.

Recommendations

Optional. The place to mention the need for future work or action by others. Should be specific and clear.

References

In reports, put references in parentheses at the end of a sentence and above the line, as[1]. List references, including author, date, and publication, in the first appendix.

Tables and Figures

Brief tables may be incorporated into text. Otherwise they should be appended to report.

Data should be graphed wherever possible. "Strict attention must be paid to data analysis to avoid indications of spurious correlations."

Use photographs "intelligently," to "eliminate usual questions arising in the reader's mind." Do not use to dress up a report.

Individual tables, figures, graphs, or photographs should not exceed 7¼ by 9¼ in. (185 by 235 mm). This is the maximum size for photographing for reproduction.

Appendixes

The place for information too lengthy for the body of the report, for situations not pertinent to the central purpose of the report, and for description of techniques developed for the study.

Short Form Report Used by a Specialty Steelmaker

The format, termed a combination of letter and report style, is reserved for such missions as reporting progress on a specific point in a study, the results of a small or simple activity or test, or the presentation of data on part of a large program.

A printed form is followed. Items include distribution (including library), report number, date of report, name of preparer, references, title, summary and conclusions, action to be taken (where appropriate), background, details, signature of writer, and approvals at the end. If the report takes more than a page, report number, date, and writer are shown in the upper left-hand corner of each page; page number is in the upper right-hand corner.

Directions for the body of the report are as follows:

"Show title in top center of page. Start with a brief statement of the research problem, or a concise description of the background, sufficient to orient the reader. Follow with summary and conclusions. Describe action taken in accordance with planned program. Follow with detail sufficiently explicit to support findings and conclusions. Present recommendations and justification for additional work or alternate action if relevant. Conclude with supporting tables, figures, and references."

In this company, the writer gets a serial number for the report from the librarian, who maintains an index that identifies serial numbers for each report issued.

Aerospace Company's Standards for Technical Reports

Policy is stated thus:

"The over-all appearance of a document does more than reflect the effort put into the work. It has a positive influence on the effectiveness of communication. It is practice . . . to prepare technical publications in accordance with certain standards. They apply to the choice of basic elements, the way they appear on the printed page, and the physical appearance of the document—its size, binding, and type of paper stock. Taken together, they constitute the format."

Elements usually appear in this sequence:

> Cover
> Title page
> Frontispiece
> Foreword
> Abstract
> Summary
> Table of contents
> List of illustrations
> List of tables
> Introduction
> Body of text
> Appendixes
> Nomenclature or glossary
> Bibliography or list of references
> Index

Not all elements are used in each instance. A short report may consist of only a title page and body of text. Order may be changed to suit requirements.

Several elements merit separate comment.

Foreword

May be used to present information relating to the report, its subject, or its preparation when such information cannot be suitably included elsewhere. The foreword may include the names of individuals or organizations participating in or contributing to the work; it may indicate that the document is one of a series or has been prepared in several parts. The section usually gives the contract authorization and administrative aspects of the work. When the author is credited, a sentence to this effect is included in the foreword.

Introduction

The purpose is to prepare the reader for the information that follows. The introduction should acquaint the reader with the background and history of the material in the text. It may state the scope of the project and any limitations that may be imposed. Although the introduction should not describe the contents of the document, it may define terms or explain the organization of the material presented.

Figures and Tables

Drawings, graphs, photographs, and charts are termed figures; a tabulated summary of data is considered a table. Each figure is identified by a number and title, and each is called out by a figure number in text. A short, simple table of only a few columns may be inserted into text without a table number and title. However, a multicolumn or full-page table is assigned a table number and title, and is referred to by table number in the text. Figure titles are centered beneath the illustration; table titles are centered above the table.

Appendixes

Supplementary material having some contributory value but not essential to completeness. Examples include portions of other reports, specification requirements, conversion tables, detailed analyses, and charts or graphs.

An appendix should be at the end of the report. A large appendix may be separately bound. Style and format should be consistent with that of the body of the text except when the appendix is a published report reproduced in its original form.

A single appendix carries only the designation "Appendix." Two or more are designated alphabetically, as "Appendix B." Titles may follow the word "Appendix" or appear beneath it.

Nomenclature

If there is a large number of mathematical symbols or uncommon technical terms, use a nomenclature or glossary section to avoid definitions in texts and footnotes. Three different placements are possible: before text, within text as a separate section, or following text.

Use a two-column format with symbols aligned on the left-hand margin and definitions in a wide-reading column on the right. Units of measure may be listed in a third column or made a part of definitions.

Index

Placed at the end of some documents to help the reader find specifics in text. Differs from a table of contents. Information is listed alphabetically rather than by order of appearance. Topics rather than generalized subject headings are listed.

Pagination

Pages in a report are usually numbered consecutively with an Arabic number. Hyphens may be used on either side of the number to set it off (as -21-). May be placed at the bottom or top of the page. When both sides of the page are used, all right-hand pages have odd numbers. All left-hand pages have even numbers. Major sections normally start on right-hand pages. A blank left-hand page that precedes a section may be assigned a number even though it is not printed.

A Government Agency's Standards for Technical Reports

Two general types are used: topical and progress reports. The topical report is usually shorter, treats a single subject, and is authored by not more than a few persons. Progress reports represent the effort of most of the technical personnel of a division, project, or section. Numerous topics and phases of the work may be covered. The main purpose is to present a current view of the status and progress of research and development work. These reports may contain many abstracts of papers published in the open literature, with only a limited amount of original research discussed in the report itself. Details that follow are limited to the topical report.

Foreword-Preface

A foreword is a statement by someone other than the author. A preface is an author's own statement about his work.

The preface usually consists of the author's reasons for undertaking the work, his method of research, acknowledgments, and, sometimes, permission granted for the use of previously published materials. If a preface is limited to acknowledgment, it should be retitled: "Acknowledgments," and placed at the end of text, preceding the "References" section.

Some Rules

A heading should not be the last line on a page.

A heading must have at least two lines of text with it or be carried over to the next page.

Do not start the first sentence of a paragraph on the last line of a page.

Do not hyphenate the last word on a page.

Footnotes

A note of reference, or explanation, or comment is usually placed below the text. When symbols are used for footnotes, the sequence should be:

* (asterisk or star)
† (dagger)
‡ (double dagger)
§ (section mark)
‖ (parallel)
(number sign)

When more symbols are needed, double or triple those above, following the same sequence, as ** and ***.

A Major Steel Company's Requirements for Technical Reports

The standard format has the following elements:

        Cover letter (letter of transmittal)
        Title page
        Subject-matter gist for mechanized information retrieval
        Abstract (when needed)
        Table of contents (if report is long)
        Summary (termed report of report)
        Introduction (optional)
        Body of report
        Conclusions, or conclusions and recommendations
        Future work (if required)
        Bibliography (if required)
        Appendix (if required).

Cover Letter

Provides a general orientation for the reader but is not supposed to do double duty as an additional summary. Use to state the chief results and their significance.

Tips on Writing Titles

Be explicit and specific. A title should not be unreasonably long, yet clarity should not be sacrificed in the name of brevity. Neither readers looking for a specific subject nor librarians and information-retrieval people can operate effectively with uninformative titles. In effect, the title is often a shorter, undetailed version of an objective or problem.

Subject-Matter Gist

Item is used in mechanized information retrieval and is not considered part of the report. Usual length is one to five paragraphs. Key words— technical terms that describe and identify the subject matter—are used even if some of the terms do not appear in the text.

Abbreviations for organizations, techniques, methods, processes, equipment, and materials should be spelled out at least once unless they are well known. Negative and speculative statements are to be omitted. If possible, equations, formulas, Greek letters, subscripts and superscripts should be available.

Some Tips on Figures and Tables

Do not overload figures and tables with content to the point where it is hard to follow the main message.

Titles of tables and captions for figures should be complete enough to stand alone. But completeness should not be mistaken for wordiness. "Rewrite each title or subtitle until you hit on the shortest form that still has all the units of meaning required for the table or figure to be understood without a rereading of the corresponding material in the text."

Never present data in both a table and a figure. Use one or the other but not both.

"If five figures can be eliminated in favor of one figure, or if several figures can be generalized into a single figure, the message will be grasped more quickly and easily. The same is true for tables. An excess of tables and figures can be relegated to appendixes if the writer cannot bear to dispense with them entirely."

Future Work

Engineers working for this company are advised: "If a report closes a project, there is no future work to propose. If such a section is required and runs more than a few sentences, it deserves its own section and a title. Also, the summary section should still include a brief, basic statement about future work." Examples of section titles include "Future Work" and "Proposed Test Program." If proposals for future work are elaborate, they should be written up separately as proposals.

Appendix

These supplementary materials must be referred to in text or in footnotes. Examples of supplementary materials: a theory section; the derivation of an equation; lists of symbols, figures, tables, or sketches.

A caveat is given: "Sometimes an engineer uses an appendix to give a fuller, more detailed treatment of a problem he has treated briefly but clearly in the body of the report. He is right not to burden the body of the report with such detail. However, by the same token, there is generally no really valid reason to tack on this kind of appendix. As a matter of fact, here are some reasons not to:

- Very few people either need or have the patience to read such an appendix.
- The argument often used by engineers is that they include such appendixes because they want people to know that all the facts are available for their inspection and that everything can be backed up. On the contrary, if a writer cannot make his points logically in the report proper, no amount of tacking-on of appendixes is going to remedy the situation. Conversely, the reader is grateful for a succinct, factual treatment of a problem and credits the author for scientific integrity."

A tip: "Reference to one's laboratory record book (by number and pages) at an appropriate point in the report is as good a way as any of noting the availability of all data."

A tip: "Once in a while an appendix can be quite functional. Take the case of a report written for two different groups of readers: one little interested in technical proofs except in a general way; the other interested in even second-order variations. In such a case, the body of the report would have only three or four master figures. An appendix would carry the dozen or so figures summed up in the master figures."

An Aerospace Company's Report Format and Guidelines

Writing practices and use of graphics receive special attention in this company's document for writers of technical reports.

"A useful report," it is stated, "must be understandable, brief, and well organized. An acceptable style is obtained by writing simply, naturally, and coherently. Abbreviations should not be used in the report. However, abbreviations for units of measurement or symbols may be used in tabular matter or in illustrations.

"The design of a good report has a varied audience in mind. The subject matter is analyzed to determine a logical sequence of events and ideas. An outline or flow chart is a useful tool in report organization. The busy manager and non-technical reader are apt to scan summaries and lead-in sentences. The engineer goes beyond lead-in sentences to determine the findings within the report. The manager-nontechnical reader must make use of conclusions or recommendations. The engineer is concerned with the details."

One section is devoted to graphics for technical reports. It is stated, "Initial report preparations should include advance planning for illustrative matter, such as photographs, that could lose effectiveness if the reproduction process for the report (such as printing versus copying) or if the change in size of the photograph (it may have to be reduced) is not considered.

"Because photographs are usually taken during an investigation, a system should be used for retrieval of the negatives during final report preparation. The system must provide for the identification of the content of photographs and, if necessary, a record of the sequence the photographs were taken. Consistency from photograph to photograph should be maintained when different cameras or photographers are used. Pay attention to such details as scale of photo, omission of distracting backgrounds, identification of top or bottom if necessary, and standard sized lettering."

Drawings and graphs, it is advised, should be simple. "Rather than attempt to superimpose all data onto one graph or chart, a set of graphs or charts can show the required information more effectively. Oversize drawings (foldout, in particular) should be simplified and redrawn specifically for the report. Callouts, labels, and legends should be as brief as possible. A series of related drawings should be proportioned and lettered for uniform appearance in the finished report. Like drawings should use like scales where possible.

"The most important precaution to observe in planning visual material is not to use too much detail or descriptive information. A cluttered, complex visual aid is no aid."

Guidelines for some individual sections of reports are of interest.

Introduction

Usually a paragraph or two to state the problem, to give a brief historical background, and to relate the problem to the company's work. If the report is long, the method of presentation of the balance should be explained. If periodic progress is being reported, earlier work should be summarized. If work is to be continued, work anticipated should be described briefly.

If there is no foreword, the introduction may be used to acknowledge assistance, material sources, the testing organization, or to identify personnel if necessary. The last item, however, is generally the exception.

The objective or purpose of the report is not stated in the introduction. It is reserved for the next section.

Objective or Purpose

States goal as presented in report. It may be an analysis, investigation, or an evaluation of a problem as presented to the reader in the introduction. The intent of the report is stated in such a manner that conclusions and recommendations will answer the problem.

Conclusions or Summary of Results

Provides an answer to the problem posed by the objective. Conclusions must be supported by data within the report—an often overlooked basic in writing. Negative results or a statement concerning the inability to obtain results may be part of the conclusion. If possible, each conclusion should be stated in a single sentence.

A summary should be used when no conclusions can be drawn from the results or when presentation of data without conclusions is the purpose of the report.

Recommendations

Should be based on the conclusions. When possible, each recommendation should be stated in a single sentence.

Materials and Equipment

Optional. Information may be put in next section on procedures. Materials tested and used in testing are described, along with procurement specifications where applicable. Description of testing equipment is optional. Special equipment built for the test should be described.

Procedures

Present in a step-by-step sequence. Be selective in how detailed steps are. Sufficient detail should be given to allow the engineer-reader to determine important aspects of the work. A factor to consider is whether the work is to be duplicated elsewhere. No results or data should be reported in this section.

Results

May be the reporting of all data obtained, selective or averaged data, or a comprehensive summary of data. When data are only summarized, complete data may be presented as an appendix. Statistical interpretation of data may require computations or sample computations. The calculations may be in this section or in an appendix.

Graphical presentation is important in this section. Visual aids such as photographs, graphs, and tables may be placed at the point of the first citation or may be presented as a group at the end of the report.

If a number of tests are reported, the order of presentation of data should be the same as that in "Procedures."

Discussion

An optional section to give the responsible engineer the opportunity to analyze the significance of the results. Facts or theories, observations made during the procedures, and unique variations within the data may be appropriate subject matter for discussion and may be of interest to both writer and reader. The engineer may discuss, or make comparisons with, other work of a similar nature. The engineer may explore directions that future work may take. Some unique approach to calculations also may be discussed.

CHAPTER 11

Guidelines for Writing Abstracts and Summaries

Abstracts and summaries are treated separately because of their unique roles. They may be the only parts of reports or other technical literature read by a variety of interested spectators. Busy technical managers want the gist of a report in a hurry. Semitechnical and nontechnical people ranging from librarians and public relations people to sales executives and production managers want a clear translation of what the report is about and its possible practical significance. Peers searching the literature also look to abstracts and summaries as a way to get a quick fix on what has been written on a subject of interest.

As in the last chapter, I report on common faults cited by readers, along with views on "what are the main requirements for well-written abstracts and summaries?" Excerpts from company documents are given to indicate current practice.

An aerospace company's definition of the abstract:

"A brief synopsis of the content of a document. Its purpose is to acquaint the reader with the subject matter so he can determine whether the information is of value to him."

The same company's definition of a summary:

"Tells a reader, in more detail than is permissible in an abstract, the significant facts presented in the text, the results and how they were obtained, and the conclusions derived. Its purpose is to give the significant information in the report to those who cannot afford the time to read the whole report."

Key phrases are (for abstracts) ". . . so they can determine whether the information is of value to them . . ."; (for summaries) ". . . give the significant information to those who cannot afford the time to read the whole report."

Typical comments on common faults found in abstracts and summaries indicate that neither objective is always achieved.

You hear, for example:

"There is a tendency to hide the outcome so one is forced to read the entire paper."

"They are not clear, so I have to read the report to find out what was done."

"Conclusions and recommendations are not adequately summarized, forcing one to locate them in the report."

"They are often too brief. Not enough information is given to the manager to determine the status of work and whether additional work is required."

"Some are too brief. Main subject is not described. The reader can't comprehend purpose without reading details."

"Too vague and general to determine the gist of a report, or too long and too detailed."

"Either too little or too much detail."

"Does not describe what is in the report."

"Many state what they plan to do and do not state clearly what was accomplished."

"Sometimes difficult to determine the bottom line. Some abstracts force you to read the paper to find results."

"There is a failure to summarize main points or to get to the meat of the report."

"Commonly abstracts do not convey information."

"Many abstracts and summaries are useless. They simply restate the introduction, and only infrequently contain major findings and conclusions."

"There is often a failure to focus on a limited number of significant accomplishments."

"Sometimes fail to give the reader an over-all perspective."

"The topic is not properly defined."

"Some are too long to be called abstracts or summaries."

"Some contain too much technical information, results, and numbers."

"At times pertinent findings are not covered."

"There is a failure to inform."

"Some are too lengthy. Some are not complete."

"Fail to make point."

Other criticisms focus on specifics. Examples follow:

> "Too lengthy . . . too general . . . not objective . . . lacking in clarity . . . too formal and technical . . . too dry . . . does not capture reader's interest . . . often written before report is written, meaning the general subject is described but not major findings and important details . . .

> "Jargon, acronyms, and convoluted sentences should not be used. There is no room to explain jargon and acronyms. Convoluted sentences are too long . . . main points of the report are diluted with redundant statements . . . conclusions are weak . . . statements of results lack specifics . . . excessive detail, and theme is not highlighted . . .

> "There is no attempt to write for an audience. The aim of the writer is to make himself look knowledgeable . . . results and conclusions

are omitted . . . many do not state the reason for the work or its objective . . . no clear statement of the problem and objectives . . . contain information that belongs in the introduction, procedures, etc. sections . . .

"Rambling, incomplete reviews . . . some data in the abstract are not in the report . . . not always tailored to the needs of the reader . . . too often guilty of false advertising, implying report contains information that is not there . . . often too many numbers, forcing the reader to draw his own conclusions . . ."

Comments on Main Requirements for Abstracts and Summaries

Definitions of these two formats include:

"They should be brief statements of purpose, approach, major findings or conclusions-recommendations."

"A precise statement of the substantive concern of the report, permitting the reader to decide whether he wishes to study the full document."

"There should be . . . a definition of topic, summary of work, and brief statement of conclusions or results."

"Should clearly summarize . . . why work was done . . . what was done."

"Summarize the content of the report and indicate the practical value of the information."

"Present major points in a logical order."

"Clearly state findings and recommendations resulting from the work."

"In an organized, clear, concise manner state . . . what was done, why, what was learned, conclusions."

"Define . . . scope, results, conclusions."

"State problem . . . summarize major results, state major recommendations."

"Tell what was done, summarize conclusions, summarize recommendations."

"Outline, briefly, type of work and nature of findings."

"A brief summary of the scope of the study."

"Concise statements of . . . objective of work, main findings in report."

"In single sentences describe scope of report and objectives. Results and conclusions should give reader sufficient information to decide if he should read the report."

"Briefly cover . . . objectives, results, conclusions."

"State objective . . . method, when pertinent . . . scope of work . . . summary of results."

"State . . . objective . . . approach (briefly) . . . conclusions."

"Highlights plus a brief descriptive statement of the main purpose or goals."

"Should contain essential points, statements, or facts."

"Abstract should not be limited by a word-count formula. It should be as long as is necessary for the reader to get the gist of the report."

"The summary should be an overview of the report, condensing background, purpose, significant results and observations, conclusions, and recommendations."

"State purpose of the work and the significance of the results."

"Subject matter, findings, or information should be summarized with sufficient precision for the reader to determine whether he should read the full report."

"It should be a brief, almost nontechnical description of method, results, and conclusions."

"Should be detailed enough to report main points or findings."

"Should answer the reader's question, 'Should I read this report?' "

Other attributes of a proper abstract or summary are spelled out in the following comments:

> "Conciseness . . . state main point without being too general . . . precise use of language . . . complete objectivity . . . focus on a limited number of significant accomplishments . . . short sentences . . . clarity . . . accuracy . . . contains key words . . . informative . . . highlight theme . . . avoids excessive detail . . . use simple language . . . give key results . . . limit to one paragraph."

Practices Recommended by a Maker of Specialty Steels

Informative abstracts, it is stated, should contain the following elements: objective and/or purpose of the investigation, methods employed, results, conclusions, and recommendations.

Objective and/or Purpose

By describing the nature and scope of the problem, the abstract should tell the reader whether the research concerns him. One brief sentence should be sufficient.

Methods Employed

Without going into detail, specifically and selectively name kind of treatment given, techniques and instruments used, materials, conditions, restrictions, and limits. Methods are especially important if they are new or unique.

Results, Conclusions, and Recommendations

The essential part of the work described in the report. Most of the abstract should be devoted to this element. If results include numerous specific data, a general statement should be made; or if the scattered nature of data allows no such statement, this fact should be clearly stated, including any important numerical values. Conclusions should interpret results.

Practices Recommended by a Maker of Automobiles

An abstract, it is stated, is optional but highly desirable. "It should capture the essence of your report. Because it draws from all the other important parts of your report, it is read by more readers than any other parts."

Most abstracts written for this company run 100 to 200 words. They may be shorter for brief reports and up to a page for longer ones. All abstracts should cover:

- The specific subject of the report (what you were trying to do, and if this is not obvious, why).
- The scope of your effort (what was involved in getting the job done).
- Significant findings or summary of thoughts.
- Major conclusions, if appropriate.
- Major recommendations, if appropriate.

"If you supply most or all of these details," it is stated, "rather than merely state such and such will be described, you are more likely to provide the kind of information your reader needs. In short, strive for the informative type of abstract, rather than the purely descriptive." The descriptive abstract, it should be noted, has a strong resemblance to a table of contents. It makes life easier for the writer, but the reader is shortchanged.

Abstracts of more than 200 words or less than a full page, the company suggests, could benefit from being split into two paragraphs. Those one page or more should be divided into at least three paragraphs.

In this company, the abstract for a research report is placed on the back of the title page, before the table of contents, if one is used.

An example of an abstract from this company:

> "The direct fabrication of automotive parts from low-cost monomers offers considerable economic advantages over the conventionally processed plastics. This work involves the investigation of the chemistry and technology of casting Nylon-6. Σ-Caprolactam may be directly polymerized to Nylon-6 plastic in a mold below the melting point of the nylon, using an alkali catalyst and an organic co-catalyst. Of the co-catalysts examined so far, polymethylene-polyphenylene isocyanate (PAPI) has been found to be the most suitable. A number of mineral fillers have been examined, and calcium carbonate has been found to be most satisfactory.

Practices Recommended by a Maker of Steel

The abstract is an optional item "but may be useful if the report is going to other divisions." A paragraph or two long, the abstract may be placed on a page by itself, following the title page and preceding the table of contents page.

The summary, or summary and conclusions, is regarded as the most important section of the technical report. In a page or two it summarizes:

1. Problem and objective.
2. What you did to achieve the objective.
3. Main results and conclusions.
4. Significance of the results.
5. Recommendations, if justified or requested, and continuing or future work, if any.

The objective may concern a problem; an area of knowledge to be added to; an existing market, program, or method; an evaluation, forecast, or extrapolation for a proposed or projected plant; or a proposal.

The objective should be stated explicitly and specifically, and the scope sharply defined. For example, instead of writing, "Tested the MN method of determining aluminum in steel to find out if the method is suitable for plant use," say something like, "Tested the MN method of determining aluminum in the p-q% range and for steels X, Y, and Z, to find out if the method is suitable for plant use."

Three common faults:

- An objective that is too general, meaning it encompasses more than the actual objective of the work.
- An objective that covers only part of the actual objective.
- An objective that is not clear, forcing the reader to go back and forth through the entire report to determine what the target is.

In describing what was done to achieve the objective, the main step or steps in the program should be stated briefly. Do not expand on methodology, test procedures, materials, or conditions, for example. Reserve for body of report.

Main results and conclusions should be limited to key results. When an extrapolation or forecast is called for, make sure facts, personal observation, estimate, trend, or guess is clearly identified.

The section on significance of results requires interpretation. Specific results do not do the job. The practical value of a result should be amplified.

The recommendation may be nothing more than a suggestion, as: "These results point to the advisability of decreasing the intake volume at least 15%." Or it may be a statement about the need for continuing or future work, as: "These results justify proceeding to pilot plant testing of 6-ton lots of material GH at X temperature conditions." In some instances, future work may require a separate proposal.

Practices in Government, an Association, and Aerospace and Materials Companies

Military Standard 847A defines a summary: "A summary may be included to provide a digest of the report, to explain the reason for the initiation of the work, and to outline principal conclusions and recommendations . . ."

An association has the following requirements for a summary. " . . . a two-to-four page summary, written by the contractor, must be included in each report. It gives a full synopsis of the report, including the purpose and scope of the research project, the method or process involved in the study, the findings, and the conclusions.

"The summary is also distributed as an independent document to executives, government officials, and individuals interested in the information who do not necessarily have a technical background. Because the summary is the only record of a project that many of these people will read, it is an extremely important element of the report and must be complete in itself. It should not be combined with the introduction or any other text section; it should not refer to tables or figures found elsewhere in the report or cite reference works without providing full bibliographic information."

An aerospace company advises that "an abstract should be a brief explanation of the content of the report. An informative abstract highlights important points in the body of the report and may include data, conclusions, or recommendations."

A summary is defined as a brief survey of the purpose, methods, results, and conclusions covered in the report. It is usually longer and more detailed than an informative abstract, and is primarily intended for the person who does not have time to study the full report.

The materials company requires an abstract for each report for use in status reports, for library reference, and for the benefit of reviewers within the company who wish to determine, rather quickly, whether they need to read the entire report. The abstract is to contain one sentence which defines the scope of the investigation, a sentence or list of the materials investigated, and a brief summary of results. In most reports, the abstract should be one short paragraph. In no case should it exceed one page.

In this instance, the summary is an alternative to conclusions. The rules: "If the objective of the assignment has been met or is incapable of being met, conclusions are appropriate. In some cases where research results are inconclusive or tentative, conclusions are not appropriate. Use a summary, which briefly tabulates results and states progress toward the objective. This may be followed by a paragraph in which recommendations for a revised approach to the objective may be given."

CHAPTER 12

Guidelines for Writing Technical Papers

"One of the greatest contributions that any technical or professional group can make is the publication and distribution of information," declares the American Foundrymen's Society (AFS) in the introduction to its guide for authors.

The statement introduces a new, if not sobering, consideration for the technical writer. When he writes for publication, he is representing his profession and writing to his peers. He is also writing for a publisher, which means he must conform to publication preferences and practices of societies, associations, and independent publishers. In other respects, the technical report and technical paper have things in common, many of which relate to format.

The administrative aspects of getting a paper published cover topics ranging from items to be included on the title page and the maximum length of abstracts to procedures for submitting manuscripts and rules for the rejection of manuscripts.

AFS summarizes its requirements thus: "The effectiveness of a technical publication, apart from actual information, is influenced by three important factors: 1. General manuscript format. 2. Illustration quality. 3. Presentation style."

AFS will reject papers because:

"Subject matter does not fall within the field of the Society's activities.

"Contents are of an advertising character.

"Subject matter is essentially of a speculative nature.

"Treatment is seriously defective in literary form and structure, continuity of thought, clarity of expression etc.

"Material is trivial or overexpanded."

In the remainder of this chapter I review practices and guidelines of a number of technical organizations and publications and present the views of readers on what they consider common faults and main requirements of technical papers. The overview is intended to put the reader of this book on the same level as the experienced technical writer.

Responsibilities of Writing for Publication

"Technical writing," advises the American Society for Testing and Materials (ASTM) in its style manual, "carries with it the expectation of new ideas, new information, or a deeper understanding of that which is known already.

"The primary requirement of good technical writing is that it be understandable to the intended audience, which carries the expectation that the ideas presented will be clear, concise, and honest. To achieve this goal, the author must consider:

"1. What is the problem that the experiment (or, in general, the work) has solved (or clarified) and, hence, the paper must explain?

"2. What approach was used to solve the problem: namely, an all-encompassing study, a specialized study, a particular bias, a critical review of previous work, etc.? The author should have, in most cases, the approach firmly in mind before the actual experimental work is begun. However, unexpected findings, invalidation of results, or any number of things can destroy such planning; in this case, he should choose one method for the presentation and adhere to its specifics.

"3. To whom shall the report be addressed? Whom will it try to convince? Will it be general in scope, application, and audience, or esoteric?"

Another guideline is offered by the American National Standards Institute Inc. in its standard for the preparation of scientific papers for written or oral presentation (ANSI Z39.16-1979). Portions of Sections 2, 4, 5, and 11 from this standard are reproduced in this book with permission from ANSI. Copies may be purchased from the American National Standards Institute at 1430 Broadway, New York, N.Y. 10018.

ANSI states, "Before preparing scientific or technical papers, authors should consider whether they have meaningful advances in knowledge to report. They should define for themselves, in a few words, what these advances are and should attempt to judge their significance objectively. Colleagues who have not worked with them on their research but who know the subject may be able to help them to decide. If the proposed contribution to knowledge is minor, the authors should postpone publication until they have more significant contributions to offer."

An added obligation is suggested. "It is not sufficient for a scientific or technical paper merely to report what was done and what was found. Authors should define for themselves before planning the paper, and for their readers when they are writing the paper, how their work fits into what is already known, and what significance the conclusions have in the context of a wider field of investigation or technology."

ANSI also has statements on what it calls awareness of primary and secondary audience and selection of journal by the author.

"Authors," the institute says, "should consider what audience the paper is to address and select the journal accordingly. Before deciding finally that a particular journal is most appropriate for their purpose, however, authors should obtain and study the journal's most recent instructions to authors and should compare the content of their proposed contribution with the journal's announced purpose and scope and with the contents of a recent issue. The proposed paper should then be planned specifically for that journal."

The secondary audience is defined in this manner, "It is important for authors to bear in mind the needs not only of the 'immediate' readers—the specialists in the same field who will encounter the article soon after publication in the course of their usual reading practices—but also of the less specialized readers who have been guided to the papers by access services of various kinds. Thinking of such less specialized readers will help authors to avoid jargon, abbreviations, and shortcut thinking intelligible only to a few."

The American Society of Mechanical Engineers (ASME) puts similar guidelines in a section titled "Style." The society states, "It is well to remember that the chief purpose of a paper is to convey information to others, many of whom will be far less familiar with the general subject than the author. Care should be exercised, therefore, to use simple terms and expressions and to make statements as concise as possible. If highly technical or unusual terms or phraseology are necessary, they should be adequately explained and defined."

The American Welding Society (AWS) proposes, "All authors should address themselves to the following five questions when writing their papers:
"Why was the work done?
"What was done?
"What was found?
"What is the significance of your results?
"What are your most important conclusions?"

The Association of Iron and Steel Engineers (AISE) prefaces its instructions to authors with these remarks:

"The subject matter of the paper should be of interest to engineers, operators, and executives involved with the various phases of the iron and steel producing industry. Whenever applicable, maintenance and cost considered should be included. The text of the paper should present new developments, processes, techniques, applications, experimental results, or important substantial conclusions relative to the design, construction, operation, maintenance or management of steel plants or their equipment. The information used to support the thesis of the paper should be specific, detailed, and accurate."

The Society of Automotive Engineers Inc. (SAE) combines its concept of an acceptable paper with writing techniques. "Good organization of thought and careful revisions," it is declared, "are two ingredients common to good papers. Good organization reflects your thinking through the facts and conclusions you wish to pass along to your readers. Careful revision improves the reader's chance of fully understanding what you are saying.

"A written outline is the most common tool used for thought-organization. It is easy to assemble and reassemble your ideas in this form. Also, the outline can serve as a source of headings for the sections in your paper. Generally, a simple outline is all that is necessary. If you find you have multiple sub-subheadings, this may be a sign that your reader will get lost along the way.

"Revisions are best generated by your colleagues. After you have written a manuscript that satisfies you, try it out on a knowledgeable associate. If portions are not clear to this person, you can be sure many others will have trouble. However, don't ask your colleague to rewrite for you. Have your critic tell you about sentences that don't read right, unnecessary statements, unsupported conclusions, or faulty logic. Make note of places where more depth would be of interest . . ."

Specifications for Getting a Paper Published

Publishers of technical papers have common items on their laundry lists. They give numbers: how many copies of the manuscript to send in, copies of photos, of drawings, of tables. They give directions: specifications for manuscript preparation directed to typists; for selecting and preparing photos, drawings, and tables. They describe preferred formats and give guidelines for writing specific parts of formats. They announce limits on manuscript length. They recommend styles for writing. Define metric practice. Set rules for copyright transfer from author to publisher. They may spell out review procedures. Cite page charges, if any. All these rules or specifications are often reduced to writing and are available from the publisher, upon request.

Instructions to authors are explicit and detailed where for economic reasons the publisher chooses to minimize or avoid expenses incident to editing. In some instances, authors are expected to supply papers ready for the printer in all respects. No changes are made after the author puts his manuscript into the mail. Traditional editorial chores like proofreading, styling, and editing are the sole responsibility of the author. Forms are supplied for the typing of manuscripts. Page layout, including the placement of figures and tables, is taken care of by the author. Pages are supplied ready for photographing and conversion to film for printing.

Some publishers provide authors with hefty do-it-yourself stylebooks covering all facets of editing, copyreading, and the presentation processes. At the other extreme, instructions are brief and cover only the essentials like selection of photographs, standards for preparing drawings, and preferences for the typing of manuscripts. In such cases, editorial staffs take care of traditional housekeeping chores. Art directors, working with editors, select graphics and tables supplied by authors, modify and otherwise improve them if necessary, then prepare page layouts. Authors may receive review copies of edited manuscripts, or in some instances, galleys. Manuscripts may or may not be reviewed by a peer group.

Whichever practice prevails, it behooves the author to find out what his prospective publisher wants. Rejection could turn on something as simple as supplying a single-spaced instead of a double-spaced manuscript. Many publishers request double spacing, which is not a frivolous requirement. Editors edit in the space between lines. Single-spaced copy must be retyped. Expenses are incurred, as are delays and the risk of creating new mistakes that may not be detected.

The term "publication" has several meanings: written presentation, oral presentation, and all possible combinations of the two media. Order also varies, from written followed by oral to oral followed by written. Requirements for oral presentation differ in some particulars from those for written presentation, and are the subject of Chapter 13. In general, there is more emphasis on visual communication of a paper than there is in written presentation. The requirements are closer to those for technical articles appearing in business and trade magazines where the writing may be handled by a professional industrial journalist who uses the author and his paper as resources and is backed by the services of a commercial artist. How-to specifications for the writing of technical articles and the gathering of graphics for consideration by an editor are covered in Chapter 14. Some guidelines for preparing technical books are included in Chapter 15, and overviews of copyright law in Chapter 25, libel law in Chapter 33.

A Potpourri of Instructions to Authors From Publishers

The American Nuclear Society

The American Nuclear Society requires authors to submit summaries of contributed papers to the technical program chairman for review by the National Program Committee. Accepted papers are presented orally, and summaries are published in the *Transactions of the American Nuclear Society*. This does not preclude publication of the complete paper elsewhere. The author is obliged to pay $140 per page published in *Transactions*. The original and three copies of the summary must be supplied to the publisher.

Summaries are to be between 450 and 900 words. Figures or tables count as 150 words each, and there must be at least 450 words of narrative, meaning that no more than three figures and/or tables are allowed. References are not included in the word count and must be held to a minimum. Each line of an equation counts as ten words. Titles should be limited to ten words. The publisher warns, "A summary that does not meet these criteria may be returned without review."

Requirements for content of summaries are spelled out. "Each contributed summary must present facts that are new and significant. A simple listing of material to be presented is not acceptable. Abstracts are not acceptable. They [summaries] must contain not only the work that has been performed but also the results achieved. Proper reference should be made to all closely related information that has been published. Summaries should include an introductory statement indicating the purpose of the work and a closing statement summarizing significant new results."

Other rules concern preparation of text, tables, and figures. For example, "Summaries must be typed double spaced on one side of the paper only. Figures and tables are normally printed one column wide (7.50 cm, or 2.95 in.). Letters on figures must be at least 1 mm (0.039 in.) high after reduction, and the design of tables should suit the width limitations. Each table or figure must be on a separate page. Figures for one copy of the summary must be high quality glossy photographs or reproducible black-on-white drawings." Metric units are used, and "a copy of the *Metric Guide for ANS Publications* is available on request from ANS headquarters."

The American Society of Mechanical Engineers

Authors are asked to use *Webster's Third New International Dictionary* "for the spelling and usage of words."

Length of text of an ASME paper is limited to 6000 words, which is equal to six printed pages in a journal. Further, "Long quotations should be avoided by referring to sources. Illustrations and tables, where they can help clarify meaning or are necessary to demonstrate results properly, are desirable, but they must be kept to a practicable minimum. Detailed drawings, lengthy test data and calculations, and photographs that may be interesting, but which are not necessarily important to the understanding of the subject, should be omitted. Equations should be kept to a reasonable minimum and built-up fractions should be avoided whenever possible. Manuscripts that fail to conform to these requirements will probably be returned for revision and condensation."

Clearances: It is the author's responsibility to secure such approvals and clearances, including company, government, etc., as may be required, and notify ASME of any potential conflicts.

Copyrights are assigned exclusively to the American Society of Mechanical Engineers "for its use any and all right of whatsoever kind of nature now or hereafter protected by the Copyright Laws (common or statutory) of the United States and all foreign countries in all languages in and to the subject technical paper, including all subsidiary rights. The signed statement must be received by ASME Headquarters before manuscript can be accepted for publication."

References: If there are no more than four or five, "use footnotes." If the total is greater, "use a list of references at the end of the paper. Each one should be numbered serially, and in the text these numbers should appear in brackets. Extreme care should be taken. A single error causes waste of time in locating the reference."

Tables: Those with no more than a half dozen lines may be typed as part of text, but should be so located that they do not run over onto a second page. Larger tables should be typed on separate sheets and assembled at the end of the manuscript. Each should have a suitable descriptive heading. Tables should be numbered consecutively and referred to in text as Table 1, 2, 3. "Do not say, 'the following table' or 'table on page 3,' etc. Each table should be identified with the author's name."

Captions: Type in a list on a separate sheet or sheets, with the proper figure number with each. Sheets should bear the name of the author for identification.

Photographs: Prints should be clear and sharp, with glossy finish. Scales should be included on photographs, as needed. Photostatic prints and halftones from printed reproductions do not reproduce satisfactorily. Photographs should not be mounted or pasted in the manuscript. Figure number and author's name should be marked lightly (pencil rather than ink, for example) on the back of each photo in such a manner that the face of the print will not be damaged. Photographs should be mailed flat between cardboards—not rolled or folded— and paper clips should not be used because they mark photographs.

Other illustrations: High contrast, black on white glossy prints of graphs, charts, line drawings, sketches, and diagrams are preferred. Black ink drawings on heavy white paper or tracing cloth are acceptable. Each sheet should be properly identified with figure number and author's name. "Do not draw or paste illustrations on text pages. Text goes to the printer to be set into type; illustrations go to an engraver to be made into halftones or linecuts."

Tabulations: "Where several considerations, conditions, requirements, or other qualifying items are involved in a presentation, it is often advantageous to put them in tabular form, one after the other, rather than string them out in text. This arrangement, in addition to emphasizing the items, creates a graphic impression in the mind of the reader that aids him in forming an over-all picture of the situation . . ."

Drawings: Those "made for general use are not suitable for reproduction in published papers. This is especially true of large drawings where excessive reduction (3 to 1 or greater) would be required. The size of lettering is invariably too small and the lines are too light for proper reproduction. Also, there is usually too much detail. Material of this type should be redrawn with only the essential details shown and as little lettering as possible. The weight of lines may be graduated to emphasize important parts. In general, the drawing should present more of a picture than a working drawing. The minimum size of letter in the reproduction should be the same as that for graphs, namely about 2 mm (1/16 in.). It is preferable to put descriptive information in the caption rather than to letter it on the drawing."

Mathematics: "Formulas and equations should be carefully typed or lettered in the manuscript. A list of all symbols used in the paper should appear at the beginning of the text. The distinction between capital and small letters should be clear. Mark all Greek letters, particularly those easily confused with English letters, like alpha, omega, rho, nu, etc. Care should be taken to avoid confusion between the small letter 'l' (el) and the figure one, or between zero and the letter 'o.' All subscript and superscript letters and figures should be clearly shown. In all mathematical expressions and analyses, explain what the symbols stand for and the unit in which each is measured. In a highly mathematical paper, it may often be found advisable to develop equations and formulas in appendixes, rather than in the body of the paper."

Heading and Numbering: "Headings and subheadings should appear throughout the text to divide the subject matter into logical parts and to emphasize the major elements and considerations. They assist the reader in following the trend of thought and in forming a mental picture of the points of chief importance. Parts or sections of the paper may be numbered if desired, but paragraphs should not be numbered . . ."

Return of Illustrations: "Illustrations and original graphs and drawings are not returned to the author unless requested at time of submittal."

Typing: "Manuscripts should be typed double spaced on one side of standard letter sized sheets (8½ by 11 in., or 220 by 280 mm) with approximately 25 mm (1 in.) margins on each side.

Number of Copies: "Five complete copies of the manuscript, with illustrations, should be submitted, and the author should retain another copy. Of the copies submitted, one should be the original typed copy and the original illustrative material for use by the Editorial Department. Carbon copies of manuscripts and blueprints, photostats, and the like of illustrations are satisfactory for the additional four sets. They are used by reviewers."

Author-Prepared Mats: Special copy layout sheets are provided so that final printed copies are uniform in appearance. Complete instructions for preparation are printed on sample sheets. Manuscripts are photographed directly from the author-prepared copy. Neatness and cleanliness are of paramount importance."

Journals of the American Society for Metals

Three copies of complete manuscripts are submitted for publication in the *Journal of Heat Treating, Journal of Applied Metalworking*, and *Materials for Energy Systems*. "High quality micrographs are required for review purposes. Xerox-type copies are not acceptable."

Figures (halftones and line drawings) and tables should be designed for final printing in single column width of 8.5 cm (3¼ in.). Double column treatment is used only when required by complexity of material.

Line drawings should be no larger than 22 by 28 cm (8½ by 11 in.). Lettering should be large enough to be 2 mm (1/16 in.) high after reduction. Glossy photocopies of larger drawings are satisfactory.

One mounted set of glossy prints with captions should be furnished for the printer's use. If necessary, indicate suitable framing, position, and proportion on a working copy.

Printer's copy of all figures (drawings and halftones) should be retained by the author until requested by the editor, after the paper is approved for publication.

Original drawings and photos are returned to the author after the printer no longer needs them.

The abstract "is the author's summary of a technical paper and is included in the review procedure. It should indicate newly observed facts, conclusions, and essential parts of any new theory, treatment, apparatus, technique, etc. It should be concise and informative and only in exceptional cases exceed 200 words."

On printer's copy, symbols, such as "oh" and "zero," should be clearly identified and differentiated by a marginal note to avoid ambiguity.

There are no facilities for translating or making editorial revisions in foreign contributions. All publication is in English.

Policy is to use dual units—inch-pound units and the International System of Units (SI). For guidelines, see ASTM E380-79, available from ASTM, 1916 Race St., Philadelphia, Pa. 19603.

"Avoid use of trade names and proprietary information whenever possible. Such use can be occasionally justified if this is the best way to specify a particular material or process."

A table of symbols should be included when symbols are used extensively.

References such as "submitted for publication" and "to be published" are not acceptable.

References to internal reports and other publications of limited availability (not available by subscription) are not desirable. The report should be available on request.

Instructions From the American Foundrymen's Society

"An untidy manuscript," advises the society, "creates a poor first impression for the material it contains. Likewise, if in improper form, it creates an unnecessary problem for the editors, and may lead to further work on the part of the author in rearranging the manuscript prior to its publication. The manuscript should be written with lines double spaced, on one side of plain white paper, approximately 8½ by 11 in. (215 by 280 mm), using an unworn black typewriter ribbon. A 1 in. (25 mm) margin on each side of the sheet and a 2 in. (50 mm) margin at the top of each page should be left for editorial notes in preparing the copy for the printer.

"It is suggested that the total length of the manuscript be limited to about 25 typewritten pages (about 5200 words). All pages should be numbered in sequence (Arabic numerals). If it is necessary to insert pages later, they may be numbered 5a, 5b, 5c, etc., according to the number of the preceding page. The original and three copies must be submitted."

Column headings for tables should include the unit in which the figures in the column are expressed. For example:

Sample Number	Riser Diam. (cm)	Pouring Temp. (C)	Yield Strength ($Pa \times 10^6$)	Hardness (Bhn)	Microstructure ($100\times$)
271A	2.1	2770	460	465	Bainitic
271B	3.3	2770	441	432	Bainitic

Most tables are photographically reproduced and reduced for publication. Suggested widths are 5 or 10 in. (125 or 255 mm). Contents of tables should be typewritten single spaced in pica type (10 characters per inch). Table titles should be listed on a caption sheet.

In preparing drawings, remember that lines may be much thinner and lettering considerably smaller when reduced to fit the printed page. Original drawings 7 in. (180 mm) wide may be reduced one-half to one-third that size for publication. If the original is not drawn with this in mind, there may be a loss of detail necessary to meaning. Line width and letter size are most critical. All drawings should be made with black felt pen or black ink on white paper, tracing paper, or chart paper. Do not use blue ballpoint pen or pencil.

Formulas require special attention in typesetting. Original copies of all formulas must be provided. All formulas should be typed on a separate sheet and the sheet placed at the end of the manuscript with tables and figures. All formulas should be numbered in sequence as they occur in text.

If a typewritten formula is wider than 4½ in. (115 mm), split it so it does not exceed this maximum. Hand printed formulas should be prepared with a lettering device, with numbers and individual capital letters 3/16 in. (5 mm) high, keeping superscripts and subscripts in proportion. Wherever clarity will permit, one-line formulas should be used. For example:

$$t = C \ (V/A)^2 \qquad not \qquad t = C \left(\frac{V}{A}\right)^2$$

Guidelines for Authors From the Society of Automotive Engineers Inc.

A title indicates the scope and subject of a paper. It should be short, simple, and explicit. An example:

"Applications for Ultrasonic Welding of Steel. "

Use of company or trade names in titles should be avoided. "You are writing a technical paper, not a sales pitch. Your company affiliation will be noted immediately following your name on both the cover sheet and the first page of the paper."

The introduction, the first paragraph or so of a paper, should give the reason for writing it. What is the problem that needed solving and why? What is the scope of the paper? These paragraphs should set the stage for the detailed presentation.

"Good headings and grouping of information should follow easily from a good outline. Make sure headings say something like 'Dynamic Test Procedure,' rather than 'What To Do About It.' "

Some tips on writing:

- The objectivity expected of a technical paper dictates the style of writing.
- Editorial comments are out of place.
- Personal history is out of place.
- Sweeping statements, however well based, are better made orally than in writing.
- Commercialism is completely out of place.
- Constant use of company or trade names is undesirable.

Photographs: The figure number or title should not be part of the print. High contrast is preferred. Crop out any unwanted material the camera might have caught.

Graphs: Plan so the width of the largest one will reduce to 3⅜ in. (85 mm) and still be readable. Larger graphs may be used in exceptional cases but maximum final width is 7 in. (175 mm).

Use a coarse grid. Never use closely lined colored graph paper because reproduction is poor. Keep all line weight near uniform to prevent the dropping of thin lines. If there are several curves, identify them by labeling or by different types of broken lines, as: - - - - or - . - . - . or -- - -- - --. Lines drawn in black ink are preferred. Pencil lines fade and break up when printed. Typewritten lettering is particularly poor.

Do not use gray tones because screening will be needed to reproduce them. Screen will break up all solid lines. If toning is essential, say to show a band in a graph, use a series of lines, such as hatchmarks, in the designated area.

Drawings: Rules are about the same as those for graphs. Once more, the largest one should fit into a 3⅜ in. (85 mm) column. Lettering should be large, not heavy, clear, and in ink. Heavy lettering fills in when it is reduced.

Instructions From American Society for Testing and Materials

Four complete copies of the manuscript are required. Optimum size of a paper is 5000 words, or about ten printed pages including tables and figures. If pertinent references are unpublished, furnish copies of the unpublished work or sufficient information to enable reviewers to evaluate the manuscript.

For the American Society for Testing and Materials (ASTM), a manuscript is made up of the following items: title page, biography, abstract and key words, body of paper, introduction, figures, tables, equations, discussion (optional), conclusions, acknowledgments, appendix, references or bibliography.

Title Page: "Use words that sufficiently describe and inform the reader concerning the contents of the paper. Major key words should be included so the title may be of practical use in indexing. (See *Thesaurus of Engineering and Scientific Terms*, Engineers Joint Council; 345 E. 47th St., New York, N.Y. 10017.) Avoid trade names, specialists' jargon, and 'cute' titles."

Biography: For use in an ASTM publication. About 100 words (half of a typewritten page double spaced), including author's name, title affiliation, and where work was done if it was not at the author's current location.

Abstracts and Key Words: In about 150 words, summarize objectives, results, and conclusions as specifically as possible. Key words are the indexer's and retriever's tool. Use of universally accepted key terms for automated retrieval systems is essential. The *Thesaurus of Engineering and Scientific Terms* is a good source.

Body of Paper: Use major sideheads to indicate sections. Underline sideheads. For example: <u>Results and Discussion.</u>

Within a section, use minor subheads, as follows:

"Pure Metals — As shown in the tables, the initial maximum potential of chromium after immersion in Na_2SO_4 was . . ."

Results and Discussion sections usually follow the Procedure and Experimental Work sections.

Repetition should be avoided. "Do not use footnotes for descriptive or explanatory information; include the information at an appropriate place in text.

Introduction: "Present the significance of your work in relation to previously published work, but do not repeat the conclusions of the previous work. Reference to previous work should be brief and the major references cited."

Procedure Section: Most readers are not interested in detailed descriptions of equipment. Items of interest include basic construction, principal methods of control, unique features, advantages and disadvantages. Specimen preparation or sampling techniques should be described if they differ from ASTM standards. Omit details in procedure that are common knowledge in your field. "Briefly describe features essential to reproduce your procedure; when possible, details should be left to references."

Experimental Work (Results): "One may start with a theoretical basis for the experimental work. Tables and figures are used to condense data and thus keep text readable. Nonetheless, do not use tables or graphs which duplicate each other or material already in the text. Omit straight line calibration curves—give information in tabular form or in a sentence or two in text. If there are many and complex data, it would be better to place them in an appendix."

Figures: Number all figures consecutively, using Arabic numbers in order of reference in text. Include a typed list of captions, and legends if appropriate, for all, double spaced on a separate page. Keep captions as brief as possible. Detailed information and interpretation should be included in the text.

Graphs and Diagrams: "The primary function of graphs is to show the trends of given data, usually in a more readily comprehensible form than that of tables. There is, of course, the limitation of attempting to illustrate more than three variables on the two dimensional chart."

Drawings are reproduced photographically (black and white only). "ASTM is not equipped to do the artwork in its entirety nor even correct it. So, submit figures as you wish to have them appear—clean, neat, and legible."

Further, "All lines, lettering, and numbering should be sharp and unbroken. Use only blue ruled coordinate paper, the rules of which will not be picked up in the reproduction process. Typed letters on drawings do not reproduce well; use black India ink with a lettering set or press-on (wax-back) lettering for all letters, numbers, and symbols. On an 8 by 10 in. (205 by 255 mm) chart, for example, lettering should be at least ⅛ in. (3 mm) high and have a line thickness of 0.017 in. (0.45 mm), necessitating a Leroy letter set template 120-C and pen No. 0; or press-on lettering, 16 point boldface letters. Lettering on copy of other sizes should be in proportion. Label ordinates and abscissa of graphs along the axes and outside the graph proper."

Photographs: "Photographs are invaluable in clarifying points discussed in text. Unnecessary backgrounds should be removed. Identifying labels should be included in the photograph where helpful. Photographs of encased apparatus, the black box mystery, should be eliminated."

Tables: Useful because exact numerical values can be shown under any given number of variables. Tables are limited, though, because they show discontinuous rather than continuous information readily visualized in graphs.

Discussion: "Not essential for the research report paper, but the section is left out too frequently, and the reader is faced with an inadequate presentation of results. For the technical discussion or review paper, this is the essential section where the author speaks out. The discussion should succinctly cover the significance of his proposals, his interpretations, and his views on the topic at hand."

Conclusions:"Use only to discuss and interpret what has been accomplished or to indicate valuable inferences from your work and significant implications for future work."

Acknowledgments: Authors are free to recognize the technical or scientific guidance of colleagues or to recognize those who financially supported the project. "Recognition of typists, proofreaders, or ASTM staff assistance is not to be included."

Appendixes: They "are useful to present supplementary material when inclusion in the body of the paper would disturb the logical flow of the text, or when complex and lengthy procedures, data, and analyses are not significant to the main points or results. Style is the same as that for the main body of the paper."

References or Bibliography: References are listed and numbered consecutively on a separate sheet in the order in which they are mentioned in text, not chronologically or alphabetically. If the bibliography is for general information rather than information on specific points, place entries in topical divisions and in alphabetical order.

Nomenclature: The unabridged *Webster's Third New International Dictionary* and the *Book of ASTM Standards* are the arbiters for the use of terms. If the nomenclature of a paper is specialized, a Nomenclature Section at the beginning of the paper is mandatory. Likewise, if symbols used are specialized, a Symbols Section at the beginning of the paper is mandatory.

Some Recommendations of the American National Standards Institute Inc.

ANSI Z39.16-1979, "American National Standard for the Preparation of Scientific Papers for Written or Oral Presentation," presents an overview of accepted practices for technical papers. Some excerpts from selected items follow:

Title: "Every word should contribute specific information. Words of little substance, such as 'studies on . . .' should be omitted."

Abstracts: "Every manuscript submitted for publication in a scientific or technical journal should include an abstract—an abbreviated accurate representation of the contents of the article. With the exception of review articles and tutorial papers, such abstracts should give definite, although not necessarily quantitative, information and should be written in accordance with 'American National Standard for Writing Abstracts,' ANSI Z39.14-1971. A review article or tutorial paper should begin with a brief descriptive abstract that specifies the scope of the article and names the topics discussed."

Organization: "Most scientific and technical papers should have an introductory section in which the authors state the purpose and scope of the work; a methods (or experimental) section, which may include a subsection on materials; a results section; a section (commonly titled 'Discussion') in which the authors state conclusions and discuss their significance; and perhaps a summary. Authors should make the introductory and concluding sections intelligible to as wide an audience as possible; these sections should not be directed only to a small group of specialists. The methods and results sections are usually directed to specialists.

"The most usual structure of a scientific article is, then: introduction, methods (and materials), results, discussion, and (perhaps) summary. Most journals use this format at present; readers are accustomed to it; and it forms a logical sequence in most instances."

Introduction: Section should make clear: aim of the paper, authors' purpose in undertaking the work described, including any hypothesis to be tested; and the relationship of the work to a larger field of inquiry. Readers should learn from the introduction exactly what question(s) the authors are to address. The introduction to a long paper may include an outline of the rest of the paper or an indication of its content. Authors must give due credit to others for what is already known about the subject, but in introducing an original scientific article, they should not attempt an exhaustive critical review of the literature on the whole topic . . ."

Results: Authors should present the results of their studies in a sequence that will logically support (or provide guidance against using) the hypothesis, or answer the question, stated in the introduction. They should include only the data and illustrative material that are pertinent to the subject of the article. Results are often best presented by means of tables and figures. Numerical data presented in tables usually need not be restated in detail in text.

Discussion: In this section, authors usually relate their experiments, calculations, or proofs to one another and to previous and related findings, and discuss the significance and limitations of their conclusions. They should clearly demarcate their work from that of others and be scrupulous in including contradictory evidence. Authors need not slavishly discuss each table and figure in the order given in the results section. On the contrary, they should construct their argument logically and make references to the results where appropriate.

"In criticizing the work of others, authors should confine themselves to the scientific aspects and should avoid personal attacks or arguments based on prejudice or personal bias. They should particularly guard against inadvertent misquotation or misrepresentation of published work." [See Chapter 23 in this handbook.]

Summary: Follows Discussion Section and is for the convenience of those who have read the paper, and is usually intelligible only after this has been done. This means a summary differs from an abstract.

Illustrative Material: Should be designed or selected before the text is written. Copies from office-type duplicating machines are rarely acceptable.

Tables: Should be used when the author wishes to present exact values. If possible, tables should be self-contained and intelligible without reference to the text. Values that do not contribute to the point being made should be omitted. Statistical accuracy should be clearly stated. For example, ranges of values should be shown either as such or as standard deviations or standard errors of the mean.

Graphs: Plotted curves, histograms, or scattergrams, for example, are preferable to tables when the author wishes to emphasize trends and relationships rather than present exact values. If possible, graphs should be self-contained and intelligible without reference to text. The number of curves should not be too large to prevent readily tracing each curve over its whole length. Letters and numerals must be large enough to be legible (not less than 1.5 mm high) after the figure has been reduced to fit the format of the publication.

Photographs: Figures that show photographs of equipment are rarely as useful as line drawings or verbal descriptions that elucidate the principles of operation. Features of interest should be clearly indicated by means of arrows or letters. If possible, the center of interest should coincide with the center of the field. Photographs should be cropped, or marked in margins for cropping, to eliminate unimportant features.

Line Drawings: Authors should avoid the confusion that results from crowded illustrations. Lines must be heavy enough and letters large enough to be legible when reduced for printing, but there should be as few of them as suffice to convey meanings.

Readers Cite Common Faults They Find in Technical Papers

The following verbatim remarks indicate the gist and scope of reader criticism.

"Sells rather than reports an advancement in knowledge."

"Too much time on method, too little time on meaning or results; so long [that] the paper is a burden to read."

"Too often papers are sales pieces to promote a product or process rather than a significant technical contribution. Other papers may deal with rather trivial or well-known phenomena and don't give the reader adequate value for the time he spends reading."

"Papers sometime are lacking in completeness."

"The purpose of the work is not clear; the significance of results is not stated; many papers remind me of a maze."

"Too many data are presented. The reader must correlate the data to the conclusions."

"Some writers try to cram too much data into one graph or figure, making them difficult to interpret. Also, dual units should be used at this time, not metric alone."

"Too much irrelevant detail is presented."

"Jargon is overused. The writer seems to be mainly concerned about communication with others in his specialized field."

"The validity of data is often open to question. In checking a reference, for example, one may find it is not the original source of data."

"I see references to out-of-date literature and cases where the writer has only a superficial awareness of the literature. Also, careless errors in equations, and scientific nepotism—papers from one's group rather than the entire field are referenced."

"The writing style is dull and impersonal; usage of acronyms and obscure abbreviations is excessive; grammar is poor; and references are incorrect."

"Photos are inadequate and hard to interpret."

"Papers often lack flow because their components are not presented in an organized, orderly manner."

"Too wordy. Written to impress."

"Too specific. Not general enough for practical applications."

"Tables and figures are not well prepared."

"Writers are not always sure who readers are, meaning that some papers have too much or too little detail and explanation. Some visuals are ill-chosen and difficult to understand."

"Publish or perish is not justification for the already large and increasing number of incomprehensible papers. I rarely see more than two or three significant papers per year."

"Most papers are too long, too dry, and do not suggest areas of applicability; very often the language is too technical for the average technical person; graphs and charts are overused; the reader spends too much time flipping from text to chart and back."

"Technical people are trained to provide a logical, sequential development of their topic, starting with the background, what they did about it, what they found, and what they conclude. In contrast, many readers would like to know first the new insight or understanding derived from the technical investigation."

Readers Cite Major Requirements for Good Technical Papers

Reader comments center on format. For example, "The format should cover what we did, why we did it, results of what we did, and what results mean to us (significance or implications)." Other opinions relate to quality of writing. For example, "The technical writer should make it interesting, make the subject attractive to the reader, not get on a pedestal and preach, and write as though he is talking to the person on the subject."

A sampling of more verbatim comments follows.

"Lead paragraphs should adequately summarize the subject matter and importance or implications of the paper. This section can be followed by conclusions or a summary of the paper, followed by sections on procedures, equipment, and data for the benefit of those sufficiently interested to study the paper in depth."

"The key to organization is a well-thought-out outline. Other important points are good illustrations in sufficient number and completeness with brevity."

"A summary or overview is usually needed. The shorter it is, the greater the care needed in planning. Important aspects of the investigation must be described and adequately reviewed. Data must be condensed and held to a minimum.

"A good paper is brief, has good organization, complete information, and is written to reach those not intimately acquainted with the art."

"Papers should have abstracts, containing objectives, summary of results, and conclusions. Text should cover, in detail, objectives, procedures, results, conclusions, and references."

"A lazy person should be able to understand the approach taken . . . though not necessarily the technical content of the piece . . . and also be able to explain the approach to someone else."

"State the message. Prove your conclusions. Restate the message."

"State purpose; state status of technology in question and where paper intends to take reader; give clear itemized statements of tests, sources of data, and test results in almost outline form for easy reference; state results and conclusions; provide references."

"Requirements are basically the same as those for a good technical report."

"Visuals should be chosen for the information they convey, not just aesthetics."

"Technical papers must put the results of others in perspective, and bring the reader up to date on new work reported."

"A paper should be well organized, have clear headings, be concise, have usable references, and be unbiased."

"A paper should have a good summary, have a logical progression, good artwork, present well-designed and conducted experimental work."

"A paper should have a logical organization (or story line); a viable, interesting topic; and should be written in a style that delivers the message to the reader without wandering all over the map."

"A paper should be written so any high school graduate can follow it. Don't try to impress readers with your technical vocabulary."

CHAPTER 13

Guidelines for Preparing and Presenting Technical Talks

A technical paper can't double as a technical talk. Those who sponsor symposia and conferences and the audiences who sit in on such events frown on the reading of conventional technical papers by speakers and the conversion of unedited figures and tables from papers into slides or other visuals. Further, a new element slips into the picture: the skill of the author as speaker.

An introduction to this new set of requirements is discussed in American National Standard ANSI Z39.16-1979, "Preparation of Scientific Papers for Written or Oral Presentation."

"An oral presentation," ANSI advises, "should be prepared with oral communication always in mind, never with an eye to eventual publication, because written and oral communications require different methods to be effective. For example, frequent reference to the major theme is recommended for an oral presentation but is undesirable in a published article. Simple sentence construction, unpretentious phraseology, and frequent use of the first person in the active voice are even more desirable than in written papers.

"Before beginning to prepare a talk, a speaker should ascertain the size, character, and range of interests of the expected audience. The presentation should be prepared—or revised, if previously presented to a different audience—so as to be readily understood by most of the audience.

"The introduction should make clear the purpose of the talk and, if time permits and circumstances are appropriate, should place the topic in a context of a larger field of inquiry. A detailed methods section may not be appropriate. In the interest of ready comprehension, the speaker may wish to omit all but essential technical details, and even technical data, from both methods and results. When doing this, the speaker should inform the audience of such omissions and should state how the data may be obtained, if desired.

"In constrast to the usual practice in written articles, results should never be separated from their discussion. The speaker should give the rationale, the conclusions to be drawn, and the bearing of these conclusions on the theme of the talk for each experiment, observation, or calculation before he or she begins to describe the next experiment or procedure.

"Slides and visual aids should be planned and constructed so that the image is legible at a distance and intelligible at a glance. In general, each slide should make a single major point. The number of slides used should be far fewer than the number that is physically possible to show in the time allotted, and should rarely be more than one per minute. If possible, the talk should be planned so that the slides will be shown in groups rather than scattered throughout the talk, because continuous examination of projected illustrations is tiring for the audience.

"The material presented should be clearly summarized in a few sentences at the end. The summary should contain an interpretation of the meaning, significance, and limitations of the experiments or mathematical derivations in the specialized field or in a wider context.

"The paper should be prepared and the slides and other visual aids planned in such a way that the talk can be easily delivered within the time allotted. Sufficient time for discussion and questions should be allowed. If no time period has been specified, the length of the presentation should be kept within the limits of the capacity of the audience to absorb the information given.

"When possible, the speaker should arrange for one or more rehearsals, complete with visual aids, before critical colleagues."

By and large, those guidelines represent a consensus among organizations with written guidelines for speakers.

Guidelines From the American Foundrymen's Society

Tips on what is expected of a speaker, how to prepare a talk, a suggested format, tips on giving a talk, and specifics on preparing visual aids are discussed in a *Guide for Authors* distributed by the American Foundrymen's Society (AFS).

On being a speaker, AFS observes, "When presenting a paper, the author is really giving an exhibit of his knowledge, his ability to apply it, and his personality. It is important not only to the author but also to his company and associates that he make an able presentation."

Authors basing their talks on papers are advised that the script should be an abstract, covering:

"A clear statement of the problem in which the author is dealing.

"A description of his method of attacking the problem.

"A forceful view of his conclusions."

Well conceived and skillfully prepared slides are a must. AFS states, "Well selected and properly prepared projection slides are a valuable aid in presentation of a paper. It should be remembered, however, that they are to be viewed on the screen for only a short time and are to be accompanied by an oral explanation by the speaker. It is neither necessary nor desirable that the slides be complete within themselves. The audience loses interest in the presentation if they are required to digest detailed slides or more information than can be assimilated in the brief showing time allowed."

Detailed attention is paid to such matters as size and type of lettering, how lettering is prepared, and maximum allowable information per slide.

Typed lettering is discouraged because it is rarely readable when projected on a screen. AFS allows an exception: an electric typewriter with a carbon ribbon. Pica type capital letters are the smallest that can be used.

AFS explains, "To be legible, letters and symbols must have a certain min-, imum size on the screen. Size depends upon distance from the screen to the most distant viewer in the audience."

A rule of thumb for determining slide size is given. "For typical viewing, the maximum viewing distance is about six times the width of the screen image." An example: the readability of a printed table 4½ in. (115 mm) is determined from a distance of 27 in. (685 mm). In other words, if it is readable from that distance, it is suitable for copying and projection. Another example: if a wall chart 5 ft. (1.5 m) wide is readable from 30 ft. (9 m),it is acceptable as a projected image.

For clarity, lettering on slides a minimum of ⅛ in. (3 mm) high is recommended. Minimum line size is set at the width of a No. 00 pen. "Use of color in art, as well as background, improves the communication impact on the audience."

AFS suggests using a template 8½ by 5¾ in. (215 by 145 mm) as a guide in preparing for 35 mm (2 by 2 in.) slides. Other tips: use one idea per slide. Do not use complex graphs and mathematical tables.

Some tips on getting ready to talk and presenting the talk:

"It is strongly advised," says AFS, "that the author prepare a written version of his presentation and read it over several times so that he will know it from memory and not have to read it at the session. . . .

"Check your meeting room ahead of time, and have your slides arranged in proper sequence, as well as marked, to assure correct showing position. Arrange to give your slides to the projectionist in advance of the start of the complete session.

"Try not to read your abstract. This impersonal approach to the presentation of your efforts is the surest way to lose your audience's attention.

"Speak loudly enough for everyone to hear you. If you have a friend with you, have him sit well back in the audience, talk to him, and if you drop your voice, he can cup his ear as a signal.

"Speak to the audience. If you must turn away to work at the blackboard or projection slide, raise your voice. . . .

"Do not overrun your time.

"Some speakers prepare small cards on which are written in large handwriting just a few words as a reminder, and glance at them during the talk. It is then advisable to watch the time, and arrange to eliminate certain cards or items if you are taking too much time."

Finally, a way to estimate the length of a talk: "One can usually present 2000 words, or about eight doublespaced typewritten pages, in 20 minutes."

Some Guidelines From the American Welding Society

A way of determining the readability of slides (this time 3¼ by 4 in. [82 by 100 mm]) is suggested by the American Welding Society (AWS): slides are suitable if they can be read at arm's length.

Ratios are provided to insure full usage of a slide. For the 3¼ by 4 in. size, the height-to-width ratio of the area available for slide projection material is 0.75 to 1.0. For the standard 2 by 2 in. (24 by 36 mm) size, the ratio is 1 to 1.5.

Use of large letters is strongly recommended: "the larger, the better." Also, "Use India ink, not pencil or typewriter unless a plastic ribbon is used, and all upper case (capital) letters are more visible to the audience than upper and lower case (small) letters."

Slide titles should be avoided. They cut down on the amount of space available for slide material. "The speaker invariably repeats the slide title anyway."

Lines should be thick. In charts, reduce the number of grid lines.

An inexpensive alternative to color artwork is to have black on white art (drawings with India ink lines and printing) photographed in reverse. Black lines and letters will appear as white on black. Selected parts of the white can be tinted with a transparent colored dye, or the entire negative can be mounted with a colored filter. The result will be color (such as orange or red) lines and lettering against a background of solid black.

Unusual effects can sometimes be obtained by cutting portions from black-and-white photographs and mounting them on a colored background and making a color slide from the art—for example, a view of weld samples cut from a black-and-white photo and mounted against a background of yellow.

Titles on tables, and footnotes, should be avoided, says AWS. Maximum results for tables, adds AWS, can be obtained with maximums of five rows across and ten rows down.

Tip for marking slides: "Each slide should bear the customary thumb mark for the operator. A right-handed operator holding the slide with his thumb over this mark will always place the slide correctly in the holder, eliminating upside-down and reversed screen images."

AWS limits the lengths of talks to 20–25 minutes "because this is the time which experience indicates as meeting with the most successful reaction. This length equates to a paper with a maximum length of 2500 words (about 12 typewritten pages running 210 words each).

Guidelines From the Society of Automotive Engineers Inc.

The Society of Automotive Engineers (SAE) has awards for Excellence in Oral Presentations. Criteria for judging suggest desirable qualities for a speaker. They include:

- Preparedness (delivery quality, absence of lengthy pauses, presentation length, effective use of aids).
- Poise (personal control)
- Delivery (conviction, forcefulness)
- Clarity (ease of understanding)
- Impact (ability of speaker to hold audience)
- Aids (clarity and quality of slides, models, etc.)
- Response to post-talk discussion

Such matters as planning, legibility and complexity of slides, organization, and handling of slides are covered by SAE instructions.

Planning: "Many excellent technical presentations fail to reach the audience due to poor slides. Slides deserve at least the same careful planning and preparation that went into your manuscript."

Legibility: "The key to effective slide lectures is legibility. When the person in the rear of the room has to cock his head and squint to read your slides, you may as well turn off the microphone. Your message will miss its mark."

SAE has a template for type copy slides with a usable area $3\frac{9}{16}$ in. (90 mm) wide by $2\frac{3}{8}$ in. (60 mm) high. A dashed area outside this one which measures about 4 in. (100 mm) wide by $2\frac{3}{4}$ in. (70 mm) high is a guide for lining up the camera viewfinder. Recommendations: copy should be no more than $2\frac{1}{4}$ in. (57 mm) high by $3\frac{1}{2}$ in. (89 mm) wide, or 14 typewritten spaces high and 35 pica (or 42 elite) characters wide. Only capital letters are used.

Some tips on making slides effective:

"The surest way to put your audience to sleep," says SAE, "is to cram your slides with data. The objective is to clarify, emphasize, organize, and enhance your presentation. Technical details should be left to the printed paper.

"Limit each slide to one main idea. Restrict each slide to a maximum of 15 to 20 words or 25 to 30 items of tabular material. Leave out data you do not plan to discuss. Use several slides to cover a detailed topic that cannot be logically included on one slide.

"Most kinds of data can be best represented in graph form rather than in tabular form. Keep graphs simple, with no more than two or three curves.

"If you must refer to one slide on several occasions, use duplicates instead of trying to return to the original. This will make for a much smoother presentation and also keep peace with the projectionist."

Guidelines From the American Society of Mechanical Engineers

Speakers at American Society of Mechanical Engineers (ASME) events are usually limited to 15–20 minutes.

ASME Manual MS-4, "An ASME Paper," sounds the typical caveat against reading and suggests a format. "Papers should not be read in full," it is stated. "Such presentations are usually monotonous and tiresome and require too much time. What the audience wants is to have the speaker tell his story briefly in a conversational style. Experience with many successful presentations shows that they should cover (1) a clear statement of the problem dealt with, (2) a brief description of the attack, and (3) a forceful review of conclusions."

Further comments on format: "The most important parts of a presentation from the standpoint of audience interest are the introductory and concluding remarks. Know definitely beforehand what you are going to say at the beginning and at the end. Avoid spending too much time on the introduction. State your purpose directly and briefly. Be sure you allow sufficient time for proper presentation of conclusions. This is usually the part in which the audience is most interested. Do not let your story run down at the end. Keep up the interest (and your voice) until you come to the finish—then stop!"

A point about notes: "A set of notes containing the principal points is invaluable. They should be carefully organized to give proper continuity to the presentation. Preferably, notes should consist of a list of different items to be discussed rather than a series of complete sentences. They should present a graphic picture to the speaker around which he can build his story. Where the speaker must move about the platform to refer to charts or other illustrations, it is convenient to have notes or small cards that may be held in the palm of the hand."

This know-how for speakers is offered: "When addressing an audience, pause for several seconds before you start to speak and look directly at your listeners. If you are inclined to be nervous, this will give you confidence. Stand erect, keep your head up, speak clearly and distinctly. Address your remarks to your audience, not to your notes, or the blackboard, or the projection screen."

Using the mike: "When a public address system with a fixed mike is used, keep a constant distance from it. Avoid turning your head or walking away. If you want to point to something on the screen or blackboard, do so, but return to the mike before you start to speak again. If the mike is portable, move it with you when you leave the rostrum. A lapel mike allows the speaker considerable latitude in his position, but he should at all times avoid turning his back on the audience."

ASME takes the position that illustrations, with the exception of photographs, for printed papers are, in general, not suitable for slides. Titles, if included, should be as brief as possible. Any lengthy explanations should be given orally and not crammed into a slide. Tables of figures are not suitable for slides. Graphs and simple line diagrams make good slides. In the case of graphs, a broad line should be shown. "Remember," says ASME, "the audience will not have sufficient time to examine details or to study points of fine distinction. The aim should be to present a general picture, setting forth only the characteristics of chief importance."

ASME comments on the size of original art and lettering for slides: "It has been found that a 150 by 230 mm (6 by 9 in.) critical area works out very well. Preparation of the art on a 250 by 300 mm (10 by 12 in.) piece of Color-Aid or Color-Vu type of paper leaves a good sized safe edge surrounding the art work, and this facilitates both handling and eventual copying. For good legibility, no letter or character should be less than 6 mm (¼ in.) high when working to the 150 by 230 mm (6 by 9 in.) size."

"Each slide," ASME continues, "should have a sticker on it, so located in the corner that when the projectionist places the slide in the projector with the thumb of his right hand over the sticker, the slide will be projected right side up. The order in which slides are to be shown should be indicated by numbers written on the stickers."

A tip: contact prints made from slides can be used for reference by the speaker. He need not turn his head toward the screen. Also, "by labeling points of special interest with letters or other characters, use of a pointer can be eliminated."

When there are several slides, it is best to show them in a group. "It is desirable to describe each slide briefly and to point out significant details. Give the audience a few seconds to study the projection either before or after you describe it. If you talk continuously while the projection is on the screen, the listener will be distracted. Do not read slides. If they are properly made, listeners can see for themselves."

Some Guidelines From the American Institute of Chemical Engineers

"If you can read 2 by 2 in. (50 by 50 mm) slides without a magnifier, people in rear seats can probably read them on the screen," advises the American Institute of Chemical Engineers (AIChE).

The size of the original drawing, the institute says, will determine the necessary sizes of characters and widths of lines. Letters, numbers, and symbols are often too small to be read by the audience.

Other slide know-how:

- The audience is unable to comprehend a large amount of information on a single slide. Flow and apparatus diagrams are especially subject to overcrowding. Such slides should present only the bare essentials. If detail must be shown, limit the view to a small section of the whole.
- Slides are illegible when contrast is low. If black is used on white, the greater part of the slide area should be transparent. A slide made by photographing ordinary graph paper results in low transparency. Contrast is lost and the slide is too dark to be read. If white is used on black, called a reverse or negative slide, the major part of the slide should be opaque. Of the two types of presentation, the negative slide has been found to be the more legible.
- Visual aids should be used only when needed to clarify or emphasize the verbal message. Too many slides are confusing.
- Slides should be outlined before the talk is written. Visual aids should show the main thread of the presentation. Write the talk to follow this line of thought.
- Slides should be simple. Figures and charts for reports, books, and magazines are usually too complex for screen projection. Edit statements and data for slides to be sure you are using the minimum information needed to support your verbal message.

- Visual aids should not duplicate spoken information. They should amplify what is said.
- A slide should not contain more than 20 words or more than 30 pieces of simple data.

A final word on presentation: "In addition to its valuable technical content, it is essential that your presentation be an enjoyable one if the audience is to receive or assimilate the information, ideas, and insight that you have evolved," states AIChE. "It is an injustice to the speaker, as well as to the audience, if the speaker fails to communicate the important ideas. No characteristics are more important than the style of delivery and quality of slides. If necessary, speak from an outline, or the presentation will be deadeningly dull if merely read. Confusing or illegible slides are extremely annoying and will alienate the audience."

More Guidelines for Speakers From a Variety of Sources

An Eastman Kodak Co. publication, *Speechmaking . . . More Than Words Alone*, contains tips on the use of such visual aids as slides, overhead projectors, flip charts, blackboards, and movies.

Some examples of a review from *Communication News* (American Society of Association Executives) follow.

- Avoid using charts and graphs as an easy way to get through a speech. Decide what information is absolutely necessary and show only that. Do not use copies from books or blueprints. Illustrations that look good on paper may be confusing on a screen. In general, a dark background with white or colored lines works best.
- Flip charts and chalkboards should be used only with small audiences. They are especially helpful in audience participation presentations.
- Overhead projectors can be used with large audiences and are helpful when dimming the lights is not possible or desirable. Overhead projectors have two shortcomings: they are manually operated and must be close to the screen.
- Normally, you should use five to six slides for each minute of speech.
- Most screens are designed for horizontal slides. Avoiding mixing horizontal and vertical types—"the change tends to jar the audience." With vertical slides, you may have to reduce the image to fit the screen.
- In showing a movie, make certain the quality of the movie is consistent with the quality of other visuals. Limit use of the movie to 2- to 3-minute segments. This way, you, rather than the movie, control the timing of the presentation.

Some tips for speakers, chairmen, and the audience from Professor David A. Rigney, Metallurgical Engineering, The Ohio State University.
This format is suggested:

- Tell the audience what you are going to talk about. Be brief.
- Give main part of talk.
- Summarize main points.

Some general recommendations from Professor Rigney:

- Limit talk to 15–20 minutes.
- Slides or transparencies should be visible from the back of the room. This guideline is suggested: can the original drawing on the slide be read clearly from a distance six times its width?
- Speak up.
- Do not read.
- Do not memorize extensive sections.
- Introduce, describe, summarize.
- Avoid cramming too many ideas into one short talk.

Some tips to chairmen:

- Introduce speakers, giving name, background, topic.
- Assist with lights, visual aids, etc., when appropriate.
- Notify speaker if he or she should speak louder or more clearly.
- Notify speaker if he or she speaks too long.
- Encourage questions.

Some tips for the audience:

- Sit near the front.
- Listen attentively.
- Participate in question-and-answer period.

An experienced speaker, Lester Alban of Fairfield Manufacturing Co., offers several guidelines to speakers and about facilities.

To the speaker:

- The most important visual aid is the speaker. Look sharp. Feel sharp. Be sharp.
- Use a limited number of visual aids.
- Do not distract the audience with equipment that is not working properly.
- Handouts should be distributed before or after, not during, a presentation. During a presentation, handouts divert attention from the speaker.
- Keep assistance from associates to a minimum. Use a remote control for slides if possible. If not, set up a signal with the operator. Do not say, "Next slide."

About facilities:

- The room should not be too large for the size of the audience.
- Arrangement of chairs and tables should be orderly, not cluttered.
- Audience should have an unrestricted view of the podium and visuals.
- Utilities (lights, etc.) should be checked carefully prior to the start of the session.

A tip on gaging the length of a talk from the size of the manuscript is offered by the Association of Iron and Steel Engineers (AISE): "It takes about 2 minutes to read one page of doubespaced typing on 8½ by 11 in. (215 by 280 mm) paper typed full width with standard margins. Most authors can cover a subject in 20 to 30 minutes."

About delivery, from AISE: "If possible, the paper should not be read—the audience will be much more receptive to the presentation delivered in a conversational style . . . speak loudly and directly to the audience . . . do not attempt to read information on a slide; augment it with additional or explanatory remarks. Allow each slide to remain on the screen long enough for the audience to comprehend. . . ."

Panel Enumerates Common Faults of Technical Talks

Technical people usually have experience on both sides of the podium. The reader panel for this book was particularly critical in citing common faults of technical talks, as indicated by the verbatim comments that follow.

- Talks are sales pitches for a company, product, or individual research.
- Delivery is poor.
- Often there is an amazing lack of knowledge of a subject.
- Poor preparation.
- Poor visual aids.
- Talks are not understandable to one outside speaker's field.
- They (talks) put me to sleep, but that may be my problem.
- Tables are too complex.
- The speaker did not rehearse.
- The speaker lacks enthusiasm.
- The talk is too long; there is not enough time left for questions.
- Most technical writers are poor speakers and they should be sensitive to audience reaction, but this is the exception.
- Visual aids are frequently too busy—because they were originally prepared for another audience.
- Talks should be limited to 20 minutes—the attention span of an audience.
- Speakers attempt to make too many points, rather than a few clearly and well.
- Slides are unreadable by the audience.
- Talks are often read in a dull monotone.
- Slides contain too much data to be comprehended during presentation.
- Information is not presented in a logical, organized manner.
- Too much editorializing.
- Too much padding with information not related to talk.
- Talk too long; no time left for Q & A.
- Too many "ands," "umms," and "er-ahs," indicating speaker is uptight or not properly prepared. I hate a statement preceded or followed by "you know." If I knew, I wouldn't be in the audience.
- Lack of a main theme.

- Glorified sales pitches.
- No graph, table, or diagram from a report or published paper belongs in a talk as a visual. Report typefaces are too small to project on a screen.
- Hesitancy.
- Poor organization—a rambling style.
- Too much time spent on the introduction; the speaker hurries through theory and results; no time is left for discussion.
- Slides that require considerable explanation.
- Slides not properly focused.
- Information not technically significant.
- Poor summation.
- Dull. Speaker does not seem to be proud of his information.
- Flow of talk not orderly.
- A lack of authority in the speaker's manner, reflected in his tone of voice and choice of words.
- Nonpertinent comments and statements.
- Material is out of date.
- Approach is often too mathematical.
- Insufficient background on why work was done.
- Poor command of English language.
- Weak openings that do not capture attention of audience.
- There is a reason and justification for most technical talks, but almost universally they are spoiled by poor visuals and poorer presentation.
- Speaker merely reads slides. Does not add anything.
- A lack of appreciation that the audience is not interested in details and minutiae for experts.
- A talk should have at least one major point. It should stand out. Speaker should try to convince the audience that it is his major point.
- Telling stories which have no relation to subject or are in poor taste.
- Being completely sober and humorless.
- Poor management of time. Say the speaker has three points. He spends 60% of his time on No. 1; 35% on No. 2; and has only 5% left over for No. 3.
- Trite and redundant phrases.
- Improper staging. Such as a blackboard that is too far from the speaker or audience.
- Don't thank the audience if you were invited to speak. The chairman should thank you. If you asked to speak, it is proper to thank audience.
- Room dark.
- Voice dull.
- Paper read.
- Some talks are so technical the majority can't understand.
- Some speakers rehash known information. Listeners learn nothing.
- Speaker does not take time to outline his talk mentally before giving it. He rambles.
- Poor stage presence. Speaker does not speak clearly or project his voice.

Main Requirements for Good Talk Spelled out by Reader Panel

Among listeners, there is a strong consensus on "what makes for a good technical talk." A cross section of opinion follows.

- Good, clear delivery.
- Good illustrations.
- Knowledge of subject.
- The talk should be . . . informal, relaxed and humorous, well organized, well illustrated, tailored to the level of comprehension of the audience; speaker should make sure everyone hears; questions should be invited at the end.
- Start with a dry run before co-workers. Have them critique. Rewrite accordingly.
- No table more than four lines long.
- No talk longer than 20 to 30 minutes.
- Speaker enthusiasm.
- Interesting subject.
- Impressive experimental techniques.
- Qualified speakers.
- Make only a few major points.
- Gear talk to audience; do Dale Carnegie things. Be interesting. Be entertaining. Have a central theme. Do not talk too much.
- Be to the point; logical in sequence; have good slides; do not go too heavy on abstract computations.
- Be relaxed. Liven up talk with a little frivolity. Some talks are worse than a sermon. Speaker should smile occasionally. Move around. Use gestures.
- Limit the presentation to full development of the theme.
- Start by outlining main points. Cover in a logical fashion. Recapitulate at the end.
- Talk to the audience as you would to friends not familiar with your work.
- Introduce, tell, summarize in making each point.
- Tell them what you are going to tell them; tell them; tell them what you have told them.
- Make one or two points only; plan illustrations to support those points; don't review earlier work or explain method unless the information is related to your points.
- Have enough visual aids for the audience to come to the same conclusions as the speaker.
- Have new, worthwhile information. Not just a sales pitch.
- Good illustrations with legible print. Given spontaneously.
- Stick with subject.
- Avoid personal bias.
- Speaker should know subject well enough to forego constant reference to notes.
- Visuals should remind speaker of his key points and order of presentation.

- Shoot for comprehension by at least 90% of audience.
- Paper written for publication should never be read.
- Speak loudly and slowly.
- Sell yourself. Sell your subject. Dress in good taste.
- The key is a relatively new or timely subject presented in an understandable manner.
- Speaker should know tricks of delivery that hold an audience.
- Remember audience has a limited attention span.
- Graphs should summarize. Pictures should reveal.
- Prepare. Relax.
- Explain results in lay language. Most of us are illiterate outside our fields of interest.

CHAPTER 14

Guidelines for Writing Technical Articles

The technical article is related to both the technical paper and the technical talk. In content, the article is often a condensation of a paper. But a paper need not be the starting point. In length, the article may run as long as a 20- to 25-minute talk, meaning 10 to 12 pages of text typewritten and double spaced with normal margins. Many articles are shorter. Some longer. Such comparisons, however, are of limited value to the writer. The article is a separate art form with its own set of practices and know-how. An aspiring author should contact the journal or magazine he sees as a home for his article and ask for counsel. Help is usually available in the form of instructions to authors. This chapter gives several examples. Also in this chapter, editors of ten magazines present their views on 20 different aspects of technical writing (for example, the usual reasons for rejecting an article), and a reader panel answers the questions: 1. What common faults do you find in the technical articles you read? 2. In your opinion, what are the main requirements for a good technical article? Magazines participating include *American Scientist, Chemical Engineering, International Journal of Powder Metallurgy and Powder Technology, Iron and Steel Engineer, Machine Design, Manufacturing Engineering, Metal Progress, Oil and Gas Journal, Scientific American, and 33 Metal Producing.*

First, instructions to authors provide an overall view of what they should know about writing articles that stand a chance of being published.

What Some Magazines Tell Prospective Authors

Typically, an editor wants writers to know the makeup of his magazine's readers, his magazine's mission or duty to its readers, his magazine's interpretation of the kinds of information sought by readers, and the standard of writing that should be pursued. Instructions concerning drawings, photographs, graphs, and tables tend to be specific because of the emphasis placed on appearance. Routine details cover such topics as average length of text, number of candidate graphics and tables, transfer of copyright from author to the magazine (the author often retains minor rights), and deadlines, which are difficult for the uninitiated to fathom because periodicals typically work on what seems to be unusually comfortable lead times. A monthly magazine, for example, normally works on a four-month lead time. Included in this cushion, however, is a period allowed for review by the author following editing. For a monthly magazine, this is generally two weeks.

Instructions to *Metal Progress* authors are apropos as an example because of their brevity.

Readers

Pretty largely, what and how we write is determined by our readers and what they want . . .

Metal Progress is written and edited primarily for metallurgists and materials engineers who are responsible for the selection and application of materials and manufacturing processes.

They want specific, useful engineering information on a professional level, as opposed to a commercial, biased, narrow point of view.

Manuscript Preparation

Develop your topic in depth. Generally, this means 10 to 12 pages of text, typewritten and double spaced.

Submission of Figures

Metal Progress puts an unusual amount of emphasis on appearance. In addition to conventional technical graphics, pay particular attention to color photos or drawings that lend themselves to functional color treatment.

Color photos are often used as the lead illustration, and latitude is encouraged. For example, you may be writing about gears for cars. A striking color photo of a car would be appropriate for the lead.

Finished drawings and graphs are not required. Everything is routinely redone to *Metal Progress* style. (Note: authors of technical articles are responsible for the raw materials for presentation, but not finished art as they are for technical talks and papers.)

Deadlines

Your article should be in our hands about four months ahead of an issue date. An October article, for example, is due in late June or the first part of July.

Instructions to authors vary all the way from impromptu and form letters touching the basics to printed booklets that cover the waterfront.

American Scientist, for example, presents its "very flexible guidelines" in letter form.

"Our mandate," states Michele Press, the editor, "is to publish articles that synthesize recent significant research findings, with enough background and interpretation to permit readers in other fields to understand them and appreciate their importance. Our aim is to present articles by experts that describe the latest findings, with emphasis on their own research, in a way that will enable scientists outside the field to learn something about it. We do not as a rule publish papers that are basically reports of primary research which have not already appeared in the primary journals.

"Papers should be no more than 25 doublespaced typewritten pages, and can include up to about ten illustrations. For line drawings our draftsman can work from clear accurate sketches; for black and white photographs, we need glossy prints; and for color photos, we need transparencies. Four-color illustrations are costly to print and therefore difficult to justify; however, we are always glad to consider any illustrations you would like to include and can consult about them once the manuscript has been accepted by the Board of Editors. Illustrations should be accompanied by interpretive captions.

"We prefer that references be cited in the text by author(s) and date and listed alphabetically at the end of the paper. If you prefer an enumerated system, we have no objection. References should be limited to those important papers and books that a general reader might be most interested in consulting for further information. We like to have a few short subheadings (five word maximum) throughout the text. Also, we include a 15 to 25 word subtitle and a brief biographical note of approximately 100 words."

Manufacturing Engineering uses a form letter, signed by Daniel B. Dallas, editorial director, "to summarize the process of getting an article into print." One paragraph describes the editorial review–author approval process. "Articles are reviewed by one or more members of the Editorial Staff. If accepted, an article is assigned to an editor who performs the necessary editing. Upon completion of his work, the editor returns the manuscript to the author for his approval."

Review by the author is fairly standard among technical magazines. There is an implicit understanding that content is the province of the author and organization and writing are the province of the editor. Editors are particularly receptive to any mandates or suggestions relating to quality of content (usually accuracy) and author intent. Additional information and second thoughts may also be supplied by the author and get a fair hearing.

Other specifics offered by *Manufacturing Engineering* are similar to requirements for papers: type manuscript double spaced on $8\frac{1}{2}$ by 11 in. (215 by 280 mm) paper; drawings and tabular matter should be on separate sheets—not integrated into text; 8 by 10 in. (205 by 255 mm) glossy prints of black and white photos are preferred; figure numbers should be typed on small pieces of paper and taped or pasted to the back of the prints; writing on the back of photos with pen or pencil is discouraged because the writing shows through and often defaces the print.

Chemical Engineering sends prospective authors a comprehensive booklet on such matters as: the role of editors in helping writers prepare an acceptable manuscript, editorial scope, editorial criteria for evaluating manuscripts, a preferred style of writing, clearances by the author, what happens before and after a manuscript is accepted for publication, copyright transfer, payment of an honorarium, and related matters.

What do you need to know to be an author?

"You don't have to be a professional writer," advises *Chemical Engineering*; "You need only present some information, based on experience or know-how, that will be of value to our readers. Our engineering-trained editorial staff . . . will help you to produce an article that will reflect credit on yourself and your employer."

What do the readers of this magazine want?

In two words, "practical information." The reader "wants concise factual information that aids in solving real problems. Such a person does not look to *Chemical Engineering* for abstract theoretical treatises, vague general discussions, reviews of previously published material, or papers that promote only one of several options."

Writers are advised to submit a proposal for an article on an exclusive basis. Editors judge proposals on six points:

1. Will the topic appeal to at least one-third of the magazine's readers?
2. Will the article provide useful, impartial information about chemical processes, equipment, design procedures, plant operations, economic evaluations, or general engineering techniques?
3. Is the material timely or interpretive, not merely a rehash of old material?
4. Is the piece aimed at helping someone become conversant with the subject, rather than being written for another expert?
5. Does the information help readers in decision making, in technical administration, or in policy formulation?
6. Will the article contribute to the professional development of the reader?

How do you go about writing the manuscript?

Chemical Engineering advises, in part, "The manuscript should begin with a statement defining its scope, and the benefits that the reader may expect to get from it. The bulk of the manuscript then will be devoted to answering the questions of 'how' and 'why.'

"Write in a simple, conversational style, as though you were telling a friend about your work. Short sentences are always best; short paragraphs, too.

"Remember that your main job is to inform, and that is best done via simple, direct statements. Your readers asre busy people, so make your point with the fewest words possible. Difficult or abstract points can often be brought home effectively and concisely by citing examples or giving some 'for instances.'

"As you write, keep in mind that you are trying to communicate information to a broad spectrum of readers. So make your conclusions as generally applicable as you reasonably can, avoid the jargon of the specialist, and indicate where the principles described might be used other than in your own special area."

Writers are advised not to be overly concerned about the opening, the ending, grammar, spelling, punctuation, and length. These are taken care of by editors. The following tips are given:

"*The Opening*—the best way for you to start is just to begin writing about the subject, again remembering to write as if you were discussing it with another engineer. If a more general introduction is needed, our editors will provide one.

"*The Ending*—If you have emphasized the important points in the development of your theme, there is no need to repeat them at the end. Let the nature of your material determine what type of summary, if any, is needed.

"*Grammar, Spelling, Punctuation*—Write as simply and clearly as you can. Our editors will take care of grammar, spelling, and punctuation, and they will make any other style changes necessary to conform to our format.

"*Length*—Go into as much detail as you feel is necessary for full understanding of what you are describing. If condensation is needed, the editors can handle that in consultation with you. (We are flexible about length, but articles typically run about four to six printed pages, and contain about one-third illustrations and two-thirds text. As a guideline, four double-spaced typewritten pages translate to about one printed page of solid text."

Authors must obtain clearances from their sources prior to submission of their manuscripts to the magazine, and provide a statement to this effect.

Particular care is demanded in the handling of math in manuscripts. These guidelines are set forth:

"Mathematical equations, formulas, and operations should be presented clearly—and double checked. Indicate in the margins all symbols that might be ambiguous, such as the letter 'ell,' which can be mistaken for the number 'one.' Avoid using superscripts on super-superscripts, bars over letters, and lengthy exponents that may be easy to write but hard to set into print. We find more errors in mathematics than in anything else—so please double check.

"Be sure to define every symbol used, including its units—even if the meaning is generally known or appears obvious to you.

"If you are describing a calculation methodology, provide a sample calculation, preferably in SI metric units.

"If you are submitting a calculator program, include the calculation flowchart, the complete program listing, and a copy of the magnetic card (if applicable). Also note the typical running time for the program.

"Whenever a manuscript uses a number of symbols representing various physical and chemical units, please supply a table of nomenclature. A good guide for standard signs and symbols appears in Perry's *Chemical Engineers' Handbook*, 5th edition, p. 1–40."

Review by the magazine follows typical practice. Editors may consult with outside sources. When authors receive edited manuscripts for review, they are asked to sign a standard copyright assignment form. Ownership of the copyright is transferred to the magazine, but "we normally grant you, upon request, the right of re-use."

Chemical Engineering pays an honorarium—not all magazines do. Authors also receive a copy of the issue in which their articles appear, along with a small supply of tear sheets. Reprints are usually available at a discount. Consult the editor who handled your article on this point.

In its booklet for authors, *Machine Design* defines four types of articles it publishes (trends and news feature articles, technical articles, personal and professional articles, and idea generating features), defines its readership, provides tips on writing, and answers such questions as "How long should the article be?"

Readers, we are told, have three main characteristics: they either work in or manage the design function; they may be in applied research, advanced design, development work, or design for production; they design all kinds of engineered products, not just production or industrial machinery.

"No one article," the magazine advises, "can appeal to all readers. But we urge you, as a prospective author, to handle your article so that it will appeal to as broad a cross section as possible."

The news feature or trends story "reports current technical events of interest to design engineers—and interprets their meaning and significance to engineers. The story may be a single event or a survey and interpretation of a major trend."

Knowledgeable people in engineering and science write these pieces; they are also written by *Machine Design* editors, or produced as a team effort—editors do the writing; authors supply the information.

Writers with ideas for trends and news feature articles are asked to query the magazine. A tip: "Emphasis is strongly pictorial. Good photographs and illustrations are needed—plus accurate and clear reporting of the news or trend."

A technical article is regarded as one that "helps readers solve their engineering problems or updates their knowledge in an area of engineering design. An article may be a survey of the state of the art; it may give advice on selecting a type of component material, or engineering system; or it may outline a design method or analysis technique.

"In any technical article, the important things are facts, the ideas, and the advice offered. Writing style is secondary. But we hope you'll give some thought to presentation. As one communication expert has asked, 'Does it have to be dull to be technical?' "

In idea articles, the magazine looks for "new or different ideas, quick solutions to problems, design shortcuts, or cost-saving tips."

Men in Management articles cover a variety of nontechnical topics: engineering management methods, systems, and approaches; the relationship of engineers, as people, to their jobs and to the profession of engineering; the profession of engineering itself—its development and future."

The discussion, it is suggested, should give authors a feel for article subjects acceptable to *Machine Design*. Other steps in the writing process are broken into eight categories; selecting information, making an outline, writing a rough draft, revising a rough draft, selecting illustrations and special presentations, assembling a final draft, submitting for clearance, submitting an article.

Selecting information—Decide "which material is interesting to *Machine Design* readers?"

Making an outline—The key to organization. "Make an outline to serve as your guide and road map through the article. It is usually a good idea to send us a copy of the outline at this point, along with a short description of your graphics."

Writing and revising a rough draft—You may want to have your rough draft typed. Put it aside for a couple of days or a week. Read it again. Look for rough spots.

Selecting graphics—Writers are advised, "One of the biggest problems an editor faces is finding dramatic illustrations with eye appeal that will attract readers. An article with only straight text, curves, tables, and equations is likely to look dull, even though it may be useful to many readers."

Authors are given examples of desirable raw materials for presentation: dramatic photos that can serve as lead figures; case histories of applications, including photographs and informative captions; a sidebar feature to place inside an article—an item that may be used as an appendix, for example; sketches of good and bad design; and checklists, such as the summary of a design procedure or a list of important points.

Once more, it is the author's responsibility to obtain clearances—company, military, and otherwise. A company public relations or advertising department can be helpful with this chore.

When you submit your article, keep at least one copy for yourself—just in case.

Machine Design pays an honorarium of "about $35 per printed page for short articles, somewhat less for long ones." Authors are given ten tear sheets and a copy of the issue in which the article appears.

Articles typically run two to five pages in the magazine. Roughly, each typewritten page (double spaced on 70 character lines) equates to one-third of a printed page. Estimate space requirements of photos, drawings, graphs, and tables the same way—one-third page each.

Do not submit your article to more than one magazine at a time.

Authors may re-use illustrations, tables, and similar information from their articles. "We customarily grant permission. Re-publication as a book chapter is also permitted on request. If you or your company wish to reprint the article, we will also grant your request providing no changes are made in text or illustration."

How much editing can the writer expect?

"Editing can vary from light to heavy. Occasionally, portions of your article may be rewritten or deleted. Our reasons for planning any modification are the same as yours for writing the article in the first place—to attract the biggest possible audience for your article. If your article is technically complex, we'll let you know beforehand if substantial changes are planned, and you'll have a chance to check the final text for technical accuracy before the article is published."

Why Editors Reject Articles and Related Matters

Editors were surveyed to get their views on topics ranging from "What are the usual reasons for rejecting manuscripts?" to "Is content a typical problem?"

The International Journal of Powder Metallurgy and Powder Technology usually rejects manuscripts because "nothing new is reported or supporting data are weak."

Iron and Steel Engineer: "Not appropriate for our readers. All accepted material is related to engineering, operation, maintenance, and management of steel plants and equipment—occasionally, a nonferrous operation."

American Scientist: "Inappropriate for our editorial purview."

Manufacturing Engineering: "Subject matter does not relate to our field of interest; manuscript is germane but too poorly organized and prepared; excessive commercialism and lack of objectivity."

Chemical Engineering: "Policy is well defined. Articles must provide hands-on practical guidance in a general or generic way. We reject manuscripts that are philosophical, descriptive (rather than nuts and bolts), centered on proprietary products or processes, or too specific (narrow) to appeal to a wide enough segment of readers."

Machine Design: "Subject not appropriate for design engineer readers; too narrow—too specialized; so poorly organized it can't be repaired; a pure product puff; a case history that does not include alternative design approaches that were considered."

Scientific American: "Manuscripts that fail to impart interesting information."

33 Metal Producing: "Manuscript does not specifically address the editorial content or subject matter of *33*, or does not explain or fully describe facts alleged."

Oil and Gas Journal: "Too commercial; slanted toward wrong audience; too general."

Metal Progress: "Outside editorial scope; of potential interest to a limited number of readers; information not important enough to be of feature article caliber; information shallow and incomplete but can't be upgraded through editing and contact with sources. Typical reasons: specifics needed are proprietary; source is not willing to replace commercialism with solid information, or is unwilling to admit bias."

The editors were asked 19 additional questions.

1. "Is content a typical problem?"   Yes 100%
2. "Is content ever too shallow?"   Yes 100%
3. "Is content ever lacking in specifics?"   Yes 100%
4. "Is content ever underdeveloped?"   Yes 86%
5. "Is content ever not accurate technically?"   Yes 75%
6. "Is content ever unclear and confusing?"   Yes 88%
7. "Is content ever lacking in originality?"   Yes 100%
8. "Is content ever too commercial?"   Yes 100%
9. "Are manuscripts ever too long for your needs?"   Yes 87%
10. "Are manuscripts ever cluttered with unnecessary detail?"   Yes 50%
11. "Is organization of text a typical problem?"   Yes 75%
12. "Do articles ever appear to be confusing or contradictory because of poor organization?"   Yes 72%
13. "Are rewrites often required because of poor organization?"   Yes 87%
14. "Are rewrites to fix poor organization normally shorter than the original?"   Yes 86%

15. "Does language typically present problems?" Yes 72%
16. "Do you frequently find needless repetition?" Yes 72%
17. "Do you frequently find overuse of language that does not inform?" Yes 72%
18. "Do you frequently find language that is ambiguous or otherwise confusing?" Yes 100%
19. "Do you frequently encounter verbosity and jargon?" Yes 86%

Editors were asked to comment on each of the 19 questions. What they said follows.

Question: "*Is content a typical problem?*"

Chemical Engineering: "Even articles that take an appropriate focus for us are often sent back to the author for reworking because they presume too much background knowledge on the part of the reader; authors make a point but do not provide enough guidance to help a reader understand how to apply this point in his or her situation; also authors do not put the content of the paper into appropriate perspective for our audience."

Manufacturing Engineering: " . . . far too many manuscripts are written to grind an ax instead of convey engineering information. Even with the commercialism edited out, the manuscript remains suspect."

33 Metal Producing: "Information misses readers' interest; does not get into the meat of the subject; lacks specifics."

Iron and Steel Engineer: "Not appropriate for our readers. All accepted material is related to engineering, operation, maintenance and management of steel plants and equipment. Occasionally a nonferrous operation is included."

Question: "*Is content ever too shallow?*"

Iron and Steel Engineer: "If subject is lightly treated, it is published as a news item. Many times an expanded version (in-depth look) will follow, either voluntarily or solicited."

33 Metal Producing: "Authors often gloss over specific data, facts, and matters of substance without explaining information, source, and whatever."

Oil and Gas Journal: "Material is too general, too obvious, not specific enough for our readers who are engineers and managers. They require specific information of use in solving technical-operating problems in their jobs and businesses."

Manufacturing Engineering: "Nine out of ten manuscripts submitted through public relations departments or advertising agencies are written by people who don't have the foggiest idea of what they are talking about. If the manuscript contains the germ of a story, it is necessary to go back for more information.

Chemical Engineering: "Authors often presume they can just highspot something, without providing real guidance or background. Many of the philosophical or descriptive drafts just spout truisms or state the obvious."

Question: "*Is content ever lacking in specifics?*"

Chemical Engineering: "We seek drafts that the reader can actually use—concrete guidance on how to do something; often an author will state that something should be done, but not provide enough specifics to enable a reader to really do it."

Machine Design: "Usually a highly theoretical piece without evidence that the technique has ever been applied."

Manufacturing Engineering: "Especially when the author is an expert, he tends to assume more knowledge than the reader has. Again, it is necessary to go back to the source for more information."

33 Metal Producing: "Statements or claims are made without documentation to back them up, or generalities that are impossible to prove or substantiate."

Question: "*Is content ever underdeveloped?*"

Manufacturing Engineering: "This is not a frequent problem. Periodically, however, we receive a manuscript in which the author provides information on a development without giving additional information on its implications, i.e., cost effectiveness, training problems, and so on."

33 Metal Producing: "This problem goes hand in hand with shallow content that is lacking in specifics."

Question: "*Is content ever not accurate technically?*"

Oil and Gas Journal: "This isn't a big problem."

Manufacturing Engineering: "Not much of a problem. We find that the words provided are generally technically accurate. But additional thoughts that should be there are often omitted. So while the manuscript itself is correct, the picture it presents is sometimes misleading."

Machine Design: "Not usually a problem. We have been mousetrapped, of course. Editors cannot be expert in all fields."

Chemical Engineering: "On accasion, but this is not a frequent problem."

33 Metal Producing: "Inaccuracy is fatal, especially for our type of publication."

Question: "*Is content ever unclear and confusing?*"

33 Metal Producing: "Yes, but the material can be clarified with a few phone calls or a phone interview with the author or an industry expert."

Chemical Engineering: "Authors often presume too much background information and knowledge on the part of their potential audience—not distinguishing between a specialist-expert readership and a more general one; also, poorly organized and poorly developed manuscripts can be unclear and confusing."

Machine Design: "Engineers don't organize well."

Manufacturing Engineering: "Engineers, noted for their intellectual clarity and precision, generally can't write."

Oil and Gas Journal: "This isn't a big problem."

Question: "*Is content ever lacking in originality?*"

Oil and Gas Journal: "We don't place great emphasis on originality. Our readers need specific, down to earth information on making systems and machinery work."

Manufacturing Engineering: "When you are dealing with PR or advertising agency prepared articles there is a one in four chance that you are dealing with warmed over material."

Machine Design: "No comment needed."

Chemical Engineering: "Not a major problem with us. In many cases authors simply do not seem to realize that the subject has already been covered— sometimes because it was a while before."

33 Metal Producing: "Varies from manuscript to manuscript. Some are good. Some bad."

Question: "*Is content ever too commercial?*"

33 Metal Producing: "Editorially we spell out that no commercial message will be tolerated. We reserve the right to edit out this type of data."

Chemical Engineering: "We encounter this frequently. Many drafts tout a particular product or option, often quite subtly. We generally reject these, though sometimes we ask authors to tackle the subject generically."

Machine Design: "Puff, puff, puff."

Manufacturing Engineering: "An article treating a subject that has commercial content is a valuable sales tool. Accordingly, the pressure is on to make the article commercial. The agencies know better but in many cases their clients do not."

International Journal of Powder Metallurgy and Powder Technology: "There is too much promotion of companies and their products."

Oil and Gas Journal: "The biggest single problem with unsolicited articles. They are either too product oriented (we require details on the use of the equipment in systems to solve problems, not descriptions of a particular product), written by the sales department or by an advertising agency to satisfy or impress the client or the sales-marketing manager of the client."

Question: "*Are manuscripts ever too long for your needs?*"

Oil and Gas Journal: "We can handle great length if content is on target."

Manufacturing Engineering: "An occasional problem, particularly with articles dealing with management subjects. We have received 20 page, single spaced manuscripts which, if used, would eat up half our available pages."

Machine Design: "Yes, but a good editor can condense."

Chemical Engineering: "Frequently, but we edit all articles for length."

33 Metal Producing: "When queried, we provide guidelines for length of text and number of illustrations; otherwise, we edit."

Question: "*Are manuscripts ever cluttered with unnecessary detail?*"

33 Metal Producing: "Not an ultimate problem, but such manuscripts may take longer to review and to evaluate. Problem stems from coverage of a lot of basics and repetition."

Chemical Engineering: "Occurs quite often. If the manuscript is of value otherwise, we edit out this material—subject to review by the author."

Machine Design: "A good editor can handle this."

Manufacturing Engineering: "Not too much of a problem. However, some authors build their articles around tables that are too long and too plentiful. Similarly, there may be far too many graphs for the supporting text."

Oil and Gas Journal: "A case of using the wrong information—information the reader may already have, or information not essential to understanding the point."

Question: "*Is organization of text a typical problem?*"

Oil and Gas Journal: "We often have to reorganize to state the point or theme at the start of the article."

Manufacturing Engineering: "A typical but not a major problem. If the article is there, we will handle organization through editing and rewriting."

Machine Design: "Engineers too frequently use the detective story structure, i.e., the butler did it on the last page."

Chemical Engineering: "This is not too much of a problem, except that many authors still put their key conclusions on page 11 instead of page 1."

33 Metal Producing: "Organization is usually pretty well thought out."

Question: "*Do articles ever appear to be confusing or contradictory because of poor organization?*"

33 Metal Producing: "In the more technically oriented pieces this is more common than not. Heavy editing is required if the piece is accepted."

Chemical Engineering: "We rarely get manuscripts that are contradictory internally. More often, though not too frequently, drafts are confusing because of poor organization, such as calling into play something before it has been developed in text."

Machine Design: "Question needs no comment."

Manufacturing Engineering: "Initially a poorly organized manuscript seems to be very confusing."

Oil and Gas Journal: "Mainly confusing; not so often contradictory."

Question: "*Are rewrites often required because of poor organization?*"

Oil and Gas Journal: "We can usually handle poor organization by editing rather than complete rewriting, but editing may be extensive. We edit for clarity and delete or add material to make sure the result is on target."

Iron and Steel Engineer: "Usually articles from foreign authors."

Manufacturing Engineering: "Because of poor organization, poor sentence structure, poor etc."

Machine Design: "Yes. Always."

Chemical Engineering: "Minor rewrites to handle some poor organization are common. More major rewrites to tackle over-all poor organization are rarer. Such manuscripts would be returned to the author with comments on what is required."

33 Metal Producing: "Usually the more technically oriented pieces."

Question: "*Are rewrites to fix organization generally shorter than the original?*"

33 Metal Producing: "More often than not. Once a piece is edited, it is sharper, crisper, and shorter."

Chemical Engineering: "Yes, when we feel it is prudent for us to tackle the rewrite, such as when the information is good. Prospects of an author making a piece more coherent are minimal. Our version is generally shorter than the original."

Manufacturing Engineering: "There is no pattern. Sometimes a rewrite is slightly longer than the original."

Oil and Gas Journal: "Some rewrites are longer—not due to language but because pertinent material was omitted from the original."

Question: "Does language typically present problems?"

Oil and Gas Journal: "Not too much because we don't stress writing skill beyond simple clarity."

Iron and Steel Engineer: "When authors with a language other than English think and write in their native language and have their papers translated into English grammar is usually correct; but word selection is not conventional."

Manufacturing Engineering: "Undefined technical terms can present enormous problems in horizontal books such as ours. New terminology is constantly being born. In many cases authors use it without defining it." [Note: a horizontal publication covers more than one field—say materials, processing, and fabrication. A vertical publication is in a single field, such as heat treating or welding. This is publishers' jargon.]

Machine Design: "Handling language problems is an editor's job."

Chemical Engineering: "Language poses a problem to us typically and not infrequently on two fronts: someone writing pedantically—to impress rather than to inform; authors to whom English is not the first language (we have many such would-be authors), who try but don't always succeed in writing understandable English."

33 Metal Producing: "Language not too bad; but jargon, in-terms, and so forth are usually major stumbling blocks."

Question: "Do you frequently find needless repetition?"

Chemical Engineering: "Not a major problem, and one we easily handle in editing."

Machine Design: "We remove it."

Iron and Steel Engineer: "A problem with authors from foreign countries—Japanese in particular."

Oil and Gas Journal: "Usually relates to poor command of language. But we don't often find repetition of points, just language."

Question: "Do you find overuse of language that does not inform?"

Oil and Gas Journal: "Technical jargon is one of the main problems."

International Journal of Powder Metallurgy and Powder Technology: "Yes. Long sentences written in a pompous, academic style."

Manufacturing Engineering: "Periodically, when the author begins to entertain the delusion he is as much of a writer as he is an engineer."

Chemical Engineering: "Many would-be authors feel the only way to appear knowledgeable is to write pedantically—writing to impress, rather than inform. They use passive constructions, long, if not pompous, words when they are not called for, etc."

Question: "*Do you frequently find language that is ambiguous or otherwise confusing?*"

Chemical Engineering: "Not a frequent problem. It crops up mainly in an author's use of symbols or such which are not properly defined, if at all."

Manufacturing Engineering: "Generally this takes the form of poor sentence structure."

Oil and Gas Journal: "Our authors are usually engineers writing in their fields. Their points are usually clear, although their language skills may be limited."

Question: "*Do you frequently encounter verbosity and jargon?*"

Oil and Gas Journal: "Yes, due to technical jargon and redundancies."

International Journal of Powder Metallurgy and Powder Technology: "The greatest sin of all in technical articles. Some authors seem to feel that verbosity is an indicator of how much they know. Not true."

Manufacturing Engineering: "Not a major problem. In some cases, the practices are a source of amusement to editors."

Chemical Engineering: "Verbosity is a frequent problem. Jargon is a problem mainly when the author falls into the trap of writing for other specialists, rather than a more general audience."

33 Metal Producing: "A typical problem of amateur authors. they think in terms of quantity rather than quality or brevity."

Reader Panel Comments on Common Faults of Technical Articles

What, in the opinion of readers, are the common faults of technical articles?

Key phrases include: too long or too short, commercial, lacking in specifics, nothing new, not complete, too narrow, good side only, data questionable, little useful information, overedited, poorly organized, too much jargon, not enough tables and illustrations, not written for the audience, claims not supported, technical significance overplayed, no interpretation of significance.

Some selected comments:

"Articles are longer than or shorter than their educational content dictates."

"They tend to be promotion pieces to publicize a company, its products, or services; often too general and do not contain specific data useful to the reader, or they present a product or process pretty well known to everyone in the field as being novel or innovative."

"Editing is poor. Some sentences are difficult to understand; too often the information is from a source with a commercial interest and this shows through."

"There is a tendency to exaggerate the good side with little or no mention of disadvantages or limitations."

"Validity of data is often open to question."

"Articles either too technical or not technical enough."

"Often lack support for claims and sound like sales literature."

"Inaccuracy due to generalization."

"I feel sales pitches are OK for trade journals, if documentation—including references—is given."

"Usually well written, but are often inaccurate, contain misquotes and misinterpretations."

Reader Panel Opinion on What Makes a Good Technical Article

In response to the question, "What are the requirements for a good technical article?" one reader quipped, "The author's background, address, and telephone number." The sarcasm has a basis in fact. Technical articles are not always complete, and there are occasions when readers want clarification, or amplification, or documentation of claims and other forms of selling.

Reader recommendations related to commercialism reflect a similar concern. A sampling:

"The good technical article does not present claims not supported by hard facts."

"Reporting should be unbiased even though selling advertising is the purpose of the magazine."

"Articles should be as noncommercial as possible, but should still recognize that costs and benefits have their place."

"Writers should present new, worthwhile information and not just sales pitches."

"A low-key sell and reporting style should be adopted."

"The writer should be limited to one sentence about the company and the rest of the piece should concentrate on technical advancements if any."

Other specifications for a good technical article include the following:

"Good illustrations intimately tied to the text."

"Graphs and charts should have coordinates that are quantitative and clearly understandable; they should not be biased or distorted by plotting from a baseline other than zero."

"Good grammar and clear prose."

"Good organization, good illustrations, completeness with brevity."

"Know the audience and write to it."

"New or state of the art information presented in a concise and interesting manner."

"Case histories that help to support facts."

"Practical information."

"A good balance between narrative information and graphics."

"Interpret."

"Give background information."

"Subject should be of general interest; information should be useful; a bibliography should be part of the package."

"Illustrations should be near references in text."

"Information should be accurate and presented with clarity."

"An article should not be as detailed as a technical paper."

"Organize, support conclusions, keep jargon to a minimum, know when to stop writing."

"Gear writing to the common knowledge and interests of the audience."

"Don't baffle or bore with unnecessary information or theoretical discussion."

"Be specific."

"Define technical terms."

"Answer the question, 'What will this do for me?' "

"Be short, be simple, use examples."

"Explain in easy to understand language."

"Come to some conclusion."

CHAPTER 15

Guidelines for Writing Technical Books

For the benefit of those who have not written a technical book, how hard is it? What are some of the things you should know before taking the plunge?

What common faults are cited by readers of this form of technical literature? What are their ideas on the attributes of a worthwhile book?

What are some of the format and other mechanical requirements? What is the gist of contracts publishers have writers sign? What about expectations for pay?

Each topic is explored in turn.

First, is a book a formidable undertaking one should approach with extreme caution?

Not necessarily. In the abstract, if you can write a decent technical article—in the eyes of an editor or a reader—you should be able to put together a book. Think of the job as a series of papers or articles, counting each as a chapter.

Further, you must have something worthwhile to say. As a rule of thumb, you qualify on this count if you have sold yourself on the project to the point, you have drummed up some enthusiasm, and can't wait to get started writing.

Further, you must be able to express yourself in a reasonably simple, reasonably straightforward manner. In part this comes with knowing your subject. In addition, you must have reached a point in your thinking where you have come up with a clear conception of what you want to write about your subject and a general plan—which will later change in detail—for the how-to aspects of writing. Having reached this mental plateau, you should have the self-confidence needed to solve the many unanticipated problems and make the many judgments that lie ahead without getting unduly bogged down in writer's depression.

Further, you should have the resolve to move ahead toward a goal—no matter how distant it may seem—without becoming disoriented and flitting about hither and yon. Good work habits are a part of this consideration.

Those are the nominal requirements.

Beyond this point are some intangibles one has to experience to appreciate. How dedicated are you in getting a big job done? How much are you willing to give up to write a book? How will you deal with competition for your time as a writer? Will you become disenchanted and quit after the apparent glamour of writing a book wears off? Will you continue to write when you aren't feeling up to par? When you are worried? When you are tired? When you are on holiday?

My experience in writing this book—as an after-hours project—is offered as an example. The job took about one year. My weekday routine: dinner, write from 6:30 to 7:30 p.m.; jog; write from 8 to 8:45; shower; read or watch TV, 9-10; bed, 10. My weekend routine: jog, 7 to 7:30 a.m.; breakfast, look at newspaper, 7:30 to 8:00; write, 8 to noon. When possible, I wrote during parts of afternoons and evenings on both Saturday and Sunday, including holidays.

On business trips, I wrote at airports, in planes, in hotel and motel rooms. If I rose early, I wrote before breakfast. When possible, I wrote after dinner.

I find I must be persistent because if I miss a couple of days, there is a danger my material will become cold. I lose my place and my pace mentally. I must work back into the groove by restudying my source material. If I plunge ahead prematurely, I make mistakes and lose confidence. I tend to become impatient and uptight at a time when it pays to be patient and relaxed.

When I can't maintain my schedule, I should note, my basic problem is content—deciding what I want to say next—rather than writing per se. If you do not write regularly as I do, writing could also be a problem if you do not keep at it.

I have this hang-up even though I feel I am well prepared. In fact, I spend what others may believe an inordinate amount of time studying, thinking about, sifting and sorting my raw materials. On the average, I invest 90% of my time on a given project getting ready to write; the 10% left over gives me plenty of time for writing, editing, and rewriting. I have few false starts and am considered a fast writer. Other writers do as well or better spending less time becoming involved with their materials. They are able to start writing early. In the process of writing, they become better acquainted with their materials and refine their thinking. As a general rule, these writers have more false starts and must do more editing and rewriting than I do.

Advice to Writers of Technical Books From Readers

Common problems readers of technical books encounter span a range of topics: organization, pedantry, value of information, readability, writing for an audience, indexing, editing, references, and graphics.

Some verbatim comments:

"Generally too much theory and not enough practical information on its application to real world today problems."

"The author assumes the reader knows more about the subject than is generally the case."

"Authors should avoid such statements as 'it is obvious that' or 'the reader will recall.' "

"Too many books are disjointed reports of conferences—a series of papers; none having any substance."

"Some editors seem to be more interested in eye-catching layouts than they are in technical content."

"Indexing is frequently inadequate. One should be able to find what he is looking for easily and quickly. Cross-referencing is needed. Using technical books as references can be frustrating."

"Too much obsolete information. Too little current information. Too much duplication of generally known information."

"Writing is difficult to follow, and specific information is hard to locate."

"Authors should discuss specifics but concentrate on principles. Most problems can be solved by applying basic principles."

"Technical books are often too steeped in theory, equations, graphs, etc."

"There seems to be a tendency to judge a book's value by its weight. Editors need to exercise stricter control over the breadth of a subject and instruct authors to be more concerned about pertinence. There is also a problem in trying to make 'commercial' books pseudotechnical."

"Writing styles are often pedantic. Some theories are developed ad nauseum, with page after page of equations, etc., as if the author is assuring us of his knowledge and competence."

"More editing is needed to remove repetition, improve flow, etc."

"The most common fault is that many books repeat information that is already available, cluttering an already cluttered field."

"Information often does not reflect state of the art."

"A lack of useful examples, poor use of graphics, insufficient references."

"Some information is out of date when it is published because the technology is changing so rapidly."

"Too much knowledge on the part of the reader is often assumed, even in books at the introductory level."

"Authors should recognize that once engineering data are published they have a way of gaining stature as they are quoted and republished. In this process, data may acquire more authority than justified by the quality of the original work. Somewhere, somehow authors should try to re-evaluate and critically review at least some of the data they publish so they don't perpetuate inaccurate or unreliable information. Citing sources protects an author from plagiarism, but it does not guarantee accuracy."

Readers are equally articulate in ticking off the basic specifications for an acceptable technical book.

Some verbatim comments:

"A logical arrangement of chapters and good descriptive material."

"Organization, clarity, and well-illustrated (data, examples, etc). Avoid page filling discussions of motherhood information."

"Technically lucid, complete, balanced."

"Well indexed and cross-referenced, lots of pictures and other graphics, simple explanations as far as possible."

"State the theory as clearly and understandably as possible, then explain its practical value: how and where it can be used—what industries, technologies use this theoretical information in making a product . . . and how it is applied."

"The author should provide a definitive statement of the scope of his information; decide on an audience and write to it."

"Have a table of contents and an index."

"The goals are good organization, a pleasant writing style, good supporting visuals, and a minimum of self-aggrandizement."

"A technical book should have good development, be well illustrated, and be accurate."

"It should provide new information."

"Each chapter should follow the format of a good technical article: a strong introduction and strong conclusions."

"Clarity, completeness of subject matter, a systematic approach, a glossary . . . are the main requirements."

"The secret is good professional editing."

"Information should be digestible and retrievable."

"A book should be written with an objective in mind—such as aiming for a level of understanding and structuring to achieve that goal; specific terminology should be covered in a glossary."

"Ample references should be provided at the end of each chapter or section."

"Knowing what's going to be covered and how much coverage is necessary are elements of good organization."

"Develop each topic before going into detail; provide appropriate examples, so reader will see all explained relationships; have a specific audience in mind and write to it."

Finally, a warning is sounded by a reader who is alarmed by the much publicized declines in reading comprehension and writing skills among students today. He asks, "Will tomorrow's technical books have to be written in rock English?"

A Little Bit About Format and Related Matters

A Guide for Preparing Manuscripts, available from the Printing and Publishing Office, National Academy of Sciences, 2101 Constitution Ave., Washington, D.C. 20418, provides an excellent introduction to book format and is the source of information for this section.

Topics of interest in the Academy booklet are in chapters entitled, "Cover and Front Matter," "The Text," "References, Notes, and Bibliographies," and "Back Matter."

What could be called a standard of acceptance, presented in the preface, is instructive. "As a general rule," it is stated, "the Academies will approve and accept manuscripts for publication that are complete in all their parts, organized appropriately to serve their purposes, correct in matters of fact and documentation, and edited for basic uniformities of style and usage appropriate to their subject matter and purposes."

Front matter is defined as all the material preceding the text. Standard elements, which may or may not be included in all instances, include some form of title page, foreword, preface, acknowledgments, contents, and lists of tables and figures.

A title page may include a title, subtitle (if any), name of author, publisher, copyright notice, and International Standard Book Number (ISBN) and Library of Congress Cataloging in Publication Data—the publisher arranges for the last two items.

For Academy books, the foreword is usually written by an authority in the field (someone other than the author). "It may present background and other material that introduces the activity and report to its readers. If the book is part of a series, the general editor of the series may write the foreword to place the book in its proper context."

The preface "is the author's statement of purpose, a brief explanation of the background, research history, and circumstances leading to the publication of his book."

Acknowledgements may follow the preface. If long, they may be related to back matter as an appendix.

A table of contents may be nothing more than a list of chapter titles or paper titles and authors. Headings and subheadings may be incorporated into the table of contents if they are useful to readers using the book as a reference.

Authors are advised by the Academy to consider separate lists of tables and figures if "the reader is likely to want to refer to them without regard to the text."

"Text," the Academy advises, "often begins with an introduction that refers in some detail to the subject matter of the book, outlining the content and structure of what is to follow. In some cases, the introductory speech to a symposium serves as the introduction. If chapters are numbered, the introduction should be counted as chapter 1." Pages are numbered consecutively with Arabic numbers, at the top or bottom of a page. The former is probably preferred because most editors seem to be programmed to expect the numbers at the top.

The body of the book is divided into chapters or papers. Within a chapter, headings for sections, subsections, and sometimes sub-subsections flag significant divisions of subject matter. The Academy discourages use of more than four levels of headings because they "are difficult for both editors and readers to follow." The same style should be followed for all headings within a given class. For example, type all section headings in all capitals (TYPE IN ALL CAPITALS) and all subsection headings in capital and lower case letters (Type in Capital and Lower Case Letters). The system helps the editor determine the levels of heading the author intended, explains the Academy.

Use of footnotes in text is discouraged "because they are expensive to set into type and often cause difficulties in the make-up of a book. They should be used sparingly." Footnotes are used to cite a bibliographic source, give a cross-reference to another part of the book, or develop a point without disrupting the continuity of text.

A numerical or symbolic system is used for footnotes. The Academy advises, "In a publication divided into chapters, sections, or papers by individual authors, each beginning on a new page, text footnotes should be numbered consecutively, beginning with 1 in each such division. If a symbolic system is used, however, the system should begin anew on each page. The sequence of symbols conventionally used is (*) asterisk, (†) dagger, (‡) double dagger, (§), section mark, and (|) parallel. If more symbols are needed, they may be doubled or tripled, or additional single-character symbols may be used." Numbered footnotes in tabular matter can cause confusion. Lower case italic superscript letters or a symbolic system may be used.

References are listed at the end of each chapter or at the end of an entire manuscript. If only a few references are used, advises the Academy, "it may be preferable to set them as footnotes, placing them at the foot of the page or, in manuscript form only, immediately following the paragraph to which they refer and numbering them consecutively throughout the paper or report."

References must contain enough information to enable an interested reader to locate source material with a minimum of effort.

Examples of several ways of citing references in text are given by the Academy.

Numbers are used consecutively in the following manner:

" . . . to more precise measures of animal response to experimental treatments.[11,25,27]"

The author's name may be cited preceding the number, as:

" . . . on the basis of the work of Reece and Leonard.[51]"

If numbers are not used, the author's name and date of publication are cited in text in this manner:

" . . . to distinguish between 'normalizing' and 'canalizing' selection (Wadington, 1942, 1957, 1962)."

Lower case letters should be used when referring to two or more separate books pubished in the same year by the same author:

" . . . as described by Clarke (1961 a.b.)."

To facilitate use of lists of unnumbered references, place the date of publication immediately after the author's name in both the text and the list of references.

If there are two authors, cite both names, as (Jackson and Green). If there are more than two, name only the first one and follow with "et al.," as (Jackson et al.). Name all authors in the list of references. If the list is arranged alphabetically, the name and initials of the author should be inverted, as (Allen, J. A.) If there is more than one author, follow this style for the first name only. Use normal word order for the remainder. If references are numbered and not in alphabetical order, names and initials of authors need not be inverted.

If a book is a collection of papers, style for references should be uniform.

When should you use references? Bibliographies? Both? The Academy offers these guidelines, "If the book refers specifically to other literature, a list of references is appropriate. If several books were used in the development of the report, a bibliography is needed. In some cases, both a list of references and a bibliography are required to give proper credit to all information sources."

Bibliographic material, say the Academy, should contain the same information that references do. Bibliographies should be arranged alphabetically by author. Individual entries are not numbered. If many works of a single author are listed, they may be arranged alphabetically by title or chronologically by year of publication. If the bibliography is a list of all the works of one author, they should be arranged chronologically and, within the same year, alphabetically by title.

A sample journal reference for references and bibliographies:

"Levene, M., and C. I. Levene, Pulmonary platelet thromboembolism, *J. Clin. Path.* 10:200–203,1957."

A sample book reference for references and bibliographies:

"G. A. Korn, *Random-Process Simulation and Measurements* (McGraw-Hill, New York, 1966)."

Back matter, by definition of the Academy, follows the last page of text and may include appendixes, glossaries, and indexes. Arabic numbering used for text is followed.

An appendix supplements text. Any number may be used. Each must be titled and lettered. Text should refer to appendixes at appropriate places.

An appendix, for example, may present derivations of formulas used in text; or copies of questionnaires used to obtain data interpreted in text. In a proceedings for a symposium, pertinent papers not presented at the symposium may be presented; or participants in the symposium may be listed in an appendix. "But appendixes should not be used to present information crucial to the understanding of the subject matter," suggest the Academy. "Such information must be included in the text itself. "

A glossary is defined as a compilation of definitions of unfamiliar terms, terms used in unfamiliar ways, and terms that, for any other reason, must be precisely understood in the context of a particular document. The glossary is usually placed at the back of a book. Terms should be listed alphabetically, followed by the definition.

Three types of indexes are used: subject or key word indexes; author or name indexes; and subject-name indexes.

Ideally, the author should prepare the index. This procedure is recommended by the Academy:

"The compilation of an index should begin with the arrival of the first galley proofs. All key words and phrases to appear in the index should be underlined or written on the galleys, and then transferred to 3 by 5 library filing cards, one entry to a card. The cards should be kept in the same order as the galleys from which the items were assembled until page proofs are received. This will facilitate the transfer of page numbers from the page proofs to the cards.

"Once the page numbers for each entry have been recorded on the cards, the indexer should arrange the cards in alphabetical order. Next, he should check through the alphabetized cards, removing duplicates and combining similar entries. For example, if he is working on a book related to foods, several cards may be items beginning with the word 'Fats,' such as 'Fats, integral components of,' 'Fats, food,' and 'Fats, substances soluble in.' These entries can be combined by alphabetizing the cards according to subtopics and crossing off the word 'Fats' on all cards except the first. The entry and subentries would appear in the printed book as follows:

"Fats
    food,00
    integral components of, 00
    substances soluble in, 00

"Cross-references (such as 'see' and 'see also' references) should be supplied when necessary. The completed index should be submitted on sequentially numbered index cards."

Some Legal Aspects of Being an Author

How do you write for permission to use copyrighted material in your book? What rights do you relinquish when you, the author, sign a contract with a publisher? What are standard items in contracts? What are the usual arrangements for royalties? How are they determined? When are they paid?

Publishers generally grant permission to use copyrighted material on receipt of a simple letter like the following:

> Name of sender
> Address
> Date
>
> Publisher
> Metal Progress Magazine
> Metals Park, Ohio   44073
>
> Gentlemen:
>
> I am preparing an article for submission to the American Concrete Institute for possible publication in <u>Concrete International</u>. I would like your permission to use materials from the following article:
>
> Slaughter, E. M., "Tests on Threaded Sections Show Exact Strengthening Effects of Threads," <u>Metal Progress</u>, Vol. 23, No. 3, March 1933, pp. 18-20.
>
> I want to cite the Slaughter article . . . because it is one of the milestone articles on the strength of threaded fasteners.
>
> Please send the written permission to me at the above address.
>
> Sincerely yours,
> Signature

Publishers typically respond with a form that requires the applicant to supply certain specific information. The form in the box on the opposite page is used by the American Society for Metals, publisher of *Metal Progress*.

APPLICATION FOR PERMISSION TO USE
MATERIAL UNDER COPYRIGHT
By the American Society for Metals, Metals Park, Ohio

Applicant _________________________ Date ___________

Address _____________________________

____

____

Permission is requested to reproduce the following excerpts or illustrations from *Metal Progress*:

Selections from text (specify date of issue, page number, and paragraph(s): ____

____

____

Illustrations (specify by date of issue, page, and figure number): ___

____

In a work to be entitled: _________________________________

Author _______________________ Publisher _____________
Estimated date of publication __________ Edition ____________
or for the following purpose: _______________________________
It is agreed that credit will be given for excerpts and illustration in the following manner (indicate style, form, and position of credit): ___

____

Applicant agrees to furnish to *Metal Progress*, American Society for Metals, at no charge, one copy of work in which material is used.

___________________   ___________________
Signed   Date

Application approved by:

___________________   ___________________
For American Society for   Date
Metals

Other matters concerning copyright law are discussed in some detail in Chapter 25.

Contracts with publishers tend to be substantial legal documents. The American Society for Metals, for example, has a seven-page contract. Specific paragraphs deal with transfer of copyright from author to publisher, a standard of acceptance, royalty agreement, assignment of rights for subsequent publications, procedures for such matters as preparation of the manuscript and author responsibilities for proofreading galleys, standards for graphics, affirmation that the author is owner of the copyright and has obtained written permission to use copyrighted materials, ground rules for editing by the American Society for Metals and an outline of production procedures, agreement of the Society to supply the author with a specified number of copies of the published work, and an agreement by the author not to publish another work with another publisher that will conflict with the sale of the work.

Paragraph No. 1 conveys the author's copyright. A sample of the language: "The Author grants and assigns to the Society . . . all rights in an unpublished literary work tentatively entitled   (herein called the 'Work'), with the exclusive right to print, publish, or to cause to be published, and sell the work, during the full term of copyright and all renewals thereof."

Paragraph No. 2 states that the manuscript, in English, shall be delivered to the Society on or before   . The Society's technical and literary standards must be met, and its editorial staff will review the manuscript. Consultants may be used for review and appraisal. The manuscript will be returned to the author and the contract terminated if it is rejected.

The Society has the exclusive right to copyright the work in its name in the United States and in its name or any other name in other countries.

Royalties are based on a specified percentage of "actual cash received." As a rule, the royalty for trade books—those sold in bookstores, colleges, etc.—is 15%, while that for mail-order books (those sold exclusively by mail) is 10%. The main reason for the difference is that the latter are promoted heavily on a regular basis by publishers. Royalties are paid yearly by the Society—with the fiscal year starting on the date of publication.

Other rights assigned to the publisher include right of digest, abridgment, condensation, anthology, quotation, reprint edition, microfilm or other reproduction, translation, foreign language book publication, motion picture, television, mechanical rendition or recording, "and any and all other rights of commercial use now or hereafter existing . . ." The Society may pay the author a percentage of any compensation received after deducting any expenses, or it may not award compensation if it is believed that the sale of the author's work is benefited.

Mechanical requirements call for the manuscript to be "clearly typewritten, double spaced, 50 characters, 25 lines per page on substantial white paper, one side only, with no longhand insertions." Two copies are supplied.

When the book gets into publication, the author is responsible for reading and correcting printer's proofs. Corrections "shall be confined to the correction of errors wherein the printer has not followed the original copy." Other changes or additions in excess of 10% "shall be paid for by the Author to the Society from accrued royalties at rates charged by the printer." The author is responsible "for the completeness and accuracy of corrections."

The Society may also ask the author to supply the title page, a preface or foreword, table of contents, index, bibliography, and a list of source materials. Photographs supplied by the author must be suitable for reproduction. Drawings, charts, diagrams, and illustrations shall be "supplied by the author in a form suitable for rendering by the Society's illustrators. Illustrations, etc., shall be clear, meaningful, and properly captioned."

The Society retains the right to edit and revise, "but such editing or revision shall not materially change its [the work's] meaning." The Society also reserves the right to publish the work in one or several volumes, to choose format and typographical style, to alter the title, to establish prices, and other related matters.

The Society also determines when "public demand for the work is no longer sufficient to warrant its continued publication." At this time, the author may ask to have the Society transfer its rights in the copyright to him.

Finally, Paragraph No. 16, declares, "This Agreement shall be construed and interpreted according to the laws of the state of Ohio and shall be binding upon the Society, its successors and assigns, and the Author, his heirs, successors, personal representations, and assigns."

Both the author and the Society sign the contract.

CHAPTER 16

Guidelines for Writing Standards and Specifications

"Provide a title that in a very few words tells what the standard covers," advises the American National Standards Institute, 1430 Broadway, New York, N.Y. 10018, in its *Style Manual for Preparation of Proposed American National Standards*.

"The title [for specifications] should be concise but complete enough to identify the nature of the test, the material to which it is applicable, and to distinguish it from other titles . . . select words that lend themselves to indexing. The essential features of a title are the particular property or constituent being determined, the material to which the method is applicable, and, when pertinent, the technique or instrumentation," advises the American Society for Testing and Materials, 1916 Race St., Philadelphia, Pa. 19103, in its *Form and Style for ASTM Standards*.

"Include in the first line of the title the indication of the commodity, form(s), method of production, and type of material in this order. Inclusion of method of production applies primarily to tubing and castings; it does not apply to specifications for sheet strip, and/or plate or to melting practice for steels and alloys . . . in the second line, include the basic elements or constituents or other characteristics necessary to identify the material and differentiate it from similar materials. In specifications for metals, nominal contents of alloying elements shall be listed in the order in which they appear in the tabulated composition except as follows . . . ," advises the Society of Automotive Engineers, Inc., 400 Commonwealth Dr., Warrendale, Pa. 15096, in its *Manual on the Preparation of Aerospace Material Specifications*.

"The title should be brief, but should contain key descriptors to identify the process. When specifications cover similar processes, the titles must distinguish between the documents. The usual elements of a title are: (a) the basic noun identifying the process; (b) descriptor modifiers to define the process; (c) product designation, if required. An example for a title is 'Cleaning Prior to Heat Treatment—Aluminum.' *Cataloging Handbook H* may be referred to for item names," advises Northrop Corp., Aircraft Div., Material and Procurement

Dept. (6000/32), 2901 W. Broadway, Hawthorne, Calif. 90250, it is *Format for Process Specification.*

No instructions for the writing of titles are given by The De Havilland Aircraft of Canada Ltd., Downsview, Ont. M3K 1Y5, in its *Instructions for the Preparation of Process Standards.*

The preceding examples were chosen to illustrate the amount of agreement and kinds of differences in preferences among organizations with standards for the writing of standards and specifications. A lack of universal agreement is evident. Parts of the five different standards will be examined in some detail for the purpose of providing the information and counsel needed by the do-it-yourself writer of standards.

Differences of opinion also exist among readers of standards and specifications—the people who must consult and rely on them for the conduct of their daily business.

For example, some readers contend that standards and specifications have become too general.

"Most military, federal, and industry specifications have become so generalized (to appeal to everyone) that they have limited value," a reader comments, and adds his idea of requirements:

"Specs and standards should define: 1. Engineering requirements. 2. Quality control requirements and acceptance test verification methods. They should not define the how-to. Manufacturing and administrative details should be left out. They should not offer choices."

The following comment is at the opposite pole:

"Standards and specs are often too technical, requiring an engineer to interpret them. They should be written for understanding by a technician."

This reader believes these documents should be "performance rather than formulation oriented. They should contain only necessary requirements to insure an acceptable item or product."

Other views of users of standards and specifications will be presented in a separate section.

Standards for Writing Proposed American National Standards

In common with the other sources in this chapter, the American National Standards Institute (ANSI) sets forth a format made up of separate elements and the order in which they are to be presented. In some details, the format differs from the four others that will be discussed.

ANSI states its requirements in a section titled "Content To Be Supplied by Committee." Items include:

> Title
> Abstract (not required for single sheet standards).
> Foreword and committee list (not required for single sheet standards).
> Complete text, including scope.

Figures, if any.

Appendixes, if any.

Table of contents (not required for single sheet standards).

Index, if applicable.

Standards, it should be noted, are in one of two formats—booklet or single sheet. "Generally, unless the committee for adequate reasons requests otherwise, typewritten manuscripts of up to five pages, double-spaced, will be published in single sheet form . . ."

Each format item and related matters, including style, are discussed. The ANSI styling practice for the numbering of sections and subsections should be described at this point because it represents the consensus among standard writing groups.

ANSI explains in 5.1.1.1 Sections and Subsections. "The system for numbering sections of an American National Standard is one that uses Arabic numbers in sequence; a subsection is designated by adding a period and number to the section number (for example, 5.1). This subsection, in turn, may be divided by a second period and a second number (for example, 5.1.1). Five numbers divided by periods (5.1.1.1.1) is maximum. Subdivision beyond this point results in cumbersome cross references and unsightly indentation. The remedy is reorganization of the material.

" . . . number sections and subsections only when more than one is required. In this manual, for example, the text for Section 3, 'Form of Publication,' consists of only one paragraph that relates directly to the heading of the section. Therefore the paragraph is not numbered."

ANSI adds, "Do not use the terms section, subsection, and sub-subsection in headings or references. Cross-references are made by referring simply to the number. The number and section heading may both be used when referring to major section headings; for example, 'see Section 5, General Style.' "

ANSI style for lists within a subsection is to number items as (1), (2), (3), etc. A further breakdown: (a), (b), (c), etc. No more lists in a subsection are permitted "to avoid confusing cross references." Closing punctuation in lists of short items should be omitted.

Figures and tables should be numbered consecutively in the order of their reference in text, such as Fig. 1, Fig. 2, Table 1, and Table 2. Consecutive letters and headings identify appendixes, such as:

Appendix A

Cleaning Procedure for New Cylinders

Text in appendixes is organized and numbered in the same manner as in text. If there is only one appendix, identify it as such but do not use a letter prefix, but use the prefix letter A in numbering sections and subsections in text, as: A1. Test Specimens. A1.1 Handling.

Figures and tables in an appendix should carry the identifying letter of the appendix in which they appear, as the first figure in Appendix A is Fig. A1.

Webster's Third New International Dictionary of the English Language (Un-

abridged) is recommended for hyphenation and spelling; "the trend is to leave out the hyphen unless its omission will mislead the reader or slow his comprehension of text." The University of Chicago *Manual of Style* is recommended for capitalization; "the trend is toward use of fewer capitals."

Use of tables is recommended. "Tables provide a clear and concise way of presenting large amounts of data in a small space. A simple table can often give information that would require several paragraphs to present textually and can do so with much greater clarity."

Tables, with the exception of informal types, should have a number and caption (or title). Tables are numbered in the order they appear in text.

ANSI style for tables:

Table 1
Nomenclature for the Parts of a Table

Stub Heading	Column Caption	Column Caption	
		Subcaption	Subcaption
Line Heading	Tabulated Data		
Subheading			
Subheading			
Line Heading			
Total			

Indicate units of measure in the title or in column headings or in a note. Use same units in each column. Do not mix units, as feet with inches. State units of measure in parentheses in column headings.

Separate digits into groups of three, counting from the decimal sign toward the left and the right. Groups should be separated by a space, not a comma, point, or other means. If the magnitude is less than unit, the decimal sign should be preceded by a zero. In numbers of four digits or less the space is not necessary, unless four-digit numbers are grouped in a column with numbers of five digits or more.

Examples:

    73 722   7372   0.133 47

Align all numbers on the decimal point. Do not combine common fractions and decimal points in the same table.

To save space in tables, use abbreviations and letter symbols in column captions and line headings where possible, and use them more freely in the body of the table than you would in text. If there are no data for a particular cell (item in the body of the table), use a dash to indicate the vacancy.

Use of horizontal rules to separate tables from text and stub headings and column captions from the body of the table is recommended. Other rules, both horizontal and vertical, should be avoided where spacing can be effectively used.

Superior numbers or letters are to be used to reference footnotes to tables unless confusion would result. If so, use *, †, ‡,§,**,‡‡ etc.

Style for tables:

Table 2

Sample Table

Model Number	Maximum Assay (wt% 235_U)	Maximum Net Weight (lb UF_6)	Activity	
			Curie(s)*	Curie(s)†
1S	100.0	1.0	0.04	0.03
2S	100.0	4.9	0.18	0.17
30A	5.0‡	4 950.0	5.40	4.85
40F	Natural	27 030.0	—	5.60

* Activities shown are for . . .

† Activities shown are for . . .

‡ Moderation control equivalent to 99.5% pure UF_6 is required . . .

Tabulations not exceeding four or five lines in depth and not exceeding the width of one column can be used in text as informal tables. Table numbers and captions are not required. The example indicates a representative width (average of 32 units, counting spaces):

Example:

E_v Units	Percent Error	
	Maximum	Minimum
$\pm 1/6\,E_v$	+12 percent	−12 percent
$\pm 1/4\,E_v$	+19 percent	−16 percent

Figures are numbered with Arabic numerals in the order in which they appear in text. Each figure should have a caption.

Equations are numbered in sequence within parentheses and prefaced by the abbreviation "Eq" to avoid mistaking the identifying number as part of the equation. (See following example.)

To save time in typesetting, equations are positioned flush with the left margin. The word "where," used to introduce explanatory material, is also set flush left on a separate line and symbols defined are indented.

Example:

$$I_0 = B(U-f)^2\, tCH\, \cos^4\theta \qquad\qquad\qquad\text{(Eq 1)}$$

B = average field luminance measured by the meter based on luminance of calibration source, lumens/square meter

U = distance from lens to object
f = focal length of lens
t = lens luminance
C = camera flare correction factor
H = vignetting factor
θ = angle image point from axis of lens
A = geometric f-number of lens

Footnotes may be included in a standard only for purposes of clarification and in the use of the standard. Footnotes are informative only and are not part of the standard. Footnotes are numbered consecutively, beginning with the first page of text and continuing through the appendixes. If a footnote occurs frequently through the text, repeat the number of the first reference.

In references to American National Standards, refer to the latest edition by title and ANSI designation, and include year date. If reference is part of the standard, include section and subsection number.

In references to standards of organizations other than ANSI, provide the name and address of the organization issuing the standard, the full title of the document, its designation, and date of issue—if the date is not part of the designation. If reference is to part of the document, include part, paragraph, or section reference. Do not reference future revisions.

In references to periodicals, put the elements in this order: last name of author and his initials, title of article, title of periodical in full (do not abbreviate), volume number, issue number, date of issue, first and last pages of article.

Example:

> NELSON, C. N. Safety factors in camera exposures. *Photographic Science and Engineering*, vol 4, no. 1, Jan-Feb 1960, pp 45-48.

In references to books, put the elements in this order: last name of author and his initials, title of book, place of publication, name of publisher, year of publication, edition number, first and last pages of reference.

Example:

> MEES, C. E. K. *The Theory of the Photographic Process*. New York, The Macmillan Company, 1966, 3rd ed, pp 1-24.

ANSI has policy on such matters as calling for the use of a patented item or commercial equipment.

The word on patented items: "Although there is no objection in principle to developing an American National Standard which calls for the use of a patented item, this practice should be avoided if practicable." Special procedures are required.

The word on commercial equipment: "References to commercial equipment shall be generic and shall not include trademarks or other proprietary designations. Where a sole source exists for essential equipment or materials, it is permissible to supply the name and address of the source in a footnote. If it is necessary to refer to a particular model number, the words, 'or the equivalent,' should be added to the reference."

Other rules pertain to such matters as typing and quality of manuscript and practice for making corrections in manuscript.

Manuscripts must be typewritten and double spaced on 8½ by 11 in. paper. A margin of at least 1 inch should be left on all sides of the page. All copy must be clean and clear. All pages are numbered consecutively, starting with the title page.

All corrections or changes in manuscript should be typed or clearly written in ink. Changes should be placed between the lines and not in margins. A long addition should be typed on a separate sheet and clearly keyed to text. If corrections are extensive, the page should be retyped.

In mathematical expression, marginal notes should be used to identify obscure modifications of symbols, to distinguish beteen the letter O and zero, the letter I and number 1, the times sign and the letter x. Superscripts and subscripts should also be clearly indicated. If Greek letters are handwritten, they should be identified the first time they appear.

Standards for Writing ASTM Test Methods and Specifications

In its *Form and Style for ASTM Standards*, this organization addresses a variety of topics: formats for test methods, specifications, and other standards; use of the modified decimal numbering system, preparation and use of terminology, legal aspects of standards, a standards style manual, and use of metric (SI) units. We will concentrate on formats for test methods and specifications.

Test methods first, starting with a listing of subject headings. Those marked with an asterisk are required. Other appropriate headings may be used. The list:

*Title	Precautions (required where applicable)
Designation	Sampling, Test Specimens, Test Units
Introduction	Preparation of Apparatus
*Scope	Calibration and Standardization
Applicable Documents	Conditioning
Summary of Methods	*Procedure
*Significance and Use	Calculation or Interpretation of Results
Terminology	Report
Interferences	*Precision and Accuracy
Apparatus	References
Reagents and Materials	Annexes and Appendixes

The title, a required item, "should be concise but complete enough to identify the nature of the test . . ."

Letter designations indicate the field covered by the test method:

A —Ferrous metals and products
B —Nonferrous metals and products
C —Cementitious, ceramic, concrete, and masonry materials
D —Miscellaneous materials and products
E —Miscellaneous subjects
F —End-use materials and products
G —Corrosion, deterioration, weathering, durability, and degradation of material and products
ES —Emergency standards

An introduction is not generally used in ASTM methods. If one is needed "for proper understanding of the user," it follows immediately after the title but does not have a section number.

ASTM has this to say about scope, another required item: "Include here information related to the purpose, application, etc., of the method. State the range of application . . . as completely as possible; do not sacrifice clarity for brevity. Avoid repetition of material included in 'Significance and Use.' If the method covers several tests, each with its own scope, include a statement regarding the scope of each." Reference to alternative or companion methods should be included in a note in the scope, giving the title and designation.

In the Applicable Documents section, all documents referenced with the standards are listed, complete with designation and complete title. Sources of documents other than ASTM are listed in footnotes.

A summary of method is not required. The section provides a brief outline of the method, focusing on essential features rather than details. A brief statement of the principle of the method may be included.

A section on significance and use is required, with emphasis on the following: meaning of test as related to the manufacture and end use of the material; suitability of the test for specification acceptance, design purposes, service evaluation, regulatory statutes, manufacturing control, development and research; fundamental assumptions inherent in the method that may affect the usefulness of results; and finally, any warning needed in the interpretation of the results of the test.

The terminology section is in effect a glossary.

The need for a section on interferences is described as follows: "If the successful application of the method requires the inclusion of explanatory statement on interference effects, include such information here."

The Apparatus section is also optional. Essential features of the apparatus and equipment required are described breifly. Schematic drawings or photographs may be used where they are necessary to supplement text. Common laboratory equipment like flasks and beakers are not listed. Trademarks are to be avoided. Generic terms replace trade names, such as "borosilicate glass" instead of Pyrex or Kimax.

A section on reagents and materials is not required. "When more than one procedure is included in one standard, list the reagents and materials required for each procedure as a separate section under each subdivision."

ASTM spells out requirements for a section on precautions: "When there are hazards to personnel in performing the test, such as explosions, fire, toxicity, or radiation, a warning to this effect should be included here. Indicate at which steps in the procedure these hazards exist. At this point in the text where a precaution is important, include the word 'Caution' in bold face type, followed by the details of the protective or precautionary measures to be taken. Such hazards shall be included in a separate section or . . . included in the scope of standards involving hazardous materials and procedures, such as the following:

> 'This standard may include the use of hazardous materials, operations, and equipment. It is the responsibility of whoever uses this standard to establish appropriate safety practices and to determine the applicability of regulatory limitations prior to use.' "

A section on preparation of apparatus is required "only when detailed instructions are required for the initial assembly, conditioning, or preparation of the apparatus."

Another optional section, Calibration and Standardization, covers apparatus, reference standards and blanks, calibration curves, and tables.

If a section on conditioning is needed, it covers "the conditioning atmosphere to be used and time of exposure to the atmosphere, as well as the atmosphere required during the test."

Procedure is a required item in the format. ASTM states, "Include in proper sequence detailed directions for performing the test. Describe the procedure in the imperative mood, present tense, for example: 'Heat the test specimen . . .' rather than, 'The test specimen shall be heated . . .' State the number of samples to be taken, and also state the number of specimens to be tested from each sample. Describe in detail the successive steps of the procedure, grouping related operations into logical divisions. Subheadings may be used if they will help the organization of the material. . . . When alternative procedures are given, state their relative status, that is, which is the preferred or referee procedure."

A section on the calculation or interpretation of results is not required.

A section titled "Report," also optional, includes "detailed information required in reporting the results of the test. When two or more procedures are described in a method, the report shall indicate which procedure was used. When the method permits variations in operating or other conditions, a statement as to the particular conditions used in the text shall be incorporated in the report. . . ."

A section on precision and accuracy is required. In ASTM's words, "Variation in test results is a consequence of variation in the material tested and variation in application of the test method. With respect to a test method, statements about precision and accuracy should refer to the variation that results from application of the test method to homogeneous material in which the variation in material quality is as small as can reasonably be attained. Every ASTM test method must make a statement about any noted bias in the test method and about the precision of the test method with respect to between-laboratory variability and a specifically defined within-laboratory variability."

Should there be a References section, it should include "only references to publications supporting or providing needed supplementary information. Historical and acknowledgment references are not desirable. . . ."

The next section may cover annexes and appendixes. Each title should carry the words "mandatory information" or "nonmandatory information." Annexes come first. They cover detailed information on apparatus or materials essential to the method but too long for inclusion in the main test. Examples include a glossary of terms, list of symbols, detailed description of apparatus, and instructions for calibrating and standardizing apparatus. Appendixes are not part of the standard. Suitable topics include notes on significance and interpretation of the method, usually to amplify the statement in the text; development of equations used in the calculations; and commentary on rationale used in the development of the test method.

Footnotes referenced in text are intended only for reference and should never include any information or instructions necessary for the proper application of the method.

Notes in text "shall not include mandatory requirements. Notes are intended to set explanatory material apart from the text itself, either for emphasis or for offering informative suggestions not properly part of the standard. . . ."

The function of an optional section on adjuncts is explained by ASTM: "Occasionally it is not practicable to publish as an integral part of the standard, because of its nature, material that may be required for use of the standard. Such material is published as an adjunct. "Adjuncts may be comparison standards, reference radiographs, charts, or drawings, for example.

A concluding section on research reports is required "where numerical data have been generated during the development of an ASTM standard or numerical data are used to establish the precision of a test, or both. The research report shall include raw data, statistical treatment of data, etc."

ASTM specifications serve any or all of these functions: facilitate dealings between purchasers and suppliers, establish standards, or provide technical data.

Use of standard headings in this format depends upon the nature of the specification. Only a section on scope is required. Headings include:

Title	Number of Tests and Retests
Designation	Specimen Preparation
Scope	Test Methods or Analytical Methods
Applicable Documents	Inspection
Classification	Rejection and Rehearing
Terminology	Certification
Ordering Information	Product Marking
Materials and Manufacture	Packaging and Package Marking

Chemical Composition

Physical Properties

Mechanical Properties

Performance Requirements

Other Requirements

Dimensions, Mass, Permissible
   Variations

Workmanship, Finish, Appearance

Sampling

Supplementary Requirements

Special Government Requirements

Quality Assurance

References

Annexes and Appendixes

The title should be "as concise as possible but complete enough to identify the material, product, system, or service covered in the specification."

Designations are the same as those used in standards for test methods defined previously.

Scope (required) "should amplify the title, state the function of the specification, and note any materials, products, systems, or services excluded. Trademarks should not be used. It is also desirable to include a statement in this section whether inch-pound or SI units shall be used in referee decisions. Related documents, or those of interest but not referenced in text should be included in a note."

General requirements for materials and methods of manufacture are frequently included in a section, particularly when reference is to a particular method of manufacture, such as the open hearth or basic oxygen processes.

When necessary, "Detailed requirements should be given as to chemical composition and other chemical characteristics to which the material product, or system must conform."

When necessary, detailed requirements should be given as to characteristics to which the material, product, or system must conform. The information is frequently presented in tabular form. Key items include: name of each property or requirement; whether the requirement is a maximum, minimum, or range; whether an allowance for measurement error is incorporated into the limits; units applicable; references to notes or footnotes; appropriate test methodology.

Details on standard shapes, mass, and size ranges are usually presented in tabular form with brief references in text. Separate sections with individual tables may be needed.

Requirements covering workmanship and finish include such general items as type of finish and general appearance, or color, and general surface condition.

If a specification applies to a unit of product or materials such as a piece of cloth, coil of wire, section of plastic, or heat of steel from which specimens are to be taken for testing, the procedure for obtaining specimens should be described.

If a specification pertains to individual units of a lot and sampling inspection is likely to be the normal procedure, it is desirable for the specification to reference or include in a supplementary section a sampling procedure for determining acceptability of the lot.

The number of test units and number of test specimens or subunits required to determine conformance of the material or product to the specification shall be stated.

Where special preparation is required, such as for mold materials, a section on this topic should be included. If possible, reference should be to a standard method.

Reference should be made to ASTM methods to be used in testing the material to determine conformance with specification. Included are sampling, chemical analysis, mechanical, electrical, thermal, optical, and other testing procedures.

Where required, a standard statement for inspection is required; it reads:

> "Inspection of the material shall be agreed upon between the purchaser and supplier as part of the purchase contract."

Technical requirements for inspection, such as sampling plan or mechanical properties, should be included in the specification.

Standard wording for rejection and rehearing is available:

> "Material that fails to conform to the requirements of this specification may be rejected. Rejection should be reported to the producer or supplier promptly in writing. In case of dissatisfaction with the results of the test, the producer or supplier may make claim for a rehearing."

More standard language is available for a section on certification:

> "When specified in the purchase order or contract, a producer's or supplier's certification shall be furnished to the purchaser that the material was manufactured, sampled, tested, and inspected in accordance with this specification and has been found to meet the requirements. When specified in the purchase order or contract, a report of the test results shall be furnished."

It is customary to specify the information to be marked on the material or included on the package, or on a label or tag attached to the package. Typical information includes the name, brand, or trademark of the manufacturer, quantity, size, weight, ASTM designation, or any other information that may be desired.

When it is customary and desirable to package, box, crate, wrap, or otherwise protect the item during shipment and storage in accordance with a standard practice, it is customary to state the requirements.

Instructions for Writing Aerospace Material Specifications

This SAE group goes further than either ANSI or ASTM in specifying the writing and content of specifications. At the beginning of this chapter, requirements for AMS titles were quoted in some detail. Representative sections will be cited to indicate the intent and flavor of this organization's lengthy specifications for specifications.

An overall format is given for specifications, "except for those for dimensional tolerances and for identification of materials and parts, which have special formats. . . ." The standard format has these sections:

1. Scope
2. Applicable Documents
3. Technical Requirements
4. Quality Assurance Provisions
5. Preparation for Delivery
6. Acknowledgment
7. Rejections
8. Notes

AMS goes further and states, "If no information is to be given under any of these section titles, the phrase, 'Not Applicable,' shall follow the section title, except that the word, 'None," shall follow the titles of Sections 2 and 8."

The Scope section "shall define what the specification covers and include the following paragraphs, as applicable, in the following sequence." They cover form (or type or purpose), application, classification, and precautions.

The Form paragraph "shall show each product form covered by the specification, whether or not all forms are listed in the title. The paragraph shall begin with the words 'This specification covers (commodity) in the form of . . .' "

The Type paragraph "is usually used in specifications for parts; e.g., the 74XX series. This paragraph should begin with a statement similar to that shown above for the Form paragraph."

The Purpose paragraph "is usually used in specifications for processes. This paragraph should begin with a statement similar to that shown for the Form paragraph."

All specifications include an Application paragraph, which "shall be worded so as to indicate those applications and purposes for which the product is primarily intended, but shall not require that the product be used for such purposes. Complete sentences are not necessary if the intended application of the product can be clearly and suitably expressed otherwise."

Standard language for the Application paragraph is provided, for example, where alloys are subject to significant stress corrosion. The following is standard language for specifications covering aluminum alloys in the fully heat treated condition:

> "Certain design and processing procedures may cause these products to become susceptible to stress-corrosion cracking; ARP recommends practices to minimize such conditions."

The Applicable Documents section "shall list all specifications, standards, and recommended practices referenced within sections 3 (Technical Requirements), 4 (Quality Assurance Provisions), and 5 (Preparation for Delivery) of the AMS. . ."

The introductory statement for this section shall read:

> "*APPLICABLE DOCUMENTS*: The following publications form a part of this specification to the extent specified herein. The latest issue of Aerospace Material Specifications (AMS), Aerospace Standards (AS),

and Aerospace Recommended Practices (ARP) shall apply. The applicable issue of other documents shall be as specified in AMS 2350."

The Technical Requirements section "defines characteristics of the product required to ensure that material or parts procedured to the specification will be satisfactory for the intended use (misapplication excepted)." Further, "The fewest requirements necessary to define adequately the properties and quality of the material shall be included. Other properties which will be inherent in the material if the specified requirements are met or which may be of interest to designers but do not help to define and control the material shall not be included in this section but may be included in the Notes section."

Subjects in this section should be presented in the following order:

> Compositions (or material)
> Condition
> Casting (in specifications for castings)
> Master Heat Qualification (in specifications for investment castings)
> Properties
> Quality
> Sizes
> Tolerances (last two items may be combined)

In stating compositions, here are instructions for "aluminum, magnesium, copper, titanium, precious metals, and their alloys, and for nickel-copper alloys: the first elements to be listed shall be the intentionally added alloying elements, listed in decreasing order of the medians of the ranges or the minimum limits when no maximum limits are specified; these shall be followed by a listing of the named impurities in decreasing order of maximum permissible amounts; and then a listing of limits for individual (where necessary) unnamed impurities, total unnamed impurities, and the base metal if expressed as a 'remainder.' If the base metal is specified to a definite range or minimum, it shall be listed first."

The Quality Assurance section "shall describe the type and extent of sampling and testing and the sampling procedures for ensuring the production capability and product reliability of the material."

Subjects for the section include:

> Responsibility for inspection
> Classification of tests
> Sampling
> Approval
> Reports
> Resampling and retesting

Standard language for responsibility for inspection is as follows:

> "The vendor of the product shall supply all samples for vendor's tests and shall be responsible for performing all required tests. Results of such tests shall be reported to the purchaser as required by 4.X. Purchaser reserves the right to sample and to perform any confirmatory testing deemed nec-

essary to ensure that the product conforms to the requirements of this specification."

The Preparation for Delivery section "shall describe requirements to ensure that the product will not be damaged in shipment and can be identified at destination as being the product required." Subjects in this section are: identification, protective treatment, and packaging.

Standard language for the Acknowledgment section should read as follows: "A vendor shall mention this specification number in all quotations and when acknowledging purchase orders."

Standard language for the Rejections section shall read: "Material not conforming to this specification and the applicable detail specification or to modifications authorized by purchaser will be subject to rejection."

The Notes section "shall include information and data of a non-mandatory or descriptive nature. Such information should include precautions to be observed in processing, data to be shown in purchase orders, the meaning of the phi (ϕ) symbol (See 3), and the listing of similar specifications, if any."

A Canadian Aircraft Company's Format for Process Standards

A production process standard is defined by this company as a document containing a clear and accurate description of the technical requirements for a process, including the procedure to be adopted to meet the requirements.

Six sections are prescribed—as needs require: Scope; References; Material and Equipment; Procedure; Requirements; and other sections, which may include such titles as Other Data, Maintenance of Equipment, and Ordering Information. The first five sections named are mandatory.

The Scope section is a clear, concise abstract of the coverage of the standard. "If only one type of process is covered by the standard, a statement to this effect shall be made."

The References section summarizes specifications, standards, and other documents to which reference is made in text. A minimum of references is recommended, and a reference cannot be listed in this section unless it is in text.

Material and equipment required to carry out the process are listed in the section bearing this name.

Quality standards the finished product or process must meet are spelled out in the Requirements section. "Requirements shall be so worded as to provide a definite basis for rejection in those cases where quality or workmanship is not suitable for the purpose intended."

The company provides the writer of specifications with a diversity of styling instructions. Several examples follow.

"When it is necessary to use the phrase 'Unless otherwise specified' to indicate an alternate course of action, the phrase shall appear at the beginning of the sentence and preferably, at the beginning of a paragraph. This phrase shall be limited in its application and used sparingly."

The term "and/or" cannot be used under any circumstances.

The terms "flammable" and "non-flammable" are to be used in place of "inflammable" and "uninflammable."

The symbol "%" may be used in place of the word "percent."

All headings are underlined. But sentences or portions of sections may not be underlined for emphasis. No words, phrases, or sentences may be put in quotes for emphasis.

Tables are placed immediately following the paragraph in which they are first referenced. If space does not permit, they are put at the beginning of the next page, or if they are extensive, on a separate page. If tables are numerous and placement with related content would cause difficulty in understanding or interpretation, they may be placed in numerical sequence at the end of text.

Format for Process Specifications at a U.S. Aircraft Company

A definition: "Process specifications impose technical requirements peculiar to a given process to produce a product that will fulfill the design function . . . a process specification . . . affects contractual compliance with customer requirements. . . . "

The general outline for format has six components:

    1. Scope
    2. Applicable documents
    3. Requirements
    4. Quality assurance provisions
    5. Preparation for delivery
    6. Notes

The format is based on MIL-STD-490 and Appendix XIV modified in accordance with MIL-S-83490, Form 1b, and source material from *Defense Standardization Manual* 4120.3-M and the *U.S. Government Printing Office Style Manual*.

The scope is always Section 1. The first paragraph (1.1) is a scope-type statement and may contain general information concerning applications. If a statement of compliance to a contractual specification is required, it should be the next paragraph (1.2), citing the applicable specification number, revision letter, and date. Classification within the process (type, class, and grade) is in the next paragraph (1.3).

Type covers differences in composition, construction, application range, process response, or properties. Type is generally designated by Arabic numerals. Class provides further designations of differences in characteristics or properties and is generally designated by Arabic numerals. Grade covers differences in quality and is designated by capital letters.

Examples:

    1.   *SCOPE*
    1.1 This specification establishes requirements for cadmium
         plating by electrodeposition.

1.2 This specification meets or exceeds the requirements of QQ-P-416, Revision B, dated 4–5–67.

1.3 Cadmium coatings applied in accordance with this specification are of the follow types and classes.:

   a. *Types*

     Type 1—Cadmium plate without supplementary treatment.

     Type II—Cadmium plate with supplementary chromate treatment.

   b. *Classes*

     Class 1—0.0005 to 0.0010 inch thick.

     Class 2—0.0003 to 0.0005 inch thick.

All documents cited in sections covering Requirements (Section 3), Quality Assurance Provisions (Section 4), and Preparation for Delivery (Section 5) are listed in the Applicable Documents section (Section 2). They include those required for use in the process and those identifying materials for the process. Information sources are not listed here but may be in Notes (Section 6). Examples:

2.1.1   *Government*

2.1.1.1 FED-STD-151   Metals: Test Methods

2.1.1.2 MIL-STD-453   Inspection, Radiographic

The Requirements section (No. 3) "includes all processing information and defines requirements for acceptance of detail parts or assemblies after processing. The requirements should be complete and to the level necessary to guarantee identical repetitive processing. The order of paragraphs shall be as listed below or adjusted by additions or deletions of paragraphs as needed and renumbered accordingly.

   a. *Simple Process*

Paragraph Number	*Paragraph Headings*
3.1	Materials and Equipment
3.2	Flow Chart
3.3	Process Requirements and Procedures
3.4	Rework and Repair
3.5	Facility Requirements
3.6	Qualification

   b. *Complex Process*

Paragraph Number	*Paragraph Headings*
3.1	Materials
3.2	Equipment
3.3	Flow Chart

3.4-3.T	(Subtitled as necessary by process requirements and procedures)
3.U	Workmanship
3.V	Parts Protection
3.W	Rework and Repair
3.X	Facility Requirements
3.Y	Safety
3.Z	Qualification (e.g., supplier approval; certification)

Examples for complex process:

3.1   *Materials and Equipment*
3.1.1 *Acid Paste Cleaner Ingredients*

a.	CAB-O-SIL	Cabot Corp.	
b.	Acid, Chromic	O-C-303	(23-26-00-2010)*
c.	Acid, Sulfuric-66 Deg Be	O-S-809	(28-26-00-2060)*

*In-house codes.

In some cases, a Qualification paragraph is required: "Some processes may require product qualification, supplier approval, certification of personnel, qualification/certification of equipment, or all of these. Care must be taken to separate requirements and procedures applicable only to product qualification, supplier approval, or certification from those relating to production use of the process. Qualification of a process may be overlooked if the documentation being prepared is to provide coverage for a process already in use."

The Quality Assurance Provisions (Section 4) "are enumerated in a separate section for ease in locating sampling and testing procedures. The section serves two functions:

"a. Description of sampling plan, specimen preparation, and test methods and procedures to verify conformance of the process to the requirements of Section 3 (Requirements) and if applicable, Section 5.

"b. Definition of departmental responsibilities (such as Quality Assurance and Materials and Process Applications Departments) for quality assurance."

Paragraph Number	*Paragraph Headings*
4.1	Responsibility for Inspection
4.2	Inspection
4.3	Process Controls
4.4	Qualification
4.5	Test Methods

4.6 (as required)	Rejection and
	Retest
4.7	Reports

Examples:

4.2	*Inspection*
4.2.1	Random samples of painted parts shall be selected by Quality Assurance after a minimum of 48 hours drying at 75 F ± 5 (24 C ± 3) and examined for surface defects, thickness, and adhesion.
4.2.2	When examined visually, parts shall be free from surface defects as specified in 3.6.

The Preparation for Delivery Section (No. 5) is not always required. It establishes delivery requirements for preservation, packaging, and packing of parts processed by suppliers, and for marking of packages and containers.

Example:

5.	*Preparation for Delivery*
5.1	*Preservation and Packaging*—Parts plated by suppliers shall be wrapped in paper conforming to MIL-P-130. Each paper wrapped part shall be nested in shredded waxed paper conforming to PPP-p-150 and packaged in individual cardboard boxes.

Definitions and information sources are given in the Notes section (No. 6). "When the exact meaning of terms is required, definitions are normally placed in one paragraph (6.1) rather than in parenthetical phrases at the point of occurrence in text.

Example:

6.	*Notes*
6.1	Information pertaining to this section may be obtained from the Materials and Process Department, Aircraft Division.

Opinions of Engineers Who Use Specifications

Common faults cited reflect a grouchy consensus.

A potpourri of verbatim remarks:

"Sometimes specifications lack the information needed by the user to follow them. Bits of information are missing. Only an expert can understand."

"Specifications are long and overly verbose. Examples and exhibits are sometimes inadequate."

"There are too many specific requirements without real technical basis. Progress is prevented."

"Requirements are not relevant to the material's application or the supplier's process control techniques. They are product rather than materials oriented."

"They are written around the capabilities of suppliers."

"They are theoretical, not attainable in production."

"They are either too broad or too narrow."

"Some standards are based on incomplete data, but they tend to become permanent. Revision is needed periodically."

"Standards are often difficult to read and to understand."

"Because the writer deals with standards and specifications, he sometimes goes overboard on minutiae—charts, tables, etc., which are sometimes difficult to understand."

"They are hard to understand and have a commercial slant."

"Standards and specifications usually get so watered down, due to proprietary interests of individuals and organizations, they become a compromise."

"Specifications are so hard to follow and information on a specification item is so hard to locate they are not used in the shop. An example: MIL-I-6868 and MIL-14-6875."

"Common problems are ambiguity, a lack of standardization of terms, unnecessary length."

"They often specify things that aren't needed."

"Achievability is not taken into consideration . . . a shortage of appropriate examples . . . inconsistencies in writing style."

"Not kept simple . . . format not uniform."

"Requirements and quality assurance provisions do not jibe."

"Do not adequately consider effect upon user when all clauses are interpreted as instructions."

"Overemphasizing some procedures with extensive detail, while providing incomplete information on equally important procedures in the same document . . ."

"Can be vague, redundant, incomplete, or phrased in confusing language."

"Too often terms are qualitative and ill-defined, or undefined. Example: 'Material to be free from injurious defects and have a workmanlike finish.' "The generalities 'injurious' and 'workmanlike' are not defined. They can't be defined without knowing the application. Quality standards for materials in aircraft gas turbine engines are more critical than those for a sash weight."

"Redundancy is a problem. In a specification something rarely has to be said more than once."

User Views on How To Write Standards and Specifications

Prescriptions tend to be cures for a given technical person's diagnosis of what's wrong with standards and specifications he or she deals with.

A grab bag of opinion:

"Instructions should be thorough and precise where necessary so the reader is not forced to consult an expert."

"Use good plain English . . . a common sense . . . be brief and concise."

"Requirements must stipulate properties relating to the control of processes and use. Use standard tests (ASTM, etc.) that are specific and to the point."

"They should be pragmatic."

"State minimum requirements . . . reference national standards if applicable . . . if arbitration is required to resolve a point in dispute or a deviation, state who has the responsibility: engineering, metallurgy, quality assurance . . . keep as brief as possible."

"The purpose of the standard must be met by the standard. Deviations should be defined."

"Choose supporting or illustrative charts, etc., with care so the message is clear and not subject to misinterpretation."

"If you can't find a loophole to cover for a goof, the standard or spec is well written."

"For a model follow ASTM standards."

"Should be brief, follow a standard format, state requirements clearly."

"They need to be complete, which promotes length but usually pays off in the end."

"As far as possible, properties and attributes should be spelled out in measurable or quantitative terms. The ideal specification or standard leaves no doubt in the mind of the reader as to what is expected and what tests or inspection methods will be used to determine conformance to specified performance."

"Procedures or criteria should be chartable."

"They should . . . have continuity and logical flow . . . material should be organized . . . good grammatical construction . . . ease of readability . . . free of error and ambiguity . . . accurate . . ."

CHAPTER 17

Guidelines for Writing Proposals and Idea Submissions

The proposal is a complex document.

In the words of one reader, "A proposal is a combination of a specification, an instruction, and , to a lesser extent, a report. If it doesn't spell out exactly what is to be done, by whom, within what time frame, and at what cost, it is not satisfactory. It is also a sales document and as such must indicate the benefit accruing to the purchaser."

Further, there are proposals and proposals. Our look is confined to proposals submitted in response to a customer's request for a proposal (termed an RFD) and to proposals to do research financed by the company or by an outside sponsor. In addition, we back up a step to consider idea submittals, after the nascent form of both proposals and patent applications. As usual, a cross section of reader opinion on common complaints and recommendations is presented.

As in the case of specifications and standards, formats for proposals seem to have a number of standard components. Any differences, and they tend to be large ones, are mostly in amount of detail. The proposal, for example, is stripped to its bare essentials in the brief instructions a steel company gives to its technical people seeking permission to do research for the company in-house.

The format has these main sections:

1. Objective . . . to be attained or problem requiring solution, including scope of problem, requirements to be met, or specific application intended.

2. Expected benefits . . . (potential value or estimated practical importance) to the Corporation.

3. Background or introductory material . . . when required, e.g., a note on previous efforts in the field, key references in the literature, or the point of departure provided by a previous program.

4. Plan of investigation . . . i.e., the proposed test program and evaluation technique.

5. Test materials and equipment.

6. Time, effort, and cost estimate.

The company wants "the title of a proposal and heading of its parts to be as specific as possible." For example, instead of a stock heading like "Proposed Test Program," the writer is urged to try something like "Procedure for Testing Variant XYZ."

Another steel company also has unusually concise and clear instructions for doing the paperwork required when a technical employee has a project he hopes will benefit the corporation. Other formats will be required, says the company, "where government or private outside agencies specify their arrangements." They should be followed accordingly.

The basic format:

1. Prepare an introduction, giving the objectives of the research program. Include a statement of the problem plus information regarding its history, theory, available literature, and pertinent information in allied fields.

2. Describe the proposed research program. Give one or more possible approaches to the problem, and the reason for selection of the proposed method. List the equipment and procedures to be used. Comment on anticipated difficulties and the expected degree of success. If the problem is particularly difficult, suggest an alternate approach to be followed in the event the first approach proves unrewarding.

3. Summarize with a discussion of the factors that make the company's laboratories particularly suited to handling the project. Comment on the probability of successful results. This may relate to prior projects, policy, organization, equipment, and facilities. Outline the competence and specify qualification of specific technical personnel on staff who can perform the program. Recommend additions to staff if necessary.

4. Submit cost estimates. Include time required for engineer and technician manpower and related salary rates. Show cost of materials, supplies, services, or equipment to be purchased directly for this work. Indicate the overhead rate to be applied to salaries. If travel expense will be significant, include as a separate item.

From this point on, the plot thickens. In government and in some parts of industry, the proposal is a richly orchestrated work. In less euphemistic terms, the red tape is abundant. The following words from the steel company just cited on government proposals will give you an inkling.

"Government proposals are generally prepared in response to a governmental agency's Request for Proposal. This is in essence an invitation to bid on a contract for technical services that are described in exhibits attached to the request. The request specifies the number of copies, the submission date, and the name and address of the person to whom the proposal is to be submitted. Four exhibits are usually attached, as follows:

"Exhibit A—An abstract containing a general description of the contemplated program.

"Exhbit B—A statement of work which serves as an outline description of the intended scope and objectives of the program. A guide is provided as to methods, materials, tests, and studies to be covered in the proposal.

"Exhibit C—A description of the kinds of reports to be prepared, the number of copies required, and a schedule of issuance dates. Also general instructions for preparing technical proposals showing format and a guide to specific content. Additional information as to contractual period and rate of effort is given.

"Exhibit D—General instruction for preparation of cost estimates and presentation of required financial data. Generally, a form such as DD-633 entitled 'Cost and Price Analysis' is provided."

What the Electric Power Research Institute (EPRI) Prescribes

EPRI, supported by the electric utility industry in the United States, awards contracts for research. What follows is the gist of its preferences for unsolicited proposals to do research.

> The minimum requirement is to cover these items:
> Objective(s)
> Work to be performed
> Methodology (technical approach)
> Expected end product (and its relevance to the utility industry and
>   the public)
> Schedule
> Cost (including the details set forth in EPRI's Contract Cost Esti-
>   mate, EPRI form 112)
> Estimated level of effort
> Background of key personnel

Here are further details on some selected items:

Cost information—Each proposal should be accompanied by cost details calculated on a calendar year basis . . . the proposer should submit a break-down of prices or costs separately or by identifiable tasks . . . the proposal should show any major items of direct materials and their estimated costs separately . . . subcontracted effort includes, but is not limited to, the fabrication of parts and assemblies or the performance of services by other than the prime contractor in accordance with designs, specifications, or direction of the prime contractor and applicable only to a specific prime contractor. . . . When space on the EPRI Contract Cost Estimate form is not sufficient for this item, the proposer must attach supporting schedules, indicate types or categories of labor that would be needed, together with the number of man-hours, -weeks, or -months expected to be needed in each category and the rate of compensation per unit, exclusive of any overtime premium and/or shift differential.

Proposals must also include details of special testing (labor, material, overhead); a separate schedule should be attached if necessary, showing the basis for the estimate . . . the proposer should attach a brief description of items—facilities, special equipment, or tooling—to be acquired or fabricated solely for the performance of this particular procurement or project . . . attach a schedule indicating the purpose of any proposed travel . . . consultant service should be explained by indicating the specific area of the project in which it is to be used . . . other items of direct cost include special direct taxes, such as federal or excise or state franchise taxes, directly applicable to the project . . . if the proposed burden rate, or rates, have recently been accepted by a contracting agency of the U. S. government, a copy of the authorizing document should accompany the proposal . . . the proposer must provide information as to the amount of independent research and development (IR&D) being proposed, if any.

Five copies of the proposal are to be submitted to EPRI's proposal administrator in its contracts department, Palo Alto, Calif.

How a Company Handles Requests for Proposals

This company is in the big leagues of proposal preparation. Its manual for proposal writers is kicked off with a set of ten "guidelines for effective proposal preparation." For example:

"1. Know your customer—study his request for proposal—learn his needs and wants—reverify through Marketing and Field Sales.

"2. Maintain dialog with the customer during proposal preparation period—ask questions.

"3. Know evaluation criteria—examine key issues from the customer's point of view.

"6. Make a topical outline of the proposal to ensure that it responds to each specific customer requirement, and that it presents a complete and uncluttered picture of our approach, and the justification for it.

"7. Tell the customer what you will do—why you will do it—and how you will do it, to satisfy his requirements. Be specific—not general or vague—about your ability to accomplish program requirements. Use checklists when writing, also for proposal evaluation.

"9. Don't assume the customer is aware of our capabilities—or that he knows of past accomplishment—despite previous relationships. Tell a complete, concise story—the evaluator may be new and totally unfamiliar with the company."

As indicated, at the company it "is mandatory that every proposal be subjected to critical review. . . . "Comprehensive checklists cover all major headings in proposals: technical discussion, hardware design, manufacturing, quality assurance, maintenance, field support, project management, personnel, schedules, facilities, and cost. Examples from each follow.

Technical Discussions:

"Have you clearly defined the technical requirements the proposal is intended to fulfill, and is your proposal responsive to all of the technical requirements? In unsolicited proposals, have you clearly defined the particular area(s) involved in the proposed project?

"Do you clearly establish your technical competence to complete the proposed project?

"Do you clearly and convincingly demonstrate an understanding of the problem?

"If certain specifications are believed to threaten cost or time delays, have you identified the potential problems?"

Hardware Design:
"Do you technically substantiate the design you are proposing?

"Have you clearly described the hardware you expect to build?

"If you propose developing new components, do you explain why existing components are not suitable?

"Do you indicate if you intend to deviate from the specifications in the procurement package?"

Manufacturing:
"Have you detailed your manufacturing organization's responsibilities, plans, policies, quality assurance, and actual manufacturing methods, and built-in controls?

"Does your proposal clearly demonstrate your appreciation for the prospective customer's concern with adequate project management?

"Does your proposal indicate the presence of adequate manpower and facilities?

"Have you described your manufacturing facilities in sufficient detail?"

Quality Assurance:
"Have you described your quality control plan, including: policies, organization, operational systems, record systems, and technical capabilities?

"Does your proposal clearly indicate that the customer's quality requirements are recognized and will be met?

"Does your proposal indicate how your quality organization fits into your overall proposed project organization and existing company structure?

"Have you detailed your facilities and testing equipment?"

Maintenance:
"Do you demonstrate that you have given proper consideration to ease of maintenance and serviceability?

"Do you include an estimate of required maintenance procedures, and indicate if any special test or support equipment will be required?"

Field Support:
"Does your proposal include information on all aspects of support required by the proposed project? Have you considered: engineering, maintenance, technical data, training, installation, field support, provisioning and sustaining engineering, and product improvement support?

"Have you fully detailed your recommended support policies, procedure, and staff?"

Project Management:
"Does your proposal clearly demonstrate your appreciation for the prospective customer's concern with adequate project management?

"Do you demonstrate your capability to handle a project of the size proposed?

"Have you included a manpower build-up chart for the duration of the project?

"Have you included first and second tier subcontractor organization charts, and indicated how they are related to your efforts?"

Personnel:

"Do you clearly indicate the availability of adequate qualified technical and management staff? Have you related your staffing estimates to your man-hour estimate?

"Have you included resumes? Do they clearly describe the professional experience closely related to the proposed project?"

Schedules:

"Does your proposal provide assurance that the required (or proposed) delivery schedules will be met?"

"Have you included master scheduling, programming, and similar back up data?"

Facilities:

"Have you clearly indicated that adequate facilities are available to perform the project both efficiently and on schedule?

"Have you clearly indicated that the necessary facilities will be available when required for the proposed project?"

Cost:

"Have you calculated overhead and burden rates consistent with the project and the customer's requirements?

"Have you carefully estimated the space, facility, man-hour, and other cost factors?

"Have you considered 'make or buy' in depth?

"Have you considered life cycle cost considerations?"

Proposal writing at this level is hardly a walk through the park. Committees and squadrons of disciplines are members of the cast. The company establishes the ambience with these words:

"All proposals, whether large or small, multi-volume or single volume, must be specifically and directly responsive to the request for proposal. It must not be assumed that the customer knows anything about us. Tell him all, tell him clearly, and make it organized. There should be a topic statement at the beginning of each major proposal section that establishes and re-establishes in the reader's mind the theme of the proposal. At a minimum, all proposals should contain:

"Summary
"Technical Discussion
"Program Management
"Program Organization

"Related Experience
"Facilities
"Cost."

Instructions cover subsets of each topic in detail in three areas: technical information, management information, and cost information. First, the technical information encompassed in a proposal.

The company says, "The technical proposal contains the details of exactly what you are proposing, it establishes the theme of your response, and it details the plan you intend to use to accomplish the project." The topics discussed include an introduction, a summary, a statement of the problem, the technical discussion, the program plan, a task statement, exceptions (clarifications or interruptions). Examples of specifications for each topic follow.

Introduction
Items include:
"Summarize general information related to the requirements specified in the request for proposal (RFP), including the basis for submitting the proposal (in response to the RFP, etc.) and the RFP number and title.

"Summarize briefly the project, its purpose and objectives.

"Describe any company participation in proposal activities, previous contract, independent R&D, etc., related to the proposed program.

"Close with a statement of why you feel the company is best qualified to perform the project (re-establishing the theme of the proposal)."

Summary
This section "should demonstrate that your proposal meets or exceeds the customer's requirements."
Items include:
"A summary of the technical requirements which the technical proposal satisfies.

"A summary of all key points related to the capability, suitability, usefulness, technical feasibility, and reliability of the proposed system, item, or project.

Statement of the Problem
"This discussion," it is stated, "centers around a concise definition of the basic problem your proposal is concerned with."
A definition may cover:
"Why a particular item is required by your prospective customer.

"A government requirement for a particular piece of equipment, process, or system.

"The requirement for a solution to a particular problem.

"The necessity for research related to a specific problem."

Other topics in this section include the selected approach and alternate solutions considered. Each alternate solution should cover:
"The comparative cost of the approach.

"The time required for implementation.

"The expected performance parameters or end result, and the merits associated with each."

The Technical Discussion

Termed the "heart of the technical proposal." Topics cover: a detailed description of the proposed solution, the suitability of the proposed solution, a description of the operation environment, detailed characteristics of the solution, and the problems and problem resolution.

The objective of the section titled "A Detailed Description of the Proposed Solution" is to "clearly define your proposed solution, product, research project, service, etc. The manner in which you describe your solution depends largely upon the nature of the problem you are dealing with."

Four different options are given:

"If you are designing a system—this description will be devoted to explaining the characteristics of the proposed overall system.

"If you are designing a product—this discussion will consist of a description of how the product will function, the characteristics or features of the product, and an overview of the component parts.

"If you are providing a service—your description of the proposed approach will detail the nature of the service, and will identify the primary tasks involved, and the purpose of each.

"If you are proposing research—this discussion will detail how you intend to approach the research problem, and will also delineate the associated research and research support tasks."

The Program Plan

Topics are the work breakdown structure, a project master schedule, manpower considerations, project master plan, proposed project personnel, detailed implementation and control plans, task statement, exceptions, and appendixes.

The work breakdown structure, for example, provides:

- The basis for estimating the man-hours and staffing required for the project.
- A platform upon which to develop and base your proposed Project Master Plan.
- A framework of actual cost and progress reporting during the course of contractual performance.

Detailed implementation and control plans may cover: life cycle cost, design to unit product cost, make or buy, subcontracting, system engineering, configuration management, engineering, manufacturing, technical data, quality assurance, reliability, system testing, maintenance, logistics, training.

The appendix "is the place for almost all highly detailed technical information, equations, etc. In every case, however, be certain that materials included in the appendix actually strengthen the technical proposal. Never use such materials just to add bulk to your proposal."

In the arena of management information, sections include: an introduction, project management, project master plan and schedule, manpower planning, proposed project management, management control techniques, subcontracting, technical support information, related experience, and facilities.

The management proposal, says the company, "should present details of how we propose to manage the project, and why we are qualified to provide the proposed research, product, or service."

Discussion topics include:

A nontechnical overview of the contents of the technical and management components.

Details of how you intend to manage the proposed project.

Project personnel and commitment to the project.

Technical support information relating to the division's structure, and overall policies and procedures which enhance our ability to manage the proposed project.

An overview of experience in the same or closely related fields.

A description of the facilities which will be used on the proposed project.

In the introduction, the introduction from the technical proposal is tailored for the manager or nontechnical evaluator.

Project management topics cover: a project overview, project master plan and schedule, manpower planning, proposed project management, management control techniques, proposed subcontracting.

Manpower planning topics include: a month-by-month forecast of total projected man-hours; an analysis of total man-hours required in each major task category, including total for each subtask; the number of man-hours in each major labor category, and an analysis of the number of persons required to satisfy those man-hours; a table which summarizes man-hours by task and number of persons in each labor category required to perform the task; a figure which illustrates your plans to increase, decrease, or otherwise alter the work force over the life of the project.

Technical support information often varies with the nature of the project. Topics may include: value engineering program, logistics and training, quality assurance and control, life cycle costs, and designing to unit product cost.

The section on related experience "should discuss the company's experience as it relates to the proposed project. The information must be expressly for the proposal and should relate directly to the proposed project. Stress both past and current projects that either relate directly, or that demonstrate the company's ability to successfully manage a project of the proposed scope. Summarize the information in a table which enables the proposal evaluator to grasp your information at a glance."

The format for cost information is often similar to that required by government agencies. Topics: introduction, statement of work, project completion schedule or delivery schedule, cost summary, supporting schedules, cost breakdowns, supplemental cost data, estimating techniques used, government or customer furnished equipment, contractual terms and conditions.

The introduction for the management proposal may be repeated.

The statement of work defines the tasks which must be performed to complete the proposed project.

For the cost summary, the government has designed a number of standard forms to be used in summarizing project costs. Copies of forms are normally supplied with the procurement invitation package. The most commonly used forms are in the DD 633 series.

When such forms are used, "you should use a supporting schedule for the following line items: purchased parts, subcontracting items, other material, interdivisional transfers at cost, direct engineering labor, direct manufacturing labor, other costs, royalties, and federal excise tax."

Supplemental cost data are usually included in the cost proposal to justify certain cost summary line items, such as general and administrative costs, engineering overhead, and manufacturing.

In the section on government or customer-furnished equipment, the aim is to identify all facilities, equipment, tools, support staff, supplies, services, etc. to be supplied by the government or your customer.

The need for executive summaries varies with circumstances. The company says, "For smaller efforts, a brief summary to highlight the key points and project the basic theme might be provided at the beginning of each proposal component (technical, management, etc.). For major proposals, however, it is essential to provide a separate document which clearly, concisely, and attractively communicates the basic win strategy and theme—why our proposed solution is better than any other."

Criticisms by Technical People Who Read Proposals

A random sampling of verbatim complaints:

"Very hazy about objectives . . . no thought as to outcome, be it success or failure."

"Technical approach, statement of work, and organization must be well thought out and well presented. Often these guidelines are not observed."

"Often too wordy . . . theme or message sometimes difficult to follow or find."

"Much emphasis on the company with too little technical discussion."

"Need a management summary; there is a lack of convincing arguments indicating the benefits of doing what is proposed."

"There is sometimes a failure to state clearly and explicitly what will be done, how it will be accomplished, and what use will be made of the results."

"Incomplete . . . difficult to locate specific information . . . too much sales pitch."

"No final recommendation."

"Major issues, such as the objective or approach, often not stated explicitly. The reader cannot infer such important topics."

"Too long, apparently in the hope that length justifies cost; too general, in hope a lack of specifics will hide inaccurate accounting."

"Relevant literature is not reviewed."

"A particular thing may not be proposed . . . your problem is merely a convenient hook to hang their proposal on . . . there is no real effort to assess your needs or to show how their work will aid you."

"Either too long with too many details, or too short and not covering pertinent points."

"Writers of proposals try to prove how smart they are instead of responding to the request for proposal."

"Background and supporting data may be lacking. Example: cost-benefit data."

"Too vague . . . unrealistic economic assumptions."

"A lack of specifics."

"Too many promises. Too much 'relevancy.' "

"The proposal fails to respond to all the specific requirements stated in the client's request for a proposal . . . the proposer prefers to overwhelm the client rather than show capability to resolve the client's specific problems . . . most proposals tend to be longer than necessary, as if more words assure a successful proposal."

"Trying to determine what is proposed is a common problem."

"Short on ideas."

Reader Recommendations to Writers of Proposals

Verbatim comments:

"Have a clear outline of objectives and approach . . . describe the implications of negative as well as positive results. Most endeavors are not successful . . . set clear target goals and dates."

"A proposal should be clear, concise—not redundant . . . have a sound technical approach based on an understanding of the problem or procurement. . . . Be well organized with a road map to each item requested in the procurement from both the technical approach section and statement of work . . . and supply data, illustrations, examples."

"An author highly motivated to communicate is basic and will transcend many problems. . . . Ability to express oneself and to organize material precisely is beneficial."

"A proposal should be complete, specific, without ambiguity, well indexed, and easy to follow."

"Supply a good background and concise, accurate statements of objective, benefits, cost, and likely outcome."

"Be certain that times, dates, dollars, etc., are clearly shown. Include enough information to demonstrate your absolute competency for the job, but don't bury the message by citing unrelated feats of greatness."

"Brief but totally responsive to the request for proposal . . . lots of illustrations and graphics . . . no bullshit."

"A simple format . . . state purpose, give a brief description of action to be taken, state goals, provide economic and time frame estimates."

"From the seller's point of view, the proposal should probably incorporate some information supporting the competence of the people and organization which will perform the work or provide the material and/or services. It must also be specific as to what constitutes satisfactory performance of the contract as well as criteria for nonconformance."

"Spell out why it is needed, what will be done, cost, changes of success, and expected payoff."

A Steel Company's Guidelines for Idea Submittals

Ideas, says this company in its Standard Procedure No. 25.311, generally fall into one of three categories:

1. Preliminary incompletely thought through ideas.
2. Carefully considered scientific concepts that warrant consideration of a proposal to engage in a research program.
3. Ideas of inventions deemed to have patentability.

A procedure covers the essential points.

Each submittal has an identifying code number obtained from the secretary to the director of research. Submittals are identified by type, number and year, e.g., Prelim 1-66. "G" is added as suffix if the idea involves government sponsorship.

Other items:

"Title the idea with a brief but descriptive title.

"Present a background describing the technical problem or situation and outlining the submitter's prior knowledge of history, theory, limitations, and possible relationship to other fields.

"Describe the idea. Explain how and why it has merit. Discuss its novelty, applicability, usefulness or value to the company, and any other factor worthy of consideration.

"Recommend action to be taken. This may include not only research work but also activities such as economic evaluation or a literature or patent search. Make an estimate of the dimensions of the research program (if one is to be undertaken) including time and cost. Suggest specific considerations by persons competent to contribute in any way to clarifying, developing, or utilizing the idea.

"Sign it, date it, and have it witnessed by a person capable of understanding it. Submit it to your section manager."

Here is the format for the submittal:

Standard Procedure No. 25.311

ABC STEEL COMPANY
Research Laboratories
IDEA SUBMITTAL
(Date)

Prelim. 1-66
Author's Name
Title
Background

Description of the Idea

Recommended Action

Witnessed:　　　　Signed:

__________________　　__________________　　__________

(Date)　　　　　　(Author)　　　　(Date)

CHAPTER 18

Guidelines for Writing Technical Directions

"The abstractness of the topics and the diversity of the audience continually require the writer to make difficult decisions such as how much detail is needed and which terms require definition. The reader wants only to do a required task in the simplest way. To the reader, a computer manual must be not only accurate but easy to use."

That quote—a suitable entree to the subject of writing technical directions—is from an article, "Writing Computer Manuals," by D. M. Mitchell and M. J. Ransom of IBM in the Journal of the Society for Technical Communication, *Technical Communication*, Vol. 27, No. 4, fourth quarter 1980. This and other articles by other IBM authors on different aspects of the subject in the same issue are reviewed here. The principles for writing manuals for computers apply to writing technical manuals of all persuasions.

The IBM approach is thorough and includes planning, writing, choosing graphics, editing, reviewing, and testing of manuals. There is an article on each topic.

In "Planning Software Manuals," author Lynn Behnke of IBM starts by explaining the need for planning: "The apparent purpose of publications planning is to ensure that the users of a computer system or a computer program have the information they need in a usable form."

Planning homework includes such key items as learning the product, learning the audience, and analyzing user tasks. Detailed knowledge of the audience is termed critical.

"Knowledge of the product," says the author, "frames the publications plan; knowledge of the audience supports the plan. Knowledge of the product provides the planner with a store of facts; knowledge of the audience reveals the principles on which to organize those facts."

Mr. Behnke continues, "A technically accurate manual can fail, for example, by assuming too much special education on the part of the reader, by ignoring the job experience of the reader, or by confusing the professional responsibilities of various readers."

What are the secrets of an easy-to-use manual?

In the article on "Writing Computer Manuals," authors Mitchell and Ransom list six:

- Information is presented in its simplest form without abbreviations or technical jargon.
- Topics are presented in a logical order.
- A complete index is provided.
- Terms are defined clearly and completely.
- Examples are accurate and useful.
- Graphics and text are blended effectively.

Fact gathering at IBM starts with the written product specifications. "As soon as a working version of the product is available," reveal the authors, "the writer tries to use the product as a customer would use it. This work leads to more questions that must be asked."

At the beginning of writing, "great importance is placed on the outline." Review also starts at this point. "Developers and market planners review it to ensure that all necessary technical information is included. Writers, editors, and illustrators also review the outline; for example, an editor can identify structural weaknesses in the manual and an illustrator can identify areas where graphics would be effective."

Subsequent steps are: first draft, edit, review, second draft, edit, review, third draft, production, printing.

The writer's team has a sizable roster: publication planners, development programmers, fellow writers, illustrators, publication test coordinators, market planners, service representatives, customer instructors, editors, and lawyers.

Illustrator is a demanding position to play. "Illustrating software manuals," declares Sandy Johnson of IBM in her article "Graphics for Software Manuals," "requires an artist's imagination and a draftsman's precision."

As mentioned before, the illustrator gets his first call at the outline stage. "When the writer is working on the first draft," explains Ms. Johnson, "the illustrator provides illustrations for individual concepts at the writer's request. The writer usually has some idea of the kind of illustration he needs or has a sample of what he needs. The illustrator usually modifies or simplifies the writer's illustration. Together the illustrator and writer prepare a rough sketch for the first draft . . . after the first draft is complete, the illustrator begins to develop new art based on the entire draft . . . ."

A lead artist reviews the artwork. Consistency is one of his objectives. "Consistency," explains the author, "requires the use of the same figure to explain the same concept in more than one manual. It thus avoids the possible ambiguity that might result if a single concept were represented with a halftone in one figure, an isometric view in another, and two dimensional line art in another."

The lead artist also determines if art helps to explain concepts in text. "Sometimes," says the author, "a figure does not clarify a concept at all; it only takes up space. The artist suggests that such figures be deleted. The artist also checks to see that the words in the figures match the words in the text.

"In deciding how effective a figure is, the lead artist must keep the audience in mind. The artist may suggest, e.g., that a basic data-processing concept (such as translating a source program into an object program) be illustrated in an introductory manual for inexperienced users but not be illustrated in a reference manual for experienced programmers."

The editor is the reader's surrogate in the IBM communication scheme. "Giving the reader all needed information is a difficult task, but even more challenging is the task of presenting that information at a level that matches the assimilation skill of the reader," says IBM's J. Y. Harrington in "Editing Computer Manuals."

The editor evaluates what the writer has done and offers suggestions for improvements. Readers, adds the author, "demand that the information be: accurate, retrievable, complete, easily understood."

Three readings by the editor are proposed. "The reasons for this division," explains the author, "lie in the size of the task and in the fallibility of the editor. No one can see everything at once, especially when there are so many things to see. It is foolhardy for an editor to try to evaluate the total effectiveness of a manual during a microscopic examination of each page for comma faults. Once absorbed in detail, the editor will find it difficult if not impossible to evaluate whether the manual satisfies the information needs of the user."

The first time around, "the editor is trying to determine whether the audience can find the required information and can understand it. In this process, the editor checks how the manual is organized, looks at the level of terminology used, and tries to determine whether any information is missing or unnecessary.

"Assuming the role of the user, the editor asks whether the information the reader needs to do the task is really there. What tasks are more important? What tasks are done more frequently? What tasks are so simple the user may read the information only once or twice? Answers can strongly influence how the information is organized and how it is presented. In making this kind of evaluation, the editor has to focus on the total manual, trying to ignore misspellings, comma faults, and dangling participles, i.e., trying not to get bogged down in detail at this stage."

At the second reading stage, in the author's words, the editor does "battle with nonsense." He concentrates on paragraphs, sentences, words. How they are put together. How transitions are made from sentence to sentence, from paragraph to paragraph. Whether sentences are cluttered with multiple adjectives, and so on.

Styling is the main concern in the third reading. Topics such as how tables are presented, punctuation, capitalization, abbreviation are pursued. The aim is consistency of style. Things like spelling a word with alternative spellings the same way each time, always putting a comma before the last "and" in a series to denote last item, and making sure the single quotation mark (') is used in quotations inside of quotations.

All things considered, the audience gets top billing. "The audience," says the author, "can vary from the untrained, unskilled, inexperienced to the very trained, very skilled, very experienced. When information for the unskilled and information for the very skilled are packaged in one manual, problems ensue. If such packaging is really necessary, the editor must ensure that each audience has a clear path to only the information it requires."

The author adds, "The inexperienced—people who are not data processing professionals—are becoming a larger and larger part of our audience. That means we must redouble our efforts to avoid the jargon of the industry and to strictly control definitions of all technical terms." An example: IPL, an acronym for "initial load program."

IBM subjects the writer to five reviews: outline, first draft, editing draft, second draft, third draft. The procedure is outlined by Marilyn Morem of IBM at the beginning of her article "Reviews of Software Manuals."

She comments, "For each review, the writer distributes drafts of the manual to the reviewers, who are given from one to four weeks for their reviews. During this time, each reviewer makes comments (such as 'Incorrect—it really works this way,' or 'Needs more information') on the draft. One person from each reviewing group then incorporates all comments from that group into one copy. At the end of the review cycle, a representative from each group and the writer meet to discuss and take action on the comments."

Reviewers come from the many different groups involved in the development of the program being documented.

One more step—testing the manual—remains. 'The information must be presented clearly, concisely, and conveniently, in a format that ensures ready access. Testing helps ensure that the manual meets these standards," say authors Barbara Winbush and Glenda McDowell in their article "Testing: How to Increase the Usability of Computer Manuals."

Testing at IBM is in the hands of people called test coordinators. They have backgrounds in publications and are familiar with the computer system being documented. They find people who represent the intended audience to take tests, and they prepare tests.

"By documenting any problems the test subjects encounter while completing these tasks, test coordinators determine the extent to which the manual meets the customers' needs; that is, the extent to which the manual is usable. . . . Emphasis is on identifying and correcting usability problems," state the authors.

Common Complaints of People Who Read Technical Directions

In the reader study for this handbook, a cache of criticism was uncovered. Verbatim comments follow.

"Directions are too long"

"They lack clarity."

"They do not convey adequate detail as to what must be done, by whom, what procedures are necessary, what time restraints apply for starting or completing the task, and who must be informed about progress, completion, or problems encountered. A good instruction must give all necessary information and leave no doubt as to what is to be done."

"Not written for the comprehension of the intended audience—usually the author prepares for an audience with a higher level of education than he or she has, rather than below the level of the projected audience."

"Can be clear and incomplete."

"Written from a limited point of view."

"Rarely written for the intended audience. An instruction is not a specification related to the job."

"Other writing tends to be wordy. Instructions tend to suffer from brevity, terse to the point of being incomprehensible."

"They are too ambiguous. Tolerance limits are not given."

"Visuals are not located in relation to text . . . there can also be an information overload—more than necessary."

"Lack of vital information."

"Either unnecessarily long or unnecessarily brief. Descriptions and supporting drawings are spaced too far apart."

"Incomplete—steps missing or assumed to be obvious."

"Too complex—difficult to follow."

"Often repetitive, and sound as if they have not been tried out."

"Notoriously inaccurate. . . . Do not follow illustrations. . . . Often do not define terms or information required. . . . Try to assemble something from a factory and you'll see what I mean."

What do these readers recommend?

One offers, "An instruction is intended to facilitate the accomplishment of some task or action. If it is properly written, the task gets done as desired. If not, the desired result is not achieved. Whatever the writer expects to accomplish must be spelled out accurately, as briefly as possible and in language which cannot be misunderstood."

Other views follow:

"Organization is most important. It usually must be chronological or sequential—to deviate from the stepwise arrangement one must have a valid purpose. Write to a grade school level audience to assure total communication."

"Must be clear, thorough, leave no doubt for reader as to what is intended."

"Try them out on a few people."

"Be specific to the job and to the operator who will carry out the instructions. He should not be required to go elsewhere for additional information or clarification, and he should not be required to use his judgment."

"Same as any technical writing—accuracy, brevity, comprehensibility."

"If someone cannot find a loophole to cover for a goof, the instructions are well written."

"Establish who will be using the instructions and write for that level; make good use of visuals—if in doubt, use more than you think are necessary."

"Should be capable of being followed by a novice with only a typical background expected for the role."

"The main requirement: permit a person unfamiliar with the subject to follow the instructions. They should always be tested before the final draft on someone or group not familiar with the subject."

"Instructions should be idiotproof."

"Use precise short sentences. . . . Omit extraneous information that dilutes the impact of the instruction."

"Cover all possible mistakes in understanding."

CHAPTER 19

Guidelines for Writing Committee Reports

Technical committees prepare technical reports and/or reports of committee meetings other than standard minutes. Formats for the former do not differ in essential detail from those prepared by individuals. Differences of interest here are in rules, procedures, and practices of organizations used to guide committees in various aspects of producing a report. SAE Recommended Practice J1159-AUG 79 is representative. Selected parts are reviewed.

The purpose of J1159-AUG 79 is to establish a uniform practice for technical committees for the preparation of technical reports.

The committee chairman is responsible for seeing that committee members understand their responsibilities toward publications. In particular, accuracy of technical content and references, conformance to policies and guidelines in the report, and proper clearance.

The chairman is responsible for such things as classification of the report, checking the report against rules prepared by the Society's legal counsel, selection of publication method (*SAE Handbook, Handbook Supplement*, separate report), making sure that metric units are used, preparation of key word index for preparation of index by SAE staff, and a letter of transmittal to a council of the Technical Board.

The last item outlines:

1. Significance of the report.
2. Background information (rationale).
3. Reason for choice of classification.
4. Recommendation for method of distribution.

Reports fall into three classifications: SAE Standards, including product standards and performance standards; SAE Recommended Practice; and SAE Information Reports.

SAE standards document broadly accepted engineering practices or requirements for a material, product, process, procedure, or test method.

A product standard may be primarily a descriptive standard covering dimensions, composition, and other details, or it may be a functional or performance standard or both.

Performance standards involve requirements or levels against which the functions can be evaluated. This frequently involves the need to define test methods by which these requirements are measured. Preferably, performance standards and test procedure standards should be in separate reports.

SAE Recommended Practices document data intended as guides to standard engineering practice. Their content may be of a more general nature, or they may propound data that have not yet gained broad acceptance. Such a report may include an introductory note stating, "This SAE Recommended Practice is intended as a guide toward standard practice but may be subject to frequent change to keep pace with experience and technical advances, and this should be kept in mind when considering its use."

SAE Information Reports are compilations of engineering reference data or educational material useful to the technical community. Example: "Automotive Carburetor Flanges," SAE J623. A product standard based on dimensions.

Development of technical reports for publication by SAE is regarded as the "major effort of technical committee activity."

Initial work on a draft technical report is usually handled by a task force which presents its work to a parent group, preferably well in advance of a meeting date. Corrections to the proposal are officially recognized at the meeting of the parent group and documented in the minutes. Depending on procedures for each group, mailing of draft technical reports may be handled directly by the chairman, the SAE staff representative, or by a delegated member.

Because of the large number of technical reports produced, it is not practical for SAE staff to restyle them. SAE references are followed:

"Preparation of SAE Technical Reports—Surface Vehicles and Machines: Standards, Recommended Practices, Information Reports"—SAE J1159.

"Rules for SAE Use of SI (Metric) Units"—SAE J916.

"AMS Editorial Procedure and Form for the Preparation of Aerospace Material Specifications (AMS) and other Aerospace Material Documents."

"Aerospace Council's Organization and Operating Guide for Aerospace Cooperative Engineering Program"—for the Preparation of Aerospace Standards, Military Standards, Aerospace Recommended Practices, and Aerospace Information Reports.

Unanimous committee approval of draft reports submitted to a council for approval is preferred. If this is not possible, "approval of at least three-quarters of the responding committee members who have not waived their vote" is required. Dissenting views accompany reports.

Draft reports of committees do not normally require confirmation by letter ballot, except when they are to be submitted for final voice vote approval. In such instances, the draft is distributed to members of the voting group at least two weeks before the meeting.

Other requirements cover orientation of reports, naming of suppliers, and naming of test materials.

Reports, it is stated, "should be written in terms of performance rather than design, so as not to exclude any technically adequate equipment, product, design, material, or process. References to sources of supply of parts or products or the identity of manufacturers is to be avoided. Where a committee finds it necessary to name a particular brand or product, such specification is to be accompanied by the statement "or equivalent." A particular product or material may be identified by name when it is essential to uniformity in testing.

Report formats have some or all of the following elements in the sequence given:

- Title—Should not duplicate an existing title; be short, concise, and as descriptive as possible.
- Report classification—Previously explained.
- Approval note—Includes credit to originating committee, original approval date, and date of last revision or reaffirmation.
- Introduction—As applicable. Should provide the basis for data or information in the report, background or general description of the report, and a brief explanation of changes from the previous version of the report.
- Purpose and/or Scope—As applicable. If both used, purpose should precede scope. Purpose explains result obtained through use of the report. Scope briefly indicates extent of treatment and applicability of the report.
- Definitions, Glossary of terms, Terminology, and Designations—Should be reviewed for consistency with SAE practices.
- Test procedures.
- Dimensional data, including tables and charts.
- General specifications which augment dimensional data.
- Illustrations, photographs.
- Performance requirements.
- Component materials and mechanical and physical properties.
- Inspection requirements.
- Appendixes—May be issued as a separate SAE Information Report.
- Rationale for report—May be included as an appendix or made the subject of a separate SAE Information Report. Provides an expanded explanation of the purpose and scope of the technical report; an explanation of the reasons for decisions, conclusions, and recommendations; and a report on actual tests made which support conclusions and recommendations or which form the basis for performance criteria.
- References.
- Bibliography.

In preparing reports, paragraphs are numbered using a decimal system. The purpose is to aid organization of long or complicated reports. Decimal point indicates successive subheadings, as:1., 1.1, 1.1.1.

Artwork instructions are based on a 50% reduction for final size. Typewriter lettering and pencil drawings are not acceptable. Ink is used to get better reproduction in printing. Clear, sharp glossy prints of original article are acceptable if originals must be kept in committee files.

Instructions for line drawings—Main lines equivalent 0 Leroy pen (0.3 mm width); inside lines, dimension line leaders, phantom lines, etc., equivalent 00 Leroy pen (0.2 mm width).

Instructions for graphs—The heaviest line weight used should be the curves. Curves, equivalent 1 Leroy pen (0.4 mm width); ordinate and abscissa, equivalent 0 Leroy pen (0.3 mm width); grid lines and tic marks, equivalent 00 Leroy pen (0.2 mm width).

Lettering (excluding section and reference letters)—Use all capitals unless lowercase letters are necessary for a specific term. Lettering shall be vertical except for quantity symbols which shall be in italics. Use only Roman alphabet, except where letters are recognized standard symbols.

All lettering should be placed outside visible outline of part. Label, with line and arrowhead to area being identified, should be reasonably close to figure.

For lettering, including Greek, numbers, and fractions, use 120 Leroy—equivalent 00 pen (3.05 mm) letter height. Align column of numbers on decimal point.

Tables are numbered consecutively and referred to in text. Each shall be titled. Concise descriptions, measurement units, and letter symbols shall be included in column headings. In giving number ranges, do not overlap or leave gaps. Example of good practice: 0.75 through 1.25 mm, over 1.25 through 2.00 mm, over 2.00 through 3.25 mm.

The word "shall" is used "wherever the criterion for conformance with the specific recommendation requires that there be deviation. Its use shall not be avoided on the grounds that compliance with the report is considered voluntary."

The word "should" is used "wherever noncompliance with the specific recommendation is permissible. 'Should' shall not be substituted for 'shall' on the grounds that that compliance with the report is considered voluntary."

The words "safe" and "safety" "shall be used in SAE technical reports only when they are in whole or in part commonly used engineering terms, such as: fail-safe, factor of safety, and safety glass. To preclude any misinterpretation of the words 'safe' and 'safety,' more definite descriptive words shall be used."

Examples given include:

"Lock wiring," rather than "safety wiring."

"Lock nut," rather than "safety nut."

"Relief valve," rather than "safety valve."

"The integrity of the painted surface," rather than "the safety of the painted surface."

What Readers Have to Say About Committee Reports

Special problems encountered in group authorship of technical papers and in reporting what goes on in a meeting highlight reader views on the common faults of this technical format.

A representative comment, "I have a low regard for committees and for committee reports. They take a compromise position tolerable to committee members, a position which probably represents the lowest possible value one could choose to take."

Some how-to guidelines offered center on acceptable formats for reports written by committees. Some of the guidelines relate to writing summaries of minutes.

An example of the latter:

"The summary should observe good English. Detail should be sufficient so that two years later one could get the spirit as well as the sense of the meeting by reading the report." In addition, one would be in a position to judge the action taken then in light of any new circumstances that may have arisen."

Formats typically suggested for committee reports are similar to thcse for conventional technical reports. For example:

"Clearly state problem, clearly state objective of study, clearly state conclusions, clearly state recommendations."

In the following report of what readers say about committee reports, no attempt will be made to distinguish views on writing technical reports and on writing reports of committee meetings. Comments are self-explanatory. In most instances, the criticism or advice applies to either form of literature.

Common Faults of Committee Reports Cited by Readers

A cross section of opinion:

"Too long, not summarized, not specific, poor writing."

"Very dull reading."

Action items not clearly defined; no indication of who is responsible for what; too long; should cover only pertinent points."

Different sections often written in different styles because of different authors."

"Too much compromise to protect a competitive position."

"Too political."

"Represents what the recorder thought was said, which may not be what was said."

"Most often, conclusions are not reached in committee, so there is none to report. If action is taken, it is not stated simply and to the point; pages are filled with redundant rhetoric for the sake of filling paper and getting out a report; reports are often written in a language or jargon private to members of the committee."

"Organization is poor; sentence structure is convoluted; jargon and acronyms are overused; content is usually skimpy. The committee approach leads to platitudes to satisfy common views of members of the committee. The larger the committee, the smaller the number of common views and the longer the report."

"Excessive length. It is difficult to determine objectives and next steps to be taken by the committee."

"Stuffy. Not action oriented."

"Tenor of discussions not properly reported, and report reflects special views of the secretary and/or chairman. Action to be taken is not adequately identified."

"Committee members are misquoted."

"Some reports try to cover too much ground and too many opinions of individual members, rather than focus on joint conclusions and recommendations. Many reports are too long, indicating a lack of organization and editing."

"Reports are too late to be useful and too infrequent."

"Either too long and too many details or too short and insufficient information."

"Too much verbatim reporting. Essential points not clear."

"Reports are written to extend the life of the committee."

"Important elements are not emphasized."

"Not always accurate. Too short—lacking in vital information."

"Does nòt capture essence of the meeting."

"No recap of what was accomplished at the last meeting. No statement of what is to be accomplished before the next meeting."

"Reasons for committee action not given."

"Lack of reference to minority views."

"Fuzzy statements of problem studied and objectives of study."

"A craving for brevity results in lack of detail of the discussion and inaccurate reporting."

"Frequently lacking in conciseness because of the desire to include input from the majority of committee members; may be lacking in essential information; information on key facts may be incomplete."

"Some reports go too far in reporting word-for-word dialogs and blow-by-blow accounts of committee decisions. The problems stem from a misconception that the report should duplicate committee minutes."

"Reports tend to be full of circumlocutions and gobbledegook, especially if the report is written by the government or a government-sponsored group. If a report is full of buzzwords, acronyms, and jargon known only to a select group of insiders, I won't take the time to try to decipher the actual message. Writers should stick to plain English and not use a paragraph where one sentence (and a short one at that) will do."

"Lack of organization, ambiguous statements, stiff and stilted style, poor grammatical constructions, no point made."

"Final results are not pinpointed too well."

Main Requirements of Committee Reports Cited by Readers

A cross section of comments—generally relating to format:

"Summary first, followed by a brief background, if appropriate; decisions made or excerpts from minutes, as briefly as possible; list actions and responsibilities, if appropriate."

"State what was accomplished or decided and why."

"One author should edit the final draft to develop continuity of material and conformity in style."

"An easy to follow format with appropriate subject headings. Not chronological reporting."

"Report what was said, what will be done, and by whom."

"Be brief, concise, to the point. Include conclusions reached, actions taken, and by whom. If no actions were taken or conclusions reached, say so."

"Main requirements are clarity, organization, brevity."

"A clear statement of the results of the meeting, what is expected of members for the next meeting, deadline and goals for the year's work."

"A good summary. Specific recommendations. Names, sources, references."

"Writer of minutes should check persons quoted on significant items."

"Clearly state joint findings. Offer only enough documentation to support findings. Don't be overly concerned about reporting each different opinion. Everyone has an opinion."

"Be brief. State conclusions of the meeting."

"Reports should be clear and concise and readily understood by all who read them."

"Clearly state conclusions, future action requirements, and responsibiiities."

"In outline form, emphasize specifications and target dates. Identify who, what, where."

"Action items and assignments should be clearly identified."

"Report action taken and reasoning behind it."

"Identify areas of substantial disagreement on important matters."

"Clearly state problem, objective, conclusions, recommendations."

"Report should be a well thought out summary of important decisions and an appropriate explanation of actions taken by committee. If more information is needed one may refer to minutes."

"State mission or assignment, authorization for work, scope and limitations, methods used to collect information, state conclusions and limitations concerning their use; give names and occupations of committee members. If report does not represent consensus, indicate sources of divergent views."

"Summarize the important, ignore the trivial, ignore what is of interest to only the committee."

"Continuity and logical flow. Ease of readability. Freedom from errors and ambiguities. Accuracy. Documentation of conclusions and recommendations."

"State results in simple terms anyone can understand."

CHAPTER 20

Guidelines for Writing Memos and Letters

The following example could be called a memo or letter:

"NOTICE: Revision to ASTM News Release #8308

"Dear Editor:

"Please note a correction on ASTM News Release #8308—Papers Solicited for Second International Symposium on Corrosion and Degradation of Implant Materials to be held in May of 1983. The change is in the date of the symposium. *The symposium is being held on 9–10 May 1983* and not on 23–24 May 1983 as the release incorrectly states.

"The date was changed after the press release went out to the trade publications. We apologize for any inconvenience this error has caused you. If you have any questions, please do not hesitate to contact me at 215/299-5478.

"Respectfully yours,

"Manager of Public Relations"

The example illustrates the two critical "C's" of memo and letter writing: be clear and be concise.

Clarity is judged in terms of how well the writer succeeded in getting his message across to the reader.

Conciseness denotes economical use of language—not using one more word than necessary to deliver the message.

Concise and short are not necessarily synonymous, although this is generally the case. Some longer pieces are concise in the sense that a number of words are required to deliver the entire message. Conversely, brevity per se is not an unqualified virtue. A short memo or letter may not deliver the message because the writer stopped before he was finished. However, the reverse is generally true in this branch of writing.

Analyze the example. Why is it clear? Why is it concise?

"NOTICE" flags our attention.

The writer adds specifics, "Revision to ASTM News Release #8308." The statement serves the function of a title, giving us a shorthand statement of the subject. More information is needed. It is not possible to squeeze the full message into a title or a single sentence. A popular format for memos labels this element as the subject. I feel this is good practice because it demands clarity and conciseness (more on this in a moment). There is nothing comparable to this element in letter-writing formats. My opinion is that the equivalent of the statement of the subject should be the first sentence of the text of the letter.

Such is the case in the example. In fact, the first sentence could stand alone without the help of the overline titled "NOTICE." It is complete.

Note how the first sentence in the text complements and augments what we have been told to this point. The "revision" is a "correction." The title of news release #8308 is spelled out. We also learn that the symposium is to be held "in May of 1983."

So far we don't know whether the change involves the call for papers or the date of the symposium. The point is clarified in the next sentence. In the next sentence, we get the new date and the old date.

Now the curious reader will want to know, "Why was this mistake made?" This is explained in the first sentence of the second paragraph. Apologizing "for any inconvenience . . . " in the next sentence is good PR; and the last sentence is good practice. Editors may have more questions, and they always appreciate knowing the person to call and his or her telephone number.

Once more, the memo-letter is clear and concise. It is clear because all points concerning "Revision to ASTM News Release #8308" are covered. No questions were left unanswered, and the piece can stand alone. The reader need not go elsewhere for clarification.

The piece is concise because it is not possible to shorten it significantly without a loss in clarity. Clarity and conciseness go together in this bare-bones concept of memo-letter writing. The idea is: confine yourself to the message. Do not embellish or pile on unnecessary detail. Stop when you have taken care of the business. Other means of communication are open for those who want additional information.

Analyze the next example. Locate the message first. Then look for what you could eliminate to make the piece more concise.

A letter:

"Dear Editor:

"Managers of engineers and personnel managers concerned with engineers will have a unique opportunity to hear experts deal with the most relevant issues in engineering manpower employment today.

"The event will be an Association conference entitled 'How to Get Ready for the Next Generation of Engineers' to be held at The Hotel in Atlanta, Thursday and Friday, September 30–October 1.

"Sponsored by the Association's advisory committee, the Thursday morning portion of the conference is designed to inform employers of engineers about the present status of supply and demand, salaries and foreign competition. In an afternoon session four major corporations will explain how they are coping with the issues of the 80's. The session is designed for a brisk interchange among the participants.

"The second day will feature four or five successful engineers from the major disciplines who will deliver messages relating to their experiences and recommendations on preparing for a new generation of engineers.

"After you look over the enclosed program, let us encourage you to fill out the registration blank attached and mail it in with your check today.

"Sincerely,

"Chairman."

When you reduce the letter to its essentials, you learn it is an invitation to managers of engineers and to personnel managers to attend an Association-sponsored conference titled "How to Get Ready for the Next Generation of Engineers," September 30 and October 1 at The Hotel in Atlanta. Topics on the first day: status of supply and demand, salaries, foreign competition, and a panel on coping with issues of the 80's. Topics on the second day: experiences and recommendations of managers now dealing with the new generation of engineers.

In the preceding paragraph, I used 77 words to describe the meat of the letter. The author of the letter used 183 words to deliver his message. By the bare-bones test I suggest, the letter is not concise, although it is clear.

In this instance, it is easy to explain the overwriting. This is a sales or PR letter. Extra language is always used to sell and to promote. I do not presume to indict these practices. My only concern is "how do you write a short memo or letter that almost everyone understands?"

From the viewpoint of the Association, nothing can or should be eliminated from the letter. If we were rewriting the piece to the bare-bones style, we would take bits and pieces here and there and discard the remainder.

For example, I would salvage the definition of the intended audience (managers of engineers and personnel managers concerned with engineers) from the first paragraph. I would also drop the last paragraph for my purposes because it has nothing to do with the message. Parts of the other sections are needed to make the message complete.

Concentrating strictly on the message, I would do something like this:

"Dear Editor:

"How to Get Ready for the Next Generation of Engineers," an Association conference for managers of engineers and personnel managers concerned with engineers, will be held September 30 and October 1 at The Hotel in Atlanta.

"Session topics on the first day will be: status of supply and demand, salaries, foreign competition, and a panel on coping with issues of the 80's. Second-day topics: experiences and recommendations of managers now dealing with the new generation of engineers.

"For further information, call . . . '

The message is complete in itself as far as the essentials are concerned. We run into trouble when it is suggested "you should add this" or "you should add that." For this reason, I tacked on the "for further information" ending. As usual, after reading the letter, questions could be asked and answers could be supplied. It's hard to call a halt, but my contention is: cover the basics, make sure nothing is left hanging, stop.

The objective is to avoid unnecessary length. As I said, I am not concerned about going beyond the basic message where sales and PR are the motives. In my experience, when people complain about long memos and letters that require translation, the writer usually has a fundamental problem. He has not reached the point in his thinking where he knows exactly what he has to say. Unfortunately, many memos and letters must be written in haste and without benefit of much or any prior thinking or preparation.

Attempting to write a one-sentence statement of your message is an excellent test for finding out if you are ready to write.

Another established procedure is to write a first draft, put it aside long enough to shift your attention completely to something else; go back to the letter later and take another look. If the memo or letter is excessively long, you will probably discover that you used more words than you needed because you did not have a firm grasp of your subject. The first draft was a way of thinking through what you have to say. Tear up the first draft. Start over.

If you find that you have trouble reducing your message to a single sentence or your first draft does not help you see the light at the end of the tunnel, you need more time to think and prepare. If you are fed up with the project, put it aside if you can. Try again tomorrow. At times our subconscious minds continue to unravel our knotty thinking for us.

Of course, once the writer is satisfied he knows what he wants to say, he has another problem: delivering his message in a clear, understandable manner.

I find that what is required to complete the first step prepares one to take care of the second. There is still one more hurdle for the writer of memos and letters for purposes other than selling and PR: sticking to the message—not tacking on anything that isn't necessary. This is in part a matter of discipline. There is always a strong urge to teach, to make absolutely certain everything is covered, to add a detail here and a detail there. Judgment is also involved. You must select only the essential elements in the message and stick with them. This is particularly difficult where more than one person is involved. The usual result of such a review is added length and a loss of clarity.

CHAPTER 21

Guidelines for Writing Technical News Releases

The engineer-author of technical papers and articles is seldom called upon to write technical news releases, but he is involved in these sales–PR literary projects in a couple of ways and can profit from an overview of industry practices and preferences.

Take the case of the engineer who is the leader or member of a team that has come up with a new product and sales informs him that a news release will be prepared to publicize the event. He will be expected to work with a PR communicator or technical writer to explain the breakthrough, answer questions, and otherwise supply information needed to write the news release, including a spell-out of key benefits to the potential purchaser of the product.

Eventually, the engineer will be asked to review the final draft of the news release, paying particular attention to technical accuracy. This is important because the designated writer—typically trained in some form of journalism—may not have the background or experience needed to grasp the subject matter or write about it with confidence. In many instances, the authority of the supplier of information stops here.

The tone of the piece and the manner in which the information is presented are normally prescribed by company sales–public relations policies. When the engineer-author has the privilege of going further, he invariably opts for the no-frills style of reporting he uses when he writes *his* things. He obviously stands a chance of having his noncommercial copy laundered and starched at the behest of the director of public relations or the manager responsible for marketing the product. But the low-key style of technical news releases does exist as a minority industry practice.

An engineer-author becomes involved in another way on occasion. He is asked to review a technical news release in instances where he is an expert on the topic being written about. Once more, his purview is pretty well limited to technical understanding and technical accuracy. Attempts to edit or to rewrite on his part for other reasons are typically ignored. The usual response: "You are the explainer and supplier of information. I am the writer."

Excerpts from two technical news releases that follow are examples of majority and minority practices.

Let's start with the latter. The writer set the stage by stating that his company and other makers of steel tubing used in drilling oil and gas wells are researching a metal problem (called sulfide stress cracking) encountered when drilling holes with high concentrations of hydrogen sulfide. The tubing absorbs hydrogen, becomes brittle, then breaks. We pick up with a report taken from a paper written by company engineers of what this company found out in its tests and related matters:

"Results showed that the low alloy high strength steels tested exhibit good sulfide cracking resistance. Susceptibility of these steels to sulfide stress cracking depends primarily on microstructure. A fully tempered martensitic microstructure exhibits the best stress corrosion cracking resistance.

"The tests also showed that ABC Company's [trade name] alloy in the over-aged condition, at yield strengths near 100 000 psi, exhibited greater stress corrosion cracking resistance than [another company's trade name alloy], which is approved for this application by the National Association of Corrosion Engineers."

The writer went on to state how much steel was tested in this instance and the types of steel tested. He then cited a limitation: there is no universally accepted test specimen or test method for sulfide stress cracking resistance. In these tests, three types of specimens were used—"notched C-ring, NACE tensile, and Shell beam."

A second limitation was added: "The ability of tubing to resist sulfide stress cracking demands close control of heat treating and other processing parameters to provide uniform through-wall microstructure and properties."

Finally, the writer reveals that work on the problem is continuing and tells readers how they can obtain a copy of the paper.

The example was chosen to represent minority practice and its low-key, factual approach. By and large, a type of steel, rather than a proprietary product, is recommended. A trade magazine editor considering the release as the basis for a news story would ask for more documentation of "good sulfide cracking resistance" and "exhibits the best stress corrosion cracking resistance" in the first paragraph cited, but over-all, there is no hype, aggressive PR, or blatant selling. In addition, results are qualified. It is made clear that although some progress has been made, more work needs to be done.

Now for a more typical technical news release. The occasion: ABC Company has just sold a $400 000 machine to XYZ Company, and hopes to use this less than worldshaking news to publicize its product.

The first paragraph of the release:

XYZ Company placed a $400 000 order for a numerically controlled grinding machine with ABC Company, a subsidiary of ABC Corporation. The machine, which will be built in ABC Company's southern plant, is an ABC Company [trade name] twin wheel form grinder and it will be used to . . . "

Repeating company names and trade names as often as possible is standard PR–sales practice. In this instance, ABC gets four mentions. The trade name is established and will be repeated. The equipment and the application are generally described. However, another prominent PR–sales practice, stating unsupported claims, is not observed at this point.

In the second paragraph, XYZ company and its product line are generally described. The paragraph is factual except for a PR–sales cliché. XYZ is characterized as a leading manufacturer of a specific line.

Paragraph No. 3:

According to Joe Dokes, ABC Company's group sales manager, "ABC has created a new generation of equipment designed for form grinding. . . . The capability of the new machine to store multiple programs for different types and sizes of . . . will make a world of difference in going from one part to another. Changeovers can be made faster and without the need for running through the complete setup procedure."

Note the unsupported claims and the standard PR device used to stroke a client and executives with muscle: stating sales claims by quoting a company official. The company name is used two more times. The word "new" is used twice. But there is no proof of newness, or "will make a world of difference in going . . . ," or "faster," or "without the need for . . . "

In the next paragraph, "new" is repeated twice and the trade name three times, and for the first time (after about 250 words) the technology is described—in one sentence. If there is any technological news in the two-page piece, this is it.

Two claims open the next paragraph:

"The machine can be used for many operations on . . . and can be readily adapted to the grinding of parts for other industries."

Some documentation follows:

"Programs for up to 20 different . . . machining requirements can be stored in the control's memory and recalled on demand."

In the concluding two paragraphs, ABC's name is repeated twice, once in connection with general descriptions of European plant operations and once in giving the company's name, address, and telephone number for further information.

The two examples provide a background for talking about the unique role of the news release in reporting technological happenings.

A news release is almost always much shorter than a technical paper or a magazine article. The first release was about 575 words long, the second about 400.

There is generally no attempt to develop a subject in depth. In the first release, the news was an advance in knowledge about certain steels. Benefits (claims) of the advance and documentation for them were confined to two paragraphs. In the second release, the news concerned a product, presumably a new one. The point was not carefully documented. However, the machine was described in one sentence, and its capabilities were summed up in one sentence.

Many technical releases merely signal that something has happened or is about to happen. Interested readers usually want more information and have to go back to the source, or the magazine that printed the release, for more information.

Readers and editors prefer the minority factual style to the majority PR–sales, although they tolerate both because news releases are convenient—and often sole—sources of certain types of technological information. But if you listen only to the criticism, this form of technical literature does not enjoy an unblemished reputation.

Readers surveyed ranked the short comings of technical news releases in this order:

1. Overselling and PR.
2. Inaccuracy.
3. Incompleteness.
4. Lack of specifics and detail
5. Writer does not understand technology being written about.
6. Misrepresentation or ambiguous handling of data.
7. Not newsworthy.
8. Not writing for an audience.

A sampling of verbatim comments on "overselling and PR": "News releases are generally unproven sales pitches" . . . "they are hyped with TV type sensationalism" . . . "too commercial" . . . "lead paragraphs tend to come on too strong—perhaps because the writer is trying to catch an editor's attention" . . . "they read like advertisements" . . . "technical news releases can be overly commercial" . . . "too much emphasis on 'better,' 'new,' 'more accurate,' 'faster,' etc." . . .

Examples from news releases confirm and amplify those judgments.

Two companies merge to combine their separate know-how in the electronics industry. A news release heralding the event is distributed to the business and trade press. The writer starts with PR hype to position the two companies. Company A is billed as a "major supplier," while company B "is noted for innovative technology." Selling follows. As a result of the merger, the marketplace now has "a dependable source of unequaled plating processes." In the past, the reader is reminded, other suppliers of this technology have been "unreliable and unimaginative." The new company "will eliminate this problem." Finally, Company A, the originator of the news release, adds that it is "one of the world's leading designers and manufacturers" or this type of equipment.

Unsupported claims are common. Here is one from another technical news release: the product being touted "has made it far easier to produce excellent quality, out of position welds with minimal welder training." No evidence is offered.

In another example, a company bills itself as "the world's largest independent organization devoted exclusively to a specific type of laser." Its equipment is "versatile" and "functional" and represents what is "really needed to work in the production world." This sales pitch is used: "Anyone concerned with improving productivity, reducing costs, and turnaround times can economically utilize the advantages of" this equipment.

Another company claims, without proof, "the first major advance in industrial power transmission design in over a century." The new drive, "with its innovative design, is expected to have broad energy-conserving implications and considerable impact on future product design in a wide array of industries."

Repetition of company names and trade names anytime there is the slightest opportunity to do so is another favorite tactic of the technical news release writer, and the word "new" is overused. For example, a company with a "new" plating chemical claims those "who need a copper plating solution with superior throwing power to effectively coat inner surfaces of superfine holes . . . " can "now turn to a successful new product from ABC Company."

Next paragraph. The "newly developed product" for ABC's "production proven lines," (trade name) "delivers a void-free deposit." One specific is given: holes can be "as small as 0.015 in."

Next paragraph: (trade name) "provides fine grained deposits which meet all requirements of military standard MIL-P-55110-C. It promotes ductility and excellent uniformity of results."

When claims are unsupported, the information is incomplete as far as the reader is concerned. Recommended practice is indicated by the following example from a maker of aluminum pipe for irrigation applications. The company proposes the substitution of one alloy (5454) for another (5050) because it is "stronger, withstands higher water pressure, is easier to relocate, and lasts longer than conventional welded tubing." The company makes both alloy 5454 and alloy 5050.

Specifics are given: alloy 5454 "has a typical tensile strength of 35 000 psi compared with 26 000 psi for alloy 5050," and alloy 5454's weld zone strength is more than twice that of 5050—"17 000 psi versus 8000 psi." In addition, alloy 5454 has greater dent resistance because it is stronger.

Higher strength makes it possible to use thinner pipe. In comparison with alloy 5050 pipe, alloy 5454 pipe is "25% lighter," making the pipe "easier to relocate." Higher strength accounts for ability "to withstand higher water pressure" and "longer life."

In another example, a company is introducing some new technology for continuously casting steels. Claims are made by quoting a company official. Steel made by the process "has fewer defects on the surface, subsurface (0.4–0.8 inches deep), and the internal section." Steel so made can be further processed into seamless tubing without further processing.

A description of how the process removes defects documents the claims: "Molten steel is poured from the tundish into a water cooled mold. The stirrer coil inside the mold rotates liquid steel by electromagnetic force at 40 to 70 revolutions, per minute. Centrifugal force pushes lighter materials, such as inclusions, slag, and gas bubbles, to the center of the surface for absorption by mold powder applied during casting."

The technical news release is only one and one-quarter pages long, yet as far as the reader is concerned, it is complete. All claims were documented. The piece is complete in itself. That is all that is required.

"Lacking in specifics" is a frequent reader comment on incompleteness. Or to put it another way, "insufficient technical information," or "lacking in detail." Another commentator points out another source of incompleteness that is encountered from time to time: "In attempting to write to a lay audience, the author generalizes to the point of saying nothing."

This constructive comment is offered: "Provide numerical or quantitative data with sufficient documentation and/or attribution to source to give the reader an idea of how well claims can be substantiated. Is the item pie in the sky like the proverbial 100 mpg carburetor or has it been proven in service?"

Inaccuracy is also singled out for criticism. "Not technically correct" is the way it is often put. A practical cure is offered: "The technical person who supplied the information to the writer of the release should review it before it is released." Sometimes other people in the clearance procedure create a problem in this area. There is often too much emphasis on simplification. As one reader describes it, "Oversimplification without regard to technical implications, . . . and hence the information is in error by implication."

As indicated by statements in the last paragraph, the writer may not always understand the technology he is writing about. A reader explains, "Many times news releases are not technically accurate because they are written and edited by journalists or PR people not conversant with technology; also, they sometimes oversell results, implying capabilities not easy to obtain."

Another view: "There may be an inability to associate the technological news with the big picture, resulting in a misinterpretation of the importance of the development and inaccurate statements technically." Other related comments: "Technical news releases are often written by someone who is not technically trained, or by a PR man rather than a technical writer with an understanding of the subject."

Manipulating data to put it in the best possible light is the subject of frequent reader comment. An example shows one of the practices objected to.

A company publishes a news release to announce that a machine it is readying for the marketplace "continues to set new performance records during developmental testing." Fuel consumption, it is stated, is "0.5% better than the levels projected." Sounds good, but the projections are not given. In addition, the maximum thrust rating "was exceeded by 7%." Once more, sounds good; but the rating is not given. Finally, "efficiency was higher that predicted." As expected, the prediction is not given.

Readers are more specific on the topic of manipulating data.

For example, "There is a tendency to deal with the average result in reporting data, rather than minimum or maximum results needed for the reader to make a sound decision."

For example, "Some authors deliberately distort data and take information out of context to bias an audience."

For example, "By using statistical tricks it is possible to bias or distort data so that a small (even insignificant) difference appears to be very great."

The tendency to make something out of practically nothing applies to the motivation for publishing news releases. There is an implied promise: "Read this. It is newsworthy." This is not always the case. A company that makes machinery, for example, issues a news release in mid-May 1982 to announce that it will begin shipments of a new model "in early 1983." Aside from sales claims and a description of the machine, there is no other news to report.

In another example, PR people working for a mining machinery company took a circuitous route to peddle what they are selling. The title to the piece indicates that the company president applauded recent action of a Congressional Committee looking into an aspect of the strategic materials situation—most of which must now be imported. In the first paragraph of the release, it is revealed that the company president hailed the Committee's action as an important first step in assuring domestic sources of critical materials vital to defense and aerospace. Then we find out he testified before the Committee earlier. Then we are told the location of some highly promising but unworked reserves of low-grade ore. Finally, it is disclosed that the company has developed machinery to improve the yield from low-grade ores and substantially to improve the economics of using them. The company, it turns out, has a pilot plant in operation.

A reader complains, "Some news releases are not newsworthy, except to the sales manager who writes them." Another adds, "I have little tolerance for press agentry and puffery that seem to constitute too much of what is called news."

Not writing for an audience has many variations. One is use of jargon to a nonpeer audience, as, "a second generation fundamental parameters program for automatically performing quantitative analysis without suites of standards. Regions of interest are automatically set across each spectral peak and automatically make peak overlap corrections or background subtractions."

A reader puts it this way: "The writer sometimes assumes the reader knows too much or too little about the technology involved."

Readers offer a variety of formulas for writing technical news releases to their liking. They include:

"A brief description of a new product of process. . . . Compare it to something similar or state what it replaces. . . . Try to be objective."

"Cover who, what, where, how, and particularly why."

"Brief, but covers what, when, where, why, and how."

"Give a clear indication of the problem the product is designed to solve. . . . Indicate its application range, limitations, and cost effectiveness. . . . Provide sufficient technical background for the reader to understand the technology."

"Explain the technology succinctly but well, leaving no doubt about its importance."

"A source for additional information is always needed, so the reader can follow up."

"State why the item is newsworthy and explain its practical consequences."

"Deal in facts."

"Be brief, accurate, noncommercial, and give source for additional information."

"Be concise and technically accurate. Present results in real world terms and capabilities."

CHAPTER 22

Guidelines for Writing Theses

The thesis and the technical paper are members of the same family. What makes for acceptable technical writing applies to thesis writing. Formats vary, but the in-house audience of superiors is usually quite specific as to what is required.

Typical components include: title page, abstract or summary, text, lists of tables and figures, captions for figures, figures, tables, equations, acknowledgments, appendixes, references, and bibliography.

A format for text is made up of such items as introduction, literature search, description of test equipment, description of test procedures, discussion, conclusions, and recommendations.

Readers surveyed who have experience in reading theses have opinions on why many fall short of expectations and offer suggestions for getting back on track.

Examples of criticism:

"Organization is poor—outlining should be the first step. Sentence construction is convoluted. Jargon and acronyms are overused . . . the inexperience shows, but this not avoidable."

"I find fault with the writing style. Flow is a problem. Graphics are inadequate. There is a lack of investigation of past work. Conclusions should state what is new—the main reason for the award of a higher degree."

"Too long . . . poor English . . . dull writing."

"Style of writing is usually stilted and pedantic. Arguments and positions are drawn out excessively. An overabundance of supporting data are often thrown in for hoped-for good effort."

"Too long . . . poor grammar and spelling . . . not logical enough."

"A lack of coherent technical data . . . unwarranted conclusions based on insufficient evidence."

"Too wordy relative to the importance of the contribution. Written in too narrow of a field in too much detail. Usually written by individuals with a high educational level for like individuals in the chosen field such that comprehension by the average reader in industry interested in using the data is difficult."

"I do not see many, but they tend to be wooly and poorly written. Most statements try to cover too large a field instead of writing for a more closely defined field and providing specifics that may be of interest to the larger audience."

"Too much emphasis on making the thesis a scholarly piece, too much scholarship, and what seems to be an assumption: the thesis is a test of the size of the student's vocabulary."

"As a nonacademic person—I work in industry—I find theses tend to be overly erudite, as if they were written to impress academic colleagues of the thesis advisor, rather than convey and add to the general store of knowledge which is the ostensbile purpose of a research project. Also, I find that many research projects deal with rather trivial subjects in much greater depth than they can justify as far as the value of the results is concerned."

"They are too academic . . . writers and reviewers are obsessed with the fear of an inaccuracy or an ill-advised expression."

What do readers recommend to writers of theses?

Verbatim comments:

"Originality."

"Let them satisfy the academic proprieties if that is what academia wants, but don't burden the nonacademic community with the need to wade through pages of repetitive trivia to get what little meat there is. Put the statement of the problem and the results and conclusions up front in plain English."

"Organize concepts and observations. Provide a clear, careful definition of the hypothesis and objectives, collect data carefully, analyze data logically, provide a clear indication of results, conclusions, and recommendations, and indicate whether the hypothesis is supported or rejected."

"Main requirements are careful selection of topic, good detail in experimental procedure, a thorough literature survey, and rational explanation of results."

"There is a need to believe in the purpose of research and its significance to the community to which it is directed."

"The main requirements are accuracy, technical soundness, and completeness in terms of presenting all the data generated."

"What's needed is a logical and orderly presentation . . . one that is well illustrated . . . and a reasonable writing style."

"Organize the work well. Try to write in a pleasing, open style. Don't go on forever justifying a point. Use good, clear visuals and only enough supporting data as realistically necessary."

"A thesis should be comprehensive, provide a good review of the area (including the work of others in the literature), present all data, and have a good summary."

"Organization is the key. Include a summary. Write concise sentences."

Example of a University's Guidelines for Thesis Preparation

The faculty of graduate studies and research at McGill University, Montreal, publishes guidelines for thesis preparation. Several excerpts follow.

A section titled "Originality and Historical Introduction" states that "the candidate should clearly indicate in the preface, or by a separate statement at the beginning or end of the thesis, what elements in the thesis should be considered contributions to original knowledge."

The requirement is mandatory for a doctoral thesis and recommended for others. Another requirement is "an appropriate historical statement of previous relevant work or investigation." A caveat is added: "Appropriate does not necessarily mean long or exhaustive."

Another comment on the topic: "A thesis for Master's degree, while not necessarily requiring an exhaustive review of work in the particular field of study, or a great deal of original scholarship, must show familiarity with previous work in the field and must demonstrate the ability to carry out research and to organize results, all of which must be presented in good literary style. The thesis normally consists of 80 to 100 typescript pages."

Information on format is in a section titled "Manuscripts and Format." Part of the thesis may be an original paper the candidate has authored or coauthored and is suitable for submission to learned journals for publication. "In this case the thesis must still conform to all other requirements explained in this document, and additional material (e.g., experimental data, details of equipment, and experimental design) may need to be provided.

"In any case abstract, full introduction, and conclusion must be included, and where more than one manuscript appears, connecting texts and common abstract introduction and conclusions are required. A mere collection of manuscripts is not acceptable; nor can reprints of published papers be accepted."

In addition, "Every thesis should have an adequate table of contents and a thorough bibliography of the subject. Immediately following the title page, there should be an abstract, giving a summary of the thesis, not exceeding 150 words."

In its general specifications, McGill requires three copies of a master's thesis and seven copies of a doctoral thesis. Examples of other requirements follow:

"Diagrams should be drawn with India ink on rag paper or reproduced by photographic, offset, or xerographic processes. Diagrams, numerical or summary tables, and parts of the table of contents may appear on left-hand pages (on the back of other material) where this aids in the layout and general readability of text. Left-hand pages should not be numbered, but should be identified as being opposite to a given page.

"Footnotes may be placed either at the bottom of each page, collected at the end of each chapter, or placed all together after the last chapter.

"The thesis must be free from typographical and other errors. If copies are deficient in these respects, they will be returned for correction. If these corrections are numerous, the granting of the degree may be delayed on this account."

CHAPTER 23

What Technical Writers Should Know About Libel Law

It is conceivable that the writer of technical literature can run afoul of laws governing libel and related matters. He should have a layman's overview of the subject and a general idea of how writers become involved with this body of law.

Four topics are examined briefly, without pretense of being comprehensive:

1. Libel.
2. A related subject known by several names including trade libel.
3. An area related to consumer protection or product liability.
4. A recent case (1982) involving the American Society of Mechanical Engineers (ASME) and two members of an ASME subcommittee working with standards.

Libel

Libel is an offense against a person, a happening less likely to occur in technical literature that in the mass media. Publishers have libel in mind when they publish disclaimers stating they are not responsible for what their authors say. Disclaimers are placed—in small print—at the front or back of a trade journal, for example. They do not, however, provide any protection to publishers.

The summary of libel law that follows is taken from an article, "Protect Yourself from Libel Suits—Understand the Laws," from *Communication News*, January 1982, published by the American Society of Association Executives, Washington, D. C.

Libel is defined in that article as "the basis for legal action to recover damages caused by false written assertions of fact made either negligently or with actual malice. The elements of such a legal action or suit for libel are:

"A false and defamatory statement concerning another.

"Publication to a third person.

"Negligence or malice on the part of the publisher.

"Damages."

The person being sued, the defendant, has four main defenses:
"The statement named in the suit is true.
"The statement does not concern the plaintiff.
"The statement is privileged, meaning the publisher is protected.
"The statement consists of opinion or fair comment on the subject."
The person bringing the suit, the plaintiff, must prove that the statement published is false and that the reader understands the plaintiff to be the subject of the statement.
"In general, statements of opinion are safe from legal action because they cannot be proven true or false. Only statements presented as facts can be libelous. The distinction between fact and opinion is nebulous and is treated on a case-by-case basis."
However, the writer does not give himself complete protection when he qualifies his statement with "in my opinion." If the reader takes the statement to be fact, the device does not work.
If the plaintiff is a public figure—a person who has assumed a role of special prominence in society—he must prove that the statement in litigation was malicious and made with reckless disregard for the truth. "According to court decisions, malice must be proven with convincing clarity."
If the plaintiff is a private individual, he must show at least negligence on the part of the defendant.
The majority of lawsuits, by the way, are brought against publications.
In summary:
"1. Truth is a defense against any libel action.
"2. Statements of opinion and comments on factual statements may be as sharp and opinionated as the author desires without fear of liability. You should be careful to distinguish between assertions of fact and opinion, however. It is the false assertion of a fact that serves as the basis of a defamation action.
"3. Private figures—as opposed to public figures—may press libel actions when there is merely negligence and not malice. Therefore, when publishing articles about private figures, more fact checking and investigation should be undertaken to insure that the reported facts are accurate.
"4. When reporting events concerning public figures, you should still conduct careful fact checking and investigation. This will insure that you will not be held liable under even less stringent standards of proof of libel."

Trade Libel

In trade libel, the offense is against a product, process, or service, for example. To illustrate: say you work for Company A and write an article or a paper in which you intentionallly or unintentionally criticize a competitive product made by Company B. Under these circumstances, it is conceivable that Company B will threaten suit against Company A, and it is inevitable that Company A will contact the publisher of the magazine or journal in which the article or paper appears and do one or both of two things: Company A will cancel its advertising if it is an advertiser or will suggest the possibility of a lawsuit unless a full retraction is made. Further along in the negotiations, Company B will also demand equal time, so to speak, with an article on its product.

Generally, this unpleasantness can be avoided by not using competitive or odious comparisons. Editors always try to guard against this, but there is always a chance that an expert on a given subject can mask a comparison or claim that will slip by the quasi-expert editorial person.

No cases of this exact nature were turned up in a fairly comprehensive literature search for this handbook. However, there is enough law closely bordering the subject to make one wary. Cases cited usually involve advertising. The alleged offense usually goes under a motley assortment of names including false advertising; disparagement of goods, services, or processes; trade libel; slander of goods; and unfair competition.

It is possible that "an attack upon goods sold or manufactured by a plaintiff is strong enough to impugn the plaintiff's character, as where the plaintiff is implicitly accused of fraud or dishonesty in his business, an action for personal defamation may be maintainable. More commonly, however, an attack upon a businessman's goods or services entails pecuniary loss from lost sales or profits as a result of the statements made. This injury to business is properly asserted in a disparagement action."[1]

Disparagement of goods or services takes more proof than personal defamation. The offense is defined as "a false statement of fact about a plaintiff's product, which either identified the plaintiff or his goods by name or was made in such a manner that the public understood the statement to refer to the plaintiff."[1] Proof of deliberateness has also been required. This has ranged from "a demonstration that the communicator knew the statement was false to proof of actual malice or spite."[1] The plaintiff must also establish the amount of his loss, although there is some indication he is not required to allege special damages.

In one case (*Black and Yates Inc.* v *Mahogany Association*), the defendant "accused the plaintiff of deceitful behavior in selling Philippine mahogany, which according to the defendant was not a mahogany wood."[1] An injunction was granted to the plaintiff.

In a current case, a manufacturer of coffee makers is suing a maker of coffee filters and decanters for use with drip coffee makers. The plaintiff charges that the defendant is showing in national advertising a broken coffee decanter made by the plaintiff, implying that broken decanters should be replaced by allegedly unbreakable decanters sold by the defendant. The plaintiff seeks an injunction prohibiting the defendant from using pictures of its products in advertising.

Available defenses to trade libel include the puffing doctrine and the right of fair comment and criticism.

Under the puffing doctrine, "A businessman can advertise that his goods are the best in the market so long as he makes no specific assertions of fact unfavorable to his rivals. This rule is based on practicality: the average buyer is skeptical enough to attach little importance to a seller's boasting; moreover, it is expected that no one will rely on general statements of opinion if they are discernible as such."[1]

The right of fair comment and criticism is not limited to public officers and candidates for office. The doctrine extends to "commodities, wares, merchandise, and products offered to the public for use or consumption."[2]

Further, "A cause for action for defamation generally does not arise in favor of one whose merchandise or products are criticized, not for the purpose of obtaining a competitive advantage, but merely to express displeasure or dissatisfaction therewith,[3] nor is an advertisement actionable which does no more than state a claim that the plaintiff's goods are inferior to those of the defendant."[4]

References to Section on Trade Libel

1. Monachino, Vivienne, "Product Disparagement and False Advertising in the Common and Civil Law," *Tulane Law Review*, Vol 53, No. 1, December 1978, p. 190.
2. *Golson* v *Hearst Corp.* (DC NY) 128 F Supp 110.
3. *Marlin Firearms Co.* v *Shields*, 171 NY 384, 64 NE 163.
4. *National Ref. Co.* v *Benzo Gas Motor Fuel Co.* (CA8 Mo) 20 F2d 763, 55ALR 406, cert den 275 US 570, 72 L Ed 431, 48 S Ct 157; Nordlund Consolidated Electric Co-operative (Mo) 289 SW2d 93, 57 ALR2d 832.

Consumer Protection/Product Liability

The consumer protection and product liability concepts give publishers and writers new cause for concern. In a case against Rand McNally Co., a jury awarded $825 000 in damages to the plaintiffs, Carolyn Carter and Christine Bertrand, both 19 at the time. The jury decided that inadequate warnings in a textbook published by the defendant led to a chemistry class explosion that burned the two young women.

The girls, students at Paxton Junior High School at the time, said they were following a chemistry textbook that outlined the experiment as an end-of-chapter test. They were calibrating an alcohol thermometer in an exercise that called for them to heat the thermometer over a Bunsen burner and cool it in ice. The girls poured methyl alcohol into the beaker of ice which they had placed over the lit Bunsen burner. The methyl alcohol exploded. Rand McNally was not liable for the full amount of the award. Earlier, the classroom teacher, Paul Zendzian, agreed to pay Miss Carter $500 000 and Miss Bertrand $170 000 in an out-of-court settlement. The publisher was ordered to pay $155 000, with $100 000 going to Miss Carter and $55 000 to Miss Bertrand.

Codes and Standards

Yet another area of possible vulnerability to the law is of particular concern to organizations publishing codes and specifications and to members of committees who write such documents. In 1982, the United States Supreme Court affirmed a decision of the Court of Appeals for the Second Circuit in *Hydrolevel* v *ASME*, holding the American Society of Mechanical Engineers liable for anticompetitive acts of two volunteer members directed against Hydrolevel Corp. The following account is based on a news release distributed by the American Society of

Mechanical Engineers, New York, following the decision handed down by the Supreme Court on May 17, 1982.

By a vote of six to three, the Court of Appeals was ruled to be "correct in finding ASME responsible for the actions of two volunteers who had conspired to disparage a Hydrolevel product through misuse of ASME's safety standard interpretation process."

As a public service, ASME provides free of charge to members and non-members alike written and oral interpretations of its codes. Requests are referred to the ASME committee responsible for the code in question. The committee may respond to properly worded inquiries in writing.

The case involved low water cutoff devices used to stop fuel flow to a boiler before the water level falls to a dangerously low level. If the cutoff does not operate, the boiler may operate without sufficient water and lead to an explosion.

The suit centered on a query to the ASME Boiler and Pressure Vessel Committee from a maker of float-type cutoff devices asking if a cutoff device with a time delay would meet ASME standards. The plaintiff made an electrically activated probe-type cutoff device. "Some Hydrolevel cutoff devices used in residential boilers incorporated a time delay."

The letter was written by two vice presidents of McDonnell and Miller Inc., a company that had "long manufactured a float type, low water cutoff device that accounted for 70 to 85% of the American sales." One of the vice presidents and the executive vice president of the Hartford Steam Boiler Inspection and Insurance Co. were chairman and vice chairman of the subcommittee to which the inquiry was ultimately referred. Following ASME procedure, a reply to the inquiry was drafted and sent back up through channels—to a staff secretary and then to the chairman and vice chairman of the Boiler and Pressure Vessel Committee.

"On April 29, 1971, the draft letter was sent to McDonnell and Miller as an 'unofficial communication.' In essence, the letter stated that 'it is the intent that the low water, fuel cutoff operate immediately and positively when the water boiler level falls to the lowest visible part of the water gauge glass,' and a time delay feature which operated after the water had left the gauge glass would provide 'no positive assurance that the boiler water level would not fall to a dangerous low point during the time delay.'

"McDonnell and Miller used it [the letter] in promotional efforts linking the 'advice' of the 'unofficial communication' to the Hydrolevel device. This information was distributed to salesmen and potential clients and eventually a vice president of Hydrolevel became aware of its use."

Hydrolevel filed suit against ASME, International Telephone and Telegraph Corp. (which had purchased McDonnell and Miller), and Hartford Steam Boiler Inspection and Insurance Co. They were charged with conspiracy in restraint of trade under the Sherman Antitrust Act and with monopolizing the low water cutoff market in violation of the Sherman Act. ITT and Hartford settled out of court.

The majority of the Supreme Court, "while specifically acknowledging that the two volunteer members acted without the knowledge of ASME or with the intent to benefit ASME, nevertheless concluded that the acts of those members, which were within their apparent authority, were sufficient to render ASME liable to Hydrolevel."

Justice Blackmun, writing for the majority, stated, "When ASME's agents act in its name, they are able to affect the lives of large numbers of people and the competitive fortunes of businesses throughout the country. By holding ASME liable under the antitrust laws for antitrust violations of its agents committed with apparent authority, we recognize the important role of ASME and its agents in the economy, and we help to ensure that standard-setting organizations will act with care when they permit their agents to speak for them."

The lower court awarded the plaintiff $7.5 million in damages. The Court of Appeals found the amount excessive, and the Supreme Court upheld the Court of Appeals. This part of the case was scheduled for retrial.

ASME has also"initiated procedures to protect against similar misadventures." The society has started to publish all written technical inquiries pertaining to codes and their interpretation.

CHAPTER 24

What Technical Writers Should Know About Patent and Trademark Law

Novelty, usefulness, and nonobviousness of an invention are the keys to making points with a patent examiner. However, an equally important step precedes the filing of a patent application: searching the public file of the U. S. Patent and Trademark Office in Arlington, Va., to find out if the invention has already been invented. For this project, you will probably wind up hiring a patent agent, an experienced searcher, or a patent attorney to do the job. Their fees range from fifty to several hundred dollars, advises R. J. Joenk in an article, "Patents: Incentive to Innovate and Communicate—an Introduction," reprinted with permission from *IEEE TRANSACTIONS ON PROFESSIONAL COMMUNICATION*, June 1979. Other know-how from the article by Dr. Joenk follows.

The patent professional will probably ask you to write a disclosure for his guidance in determining the patentability of your invention. Briefly, a disclosure identifies the field of technology of the invention, describes (in word and sketch) its principal parts and functions, and explains how it differs from or is better than related prior art known to the patent seeker.

In the next step, a patent application, the inventor has the guidance of the law (35 U. S. C. 112). He is obliged to describe his invention "in such full, clear, concise, and exact terms as to enable any person skilled in the art to which it pertains, or with which it is most nearly connected, to make and use the same, and shall set forth the best mode contemplated by the inventor of carrying out his invention."

Once more, professional help is recommended. "Here there is no substitute for experience. A patent application involves technical as well as legal writing at its most demanding level. From the inventor's notes, drawings, and verbal descriptions, a technical disclosure document must be developed that satisfies the letter and the spirit of the law. . . . most patent applications are prepared by or under the guidance of registered patent agents or attorneys," states Dr. Joenk.

The principal text portion of a patent application—known as the specification—should include:

A statement relating the present application to appropriate previous applications or patents of the inventor.

Identification of the technology or technical field to which the invention belongs.

Discussion of the background (prior art) of the invention to encourage understanding and to expedite searching and examining the application by the Patent Office—the inventor and his legal representative must disclose to the Patent Office all prior art of which they are aware that might be relevant or useful or material to the Patent Office in deciding whether to issue a patent.

A brief technical discussion of the problem and how it is solved by the invention, particularly in the way it represents an advance over prior art.

Brief description of figures—if there are any.

Description of the invention sufficiently detailed to comply with the law (35 U. S. C. 112) previously cited.

An explanation of the best mode contemplated for carrying out the invention claimed, using examples and references to figures.

A claim of at least one sentence that states concisely and precisely the technical feature of the invention not found in prior art. More than one claim can be made.

The patent application must be accompanied by an abstract that summarizes the main features of the disclosure and claims without editorializing. By oath or declaration the applicant must declare he believes himself to be the original and first inventor of the claimed invention and states his citizenship. The abstract facilitates search and retrieval; it does not limit the scope of the invention.

Claims define the legal scope of a patent. "Over the years," states Dr. Joenk in his article, "some preferred and required styles, formats, and word usages have developed for the writing of claims which are not taught in technical writing courses."

Some examples:

A claim is written as a single sentence, however long or compound, beginning with "I claim," "We claim," or "What is claimed is." The phrase is used only once even if there is more than one claim. Thus, by itself, the second, third, etc., claim is not a complete sentence, but it must read as one with the claim phrase prefixed.

The first part of a claim after the introductory phrase is usually a preamble of unspecified length which introduces, names, or defines the things, combination, process, etc., to be claimed.

In proceeding from the preamble to the body of a claim, certain transitional phrases have specific interpretations:

"Comprising and which comprises" means "including the following elements but not excluding others."

"Including and having" means the same as "comprising" but the latter is preferred.

"Whereby" means "it necessarily follows that."

The body of a claim enumerates the elements of the invention with a description of the way the elements cooperate with each other to accomplish the invention.

The article by Dr. Joenk is one of 17 covering different aspects of patent law in the June 1979 issue of *IEEE TRANSACTIONS ON PROFESSIONAL COMMUNICATION*. They are listed in the references at the end of this chapter. One authored by a registered patent agent, James F. Cottone, 1911 Jefferson Davis Parkway, Arlington, Va. 22202 (703/920-6772), spells out practices in writing invention disclosures.

"Writing a disclosure," says the author, "is often the first step in communicating and protecting an invention. This is a task that almost every technical person takes on from time to time, but few really approach the task with confidence. The basic problem appears to be not so much the actual job of describing the novel features of the invention as it is deciding just how much to include, and often what not to include."

The most common types of disclosures—six in all—are reviewed by the author. He then lists and explains elements belonging in disclosures generally.

Types of disclosures include laboratory notebooks, technical reports, patentability search disclosures, patentability applications, defensive publications, and Patent Office disclosure documents.

Laboratory notebook disclosures include, for example, all preliminary and informal write-ups of inventions, including test results and rough notes; more carefully prepared technical diaries; sketches, photographs of equipment, traces from recording instruments, and computer printouts.

Technical report disclosures, for example, include in-house status reports, progress reports, and smoothed out compilations of notebook entries—reports written for technical or management people.

Patentability search disclosures describe the invention for the purpose of defining a search of prior art (issued patents) to determine novelty of the invention.

Patent application disclosures include all written efforts directed toward producing a formal application for filing in the U. S. Patent and Trademark Office. It is supposed to "include well-thought-out, smooth-reading, detailed descriptions, and is usually written by or under the direction of a registered patent practitioner."

Defensive publication disclosures generally concern "inventive concepts considered to have little commercial value per se, but which are published as a defense against patenting by a subsequent inventor; after a year they become an absolute bar to patenting by anyone. Thus defensive publications offer freedom of action—but not patent protection to the inventor."

Disclosure documents filed by inventors are kept on file at the Patent Office. "These are generally informal disclosures written by individual inventors who thereby obtain the benefit of a recorded date, which may subsequently be used as proof of date of conception of the invention disclosed."

Reasons disclosures are needed fall into these categories:

A. Proof of conception of the invention—when and by whom.

B. Proof of operativeness—description of the invention and key details of how it functions.

C. Proof of reduction to practice if actually built and tested—when and by whom, and with what results.

D. Actions affecting possible patent protection.

The author lists items needed in each category.

For "A":
1. Date of conception of invention.
2. Names of inventors and their relative contributions.
3. Overview description of the invention with sketches.
4. Witnessing of the written description.

For "B":
5. Functional description (in moderate detail).
6. Drawings showing structural parts and their interrelations.
7. Illustrative uses—best use and other possible uses.
8. Theoretical or mathematical description if available.

For "C":
9. Performance data and dates.
10. Operation witnessed.
11. Photographs.

For "D":
12. Prior art patents.
13. Related publications.
14. Disclosure in talks, papers, and meetings.

Note that items 4 and 10 require witnesses. The author explains, "No. 4 requires only that a witness see and understand the written description of the invention, whereas No. 10 requires that a witness see the invention after it has been reduced to practice (actually working)."

Of the 14 items, four are usually found in the laboratory notebook disclosure: date of conception, names of inventors, overview descriptions and sketches, and description witnessed. The author adds that items 9 and 10 (performance data and dates, and operation witnessed) are required "if the laboratory notebook description pertains to an invention already assembled, operating, and undergoing evaluation (no matter how tenative)." In this instance, the technical data written serve as proof of reduction to practice.

For the patentability search disclosure, item 2 (names of inventors) is optional; items 5 (functional description in moderate detail) and 6 (drawings) are required; items 7 (illustrative uses) is optional, as is item 11 (photographs); item 12 (prior art patents) is required; and item 13 (related publications) is optional.

The author comments, "The most common technical disclosure written on a day-to-day basis is found in the laboratory notebooks of employees of research and development oriented companies and in the notes or diaries of independent inventors. These disclosures tend to be informal documents and usually provide a time series of entries, often relating to a number of different projects underway at the same time. Technical people are at ease with these modest disclosures that can serve as very useful evidentiary documents. Early entries provide the best proof of date of conception of an invention; they fairly well specify what the invention is, and in the case of joint inventorship, they often clarify which inventor initially made what contribution. . . . It is only necessary that the invention be reasonably well described in writing, that the description be signed and dated by each participant, and that the entry be properly witnessed."

There is no standard length for a patentability search disclosure. A guideline is offered: "It can be presumed that the reader, evaluator, or searcher has a fair working knowledge of the technical subject matter involved. Thus the inventor or writer can first set forth the general field of the invention and then proceed directly to the believed point of novelty. It is usually not desirable to include a lot of elementary or introductory matter unless, for example, the writer believes such material—while basic—is not widely known or properly understood by most people in the field."

The author adds, "Almost any disclosure which grows beyond a very few pages invariably loses its focus on the single invention which is being disclosed."

The author concludes with sets of do's and don'ts for patentability search disclosures.

Do

. . . provide a short descriptive title for the invention.

. . . provide a concise description including the general field of technology for its application, major parts of the invention, and how it works.

. . . state what the inventor believes to be the novel feature and how it is an improvement or significantly different from the state of the art.

. . . provide one or more drawings, clearly labeling the major parts and where key functions are accomplished.

. . . provide references to related devices or descriptions in the form of clearly annotated supporting materials.

Don't

. . . use abbreviations, acronyms, or jargon. (Avoid in-house or trendy words and phrases; rely on widely known and accepted words which have withstood the test of time.)

. . . attempt to disclose more than one invention in a disclosure.

. . . use excessive detail describing items that would be well known to a person having reasonable knowledge of the general subject matter.

. . . provide large manufacturing drawings or overly elaborate or artistic drawings. (Select—cut out or photocopy—pertinent parts and cross out nonrelated parts or views.)

. . . use different names for the same entity in various parts of a disclosure.

. . . provide large quantities of support material for general background information. (Review all support material first and clearly mark the pertinent portions. Show the correlation between the names or terms used in the support materials and those in the enclosure.)

. . . attempt to have a patentability search disclosure serve as an all-purpose document by including extraneous material.

Company invention reports are similar in content to patentability search disclosures.

A steel company, for example, has a standard procedure "to provide guidance in the procedure to be followed when ideas are considered to be at the stage where patent proceedings should be initiated." If reviewers and management feel the invention merits patenting, a patent application is filed by the company's patent department.

The form for the application is shown in the box on the opposite page.

U. S. Patent Office as Source of Information

Browsing through a booklet, *General Information Concerning Patents*, one can pick up an overview of the subject. In the words of the Patent Office, "It [booklet] attempts to answer many of the questions commonly asked but is not intended to be a comprehensive textbook of patent law or a guide for the patent lawyer." A sampling of chapter and section headings: Functions of the Patent and Trademark Office; What Is a Patent?; Copyrights, Trademarks, Patent Laws; What Can Be Patented; Novelty and Other Conditions for Obtaining a Patent; The United States Patent and Trademark Office; Publications of the Patent and Trademark Office; Attorneys and Agents; Who May Apply for a Patent; Applications for Patent; Oath or Declaration and Signature; Filing Fees; Specification (Description and Claims); Drawing; Amendments to Application; Appeal to the Board of Appeals and to the Court; Interferences; Nature of Patent and Patent Rights; Assignments and Licenses; Joint Ownership; Infringement of Patents; Patent Marking and Patent Pending; Design Patents; Plant Patents; Treaties and Foreign Patents; Foreign Applicants for United States Patents; Fees and Payment; Forms (including examples).

General Information Concerning Patents is available from the Superintendent of Documents, U. S. Government Printing Office, Washington, D. C. 20302.

A section of the specification (patent application) is particularly in the realm of the technical writer. A brief statement of requirements is followed by a concise description of format requirements.

"The specification," states the Patent Office, "must include a written description of the invention and of the manner and process of making and using it, and is required to be in such full, clear, and exact terms as to enable any person skilled in the art to which the invention pertains, or with which it is most nearly connected, to make and use the same.

"The specification must set forth the precise invention for which a patent is solicited, in such manner as to distinguish it from other inventions and from what is old. It must describe completely a specific embodiment of the process, machine, manufacture, composition of matter or improvements invented, and must

SUBJECT   Invention Reports

EXHIBIT 1

DATE ______________________

cc: Patent Department
Director of Research
Director of Development
Assoc. Director of
Electromechanical R & D
Others as designated by a
Director

INVENTION REPORT NO. ______

Title: ____

1. Description of invention (refer to other reports where possible).

2. Discuss prior art.

3. Discuss status, significance to the company, urgency.

4. Identify, as by notebook number and page, idea submittal number, etc., all written material relating to the origin and development of the subject invention. Attach copies of key pages of such identified material (if not previously submitted).

5. Identify all publications which have been made disclosing this invention (other than the written information requested above).

6. Identify, by date and nature, any commercial use which may have been made of this invention.

7. Identify Government contracts pursuant to which work was done on this invention. Identify and submit copies of all reports submitted to the Government pursuant to any such contract.

8. Identify, by full name and home address, all persons whom you believe have had a part in this invention, giving facts pertinent to their contribution. It will then become the responsibility of the Patent Attorney to decide who are the inventors.

9. Has research and development work on this invention been completed?
        YES ___________   NO ___________

Attachments:

Prepared by ___________________________

Approved by ___________________________
             (name) Section Head

Approved by ___________________________
             (name) Director ___________________________
                  Research, Development or
                  Electromechanical R & D

explain the mode of operation or principle, whenever applicable. The best mode contemplated by the inventor of carrying out his invention must be set forth.

"In the case of an improvement, the specifications must particularly point out the part or parts of the process, machine, manufacture, or composition of matter to which the improvement relates, and the description should be confined to the specific improvement and to such parts as necessarily cooperate with it or as may be necessary to a complete understanding or description of it."

In summary, elements of a patent application (specification) include:

1. Title of the invention; or a preamble stating the name, citizenship, and residence of the applicant and title of the invention.
2. Cross-references to related applications, if any.
3. Summary of the invention.
4. Brief description of the several views of the drawing(s), if any.
5. Detailed description.
6. Claim(s).
7. Abstract of disclosure.

The title "should be as short and specific as possible; it should appear as a head on the first page of the specification, if it does not otherwise appear [as in a preamble] at the beginning of the application."

A brief abstract "of the technical disclosure . . . must be set forth in a separate page immediately following the claims in a separate paragraph under the heading, 'Abstract of the Disclosure.' "

A brief summary indicates the nature and substance of the invention, and may include a statement of the object of the invention.

The summary should precede the detailed description.

When drawings are used, "there shall be a brief description of the several views of the drawings, and the detailed description of the invention shall refer to the different views by specifying the numbers of the figures, and to the different parts by use of reference letters or numerals (preferably the latter)."

A section of claim(s) is the last. A claim must point out "and distinctly claim the subject matter which the applicant regards as his invention." By definition, "Claims are brief descriptions of the subject matter of the invention, eliminating unnecessary details and reciting all essential features necessary to distinguish the invention from what is old. Claims are the operative part of the patent. Novelty and patentability are judged by the claims, and when a patent is granted, questions of infringements are judged by the courts on the basis of the claims."

Beyond this point, one encounters the full force of patent law head on. Take, for example, the following excerpts from a part of the law (37 CFR 1.75) on examining procedure:

> "A claim may be typed with the various elements subdivided in paragraph form. There may be plural indentations to further segregate subcombinations or related steps.
>
> "Reference characters corresponding to elements recited in the detailed description and the drawings may be used in conjunction with

the recitation of the same element or group of elements in the claims. The reference characters, however, should be enclosed within parentheses so as to avoid confusion with other numbers or characters which may appear in the claims. The use of reference characters is to be considered as having no effect on the scope of the claims.

"Claims should preferably be arranged in order of scope so that the first claim presented is the broadest. When separate species are claimed, the claims of like species should be grouped together where possible and physically separated by drawing a line between claims or groups of claims . . . "

A few words on drawings from *General Information Concerning Patents.* "The applicant for a patent will usually be required by statute to furnish a drawing of his invention whenever the nature of the case admits of it; this drawing must be filed with the application. This includes practically all inventions except compositions of matter or processes, but a drawing may also be useful in the case of many processes."

Standards for drawings are spelled out in detail. For example, "Character of lines. All drawings must be made with drafting instruments or by a process which will give them satisfactory reproduction characteristics. Every line and letter must be durable, black, sufficiently dense and dark, uniformly thick and well defined; the weight of all lines and letters must be heavy enough to permit adequate reproduction. This direction applies to all lines however fine, to shading, and to lines representing cut surfaces in sectional views. All lines must be clean, sharp, and solid. Fine or crowded lines should be avoided. Solid black should not be used for section or surface shading. Freehand work should be avoided wherever it is possible to do so."

Other sections cover paper and ink, size of sheet and margins, hatching and shading, scale, reference characters, symbols and legends, views, arrangement of views, figures for *Official Gazette*, extraneous matter, and transmission of drawings.

Views of Engineers Who Write and Read Patents

Engineers on the front line of patents are not charitable in their views of the law and those who practice it. Witness the verbatim comments that follow.

"Too many words . . . hard to find significance of patent . . . need good introductory statements for well written patent applications, indicating what is being claimed."

"Written in an old-fashioned style. Too legalistic. Coverage of inventions is often too broad."

"Criticisms of applications include a lack of precise dates and well-defined evidence of the invention."

"I think patent language stinks. It is archaic, as are the drawings—100 years out of date. Applicants often do not describe the relationship of working parts and thereby limit their claims."

"Patent law rewards those skilled in the application of the law, which is highly redundant."

"I see applications that don't tell what the patent is . . . wording is so general the claims cover the whole field and prevent others from taking out a similar patent."

"The few applications I have seen were so boring and flowery they put me to sleep."

"The legal language is overly complicated."

"Legal wordiness tends to confuse (disguise) the invention and its purpose."

"Because of traditional patent jargon, which is apparently required, all of these documents are difficult to read."

"Patent applications written by lawyers are too wordy and complex—usually the inventor cannot understand what is being said."

"Patents are too long. The gist of the patent should be stated in one paragraph. A person reading a patent should not be required to go through a dozen pages of details to extract this information for himself."

"Lack of clarity due to legalese."

"Some applications are not detailed enough to allow prior art to be assessed. Others need to be broadened to provide adequate coverage."

"Quoting of prior art can be excessive . . . supporting diagrams are sometimes weak . . . there is a deliberate overuse of legalese."

"A lack of clarity and too much legalese."

Recommendations of engineers to writers of patent applications mirror information found in the Patent Office booklet, *General Information Concerning Patents*. For example, they advise:

"You need . . . good knowledge and analysis of prior patents (art) on the subject . . . a good description of the invention . . . a good description of what it will do . . . good evidence and documentation, including specific dates, notes, reports."

"Claim everything you can. Try, in spite of your patent attorney, to write in English."

"An application should be brief and simple, stating why the invention is needed, describing the invention, reporting test results (if relevant), and conclusions from tests."

"Show why your invention is better, clearly and succinctly. Use clear, well labeled diagrams. Use legalese only when necessary. The patent examiner should rule on the invention, not the patent attorney."

Writer's Overview of Trademark Law From the U.S. Patent Office

As in the case of patent law, the U.S. Patent Office has a booklet available on the subject of trademarks. Entitled *General Information Concerning Trademarks*, the booklet can be purchased from the U.S. Government Printing Office.

A pamphlet, *Q&A About Trademarks*, may be obtained without charge from the same source.

First, some background from the pamphlet. By definition, "a trademark is a word, name, symbol, or device, or any combination of these, adopted and used by a manufacturer or merchant to identify his goods and distinguish them from those manufactured or sold by others. In short, it is a brand name used on goods moving in the channels of trade." It should be noted that the term trade name is often used in this sense. Legally, though, a trade name is confined to "business names of manufacturers, merchants, and others to identify their businesses. . . ."

A trademark may be registered at the U.S. Office of Patents and Trademarks in somewhat the same manner one applies for a patent. However, "there is no provision . . . in present trademark law for the registration of trade or commercial names merely to identify a business entity."

It is not necessary to register a trademark if it is "in actual use in goods moving in trade . . . it is protected under common law." But there are advantages to registration. "It constitutes notice of the registrant's claim of ownership; and it creates certain presumptions of ownership, validity, and exclusive right to use the mark on goods recited in the registration." Registration provides protection for 20 years from the date of issue, and may be renewed every 20 years "so long as the mark is still in use in commerce." Applications are directed to the Commissioner of Patents and Trademarks, Washington, D.C. 20231. Representation by an attorney is not required. As in the case of patents, a search should precede the application. "A search may be made in the Public Search Room of Trademark Examining Operation of the Patent and Trademark Office, Crystal Plaza Building No. 2, 2011 Jefferson Davis Highway, Arlington, Va."

Representative topics in *General Information Concerning Trademarks* include trademark statutes and rules, registration of trademarks, application for registration, the written application, drawing, specimens, examination of application, renewal, fees and payment of money, forms for trademarks.

Parts of an application include:

1. A written application.
2. A drawing of the mark.
3. Five specimens or facsimiles.
4. Payment of a filing fee.

An application must specify:

1. Name of applicant.
2. Citizenship of applicant—if a partnership, the names and citizenship of general partners; or if a corporation or association, the state or nation under the laws of which it is organized.
3. Domicile and post office address of applicant.
4. Statement that the applicant has adopted and is using the mark shown in a drawing (accompanying the application).
5. The goods or manner in which the trademark is used.
6. Official classification of goods or services covered by trademark (classifications are spelled out in the booklet).

7. Date of first use of mark.
8. Manner in which the mark is used.

The application must be signed and verified (sworn to), or must include a declaration by the applicant or by a member of the firm or an officer of the corporation or association applying.

"The drawing must be a substantially exact representation of the mark as actually used in connection with the goods or services. The drawing of a service mark may be dispensed with if the mark is not capable of representation by a drawing, but in such case the written application must contain an adequate description of the mark.

If the application is for registration only of a word, letter, or numeral, or any combination thereof, not depicted in special form, the drawing may be the mark typed in capital letters on paper, otherwise complying with the requirements."

The need for trademark specimens is explained in part in this manner:

"A trademark may be placed in any manner on the goods, or their containers or displays associated therewith, or on tags or labels attached to the goods. The five specimens [required] shall be duplicates of actually used labels, tags, containers, or displays or portions thereof, when made of suitable material and capable of being arranged flat and of a size not larger than 8½ by 13 inches. Third dimensional or bulky material submitted as specimens cannot be accepted, and the submission of such material may result in a delay in receiving a filing date."

A schedule of filing fees is in *General Information Concerning Trademarks*. The basic fee for an application to register is $35; that for a renewal application $25.

Bibliography From IEEE TRANSACTIONS ON PROFESSIONAL COMMUNICATION

The special issue of this periodical mentioned previously (June 1979) contains the following articles:

R. J. Joenk, "Patents: Incentive to Innovate and Communicate—an Introduction."
H. Skolnik, "Historical Aspects of Patent Systems."
B. M. Vanderbilt, "The Plight of the Independent Inventor."
H. J. Nussbaumer, "Patents and the Engineer."
W. G. Wolber, "The Business Value of Patents."
G. R. White, "Management Criteria for Effective Innovation."
R. V. Hughson, "The Right Way to Keep Laboratory Notebooks."
W. L. Franz and J. S. Child, Jr., "Good Habits Before Filing a Patent Application."
C. M. Wright, "Publication, Public Use, and Sale as Bars to Patenting."
G. A. Hauptman, "Patenting Inventions Based on Algorithms."
K. J. Dood, "The U.S. Patent Classification System."
P. J. Terragno, "Patents as Technical Literature."
J. F. Cottone, "Writing an Invention Disclosure."
"Guide for Patent Drawings" (no author specified).

J. T. Maynard, "How to Read a Patent."
A. B. Kimball, Jr., "Patenting U.S. Inventions Abroad."
T. M. Noone, "Trade Secrets vs Patent Protection."

Copies of the issue may be obtained (if available) by writing to the IEEE Service Center, 445 Hoes Lane, Piscataway, N.J. 08854. Single copies are $5 for IEEE members and $10 for nonmembers.

CHAPTER 25

What Technical Writers Should Know About Copyright Law

New copyright law went into effect January 1, 1978. Officially called the Copyright Act of 1976, it is described and discussed in *Circular R1, Copyright Basics*, published by the Register of Copyrights, Library of Congress, Washington, D.C. 20559.

A preamble states, "This general revision of the copyright law of the United States, the first such revision since 1909, makes important changes in our copyright system and generally, but not entirely, supersedes the previous Federal copyright statute."

The summary of the new law that follows is drawn exclusively from *Circular R1* and is limited in scope to literary works.

Protection under the law (title 17, U.S. Code) "is available to both published and unpublished works." The owner of the copyright has several exclusive rights:

1. To reproduce the copyrighted work in copies.
2. To prepare derivative works based upon the copyrighted work.
3. To distribute copies to the public by sale or other transfer of ownership.
4. To display the copyrighted work publicly.

There is one general limitation to exclusive rights called the doctrine of fair use. Further explanation is found in section 107 of the new law (Public Law 94-553-Oct. 19, 1976) and in another publication of the Copyright Office, *General Guide to the Copyright Act of 1976*. Discussion of fair use in this chapter will be delayed until this review of the law is completed. It should also be noted at this point that the exclusive rights of the copyright owner come into play when the writer has his literary work published in a periodical or book. He transfers all or part of his exclusive rights to the publisher at that time.

Under the new law, copyright protection for the writer starts from the time the work is created. This is automatic and registration is not necessary. An exception is made for persons writing a paper or article as part of their jobs. They may be employed as writers for journals or magazines. This is called a work for hire and the employer owns the copyright. Generally, coauthors are joint owners of a copyright, unless they have worked out some other arrangement.

A special situation exists where a paper or article is part of an anthology or collection of literary works published in a single issue of a magazine or book. The publisher must obtain a copyright transfer from each author. All or only part of the owner's exclusive rights may be transferred. In the example shown in the box below, the owner retains some rights.

Some creations are not eligible for copyright protection. These include: titles, names, short phrases, and slogans; familiar symbols or designs; mere variations

---

Metal Progress

Copyright Transfer

Article   ____

____

Author(s) ____

____

Copyright, title, interest and all rights in the Article is hereby assigned and transferred to American Society for Metals, Metals Park, Ohio 44073.

The author(s) reserve the following:

1. All proprietary rights other than copyright, such as patent rights.
2. The right to make oral presentation of the same material in any form.
3. The right to reproduce figures and extracts from the Article with proper acknowledgement.

____

Author's Signature   Print Name   Date

____

Author's Signature   Print Name   Date

____

Author's Signature   Print Name   Date

This signed transfer document must be received by the Editor's office before the manuscript can be published. Please mail **one copy** of this document to the following address:

The Editor
Metal Progress
American Society for Metals
Metals Park, Ohio 44073

of typographic ornamentation, lettering, or coloring; mere listings of ingredients or contents; and works consisting entirely of information that is common property and contains no original authorship, as standard calendars and weight charts.

As stated before, copyright protection starts automatically upon creation. "No publication or registration or other action in the Copyright Office is required to secure copyright under the new law, unlike the old law which required either publication with the copyright notice or registration in the Copyright Office."

In other words, publication "is no longer the key to obtaining statutory copyright . . . however, publication remains important to copyright owners . . . .

"When a work is published, all published copies should bear a notice of copyright.

"Works published with notice of copyright in the United States are subject to mandatory deposit with the Library of Congress.

"Publication of a work can affect the limitations on the exclusive rights of the copyright owner.

"The year of publication is used in determining the duration of copyright protection for anonymous and pseudonymous works, and for works made for hire."

When a work is published under the authority of the copyright owner, a notice of copyright should be placed on all publicly distributed copies. Use of the notice is the exclusive responsibility of the copyright owner. Permission of or registration with the Copyright Office is not required.

The notice "for visually perceptible copies" should include three elements:

1. The symbol © (the letter C in a circle), or the word "Copyright," or the abbreviation "Copr."
2. Year of first publication of the work. In the case of collections of works or derivative works, the year date of first publication of the collection or derivative work is sufficient.
3. Name of the owner of the copyright.
Example: © 1981   John Doe.

Notice "should be affixed to copies of the work in such a manner and location as to give reasonable notice of the claim of copyright."

For example, notice is on the inside of the front cover of *Technical Communication*, the journal of the Society for Technical Communication. It reads: "All rights reserved, and reproduction without written permission is prohibited. Copyright © 1980 by the Society for Technical Communication."

Copyright notice is not required on unpublished works. But "it may be advisable for the author or other owner of the copyright to affix notices to any copies which leave his or her control."

Failure to publish the notice or errors in the notice are not necessarily fatal. Two conditions must be met: registration of the copyright must come within five years after publication without notice and "a reasonable effort [must be] made to add the notice to all copies distributed to the public in the United States after the omission has been discovered."

Once more, registration is not a condition of copyright protection, but there are advantages:

1. A public record of the copyright claim is established.
2. Registration is usually necessary before an infringement suit may be filed in court.
3. If registration is made before or within five years of publication, it is prima facie evidence in court that the copyright is valid.
4. "If registration is made within three months after publication or prior to an infringement, statutory damages and attorney's fees will be available to the copyright owner in court actions. Otherwise, only actual damages are available to the copyright owner."

Three items are required ("in the same envelope or package") in registering a copyright:

1. Completed application form.
2. $10 for application fee.
3. A deposit of the work being registered. If the work is unpublished, one copy; if published on or after January 1, 1978, two copies of the work as first published.

Applications must be submitted only on forms printed and issued by the Copyright Office and should be completed legibly in dark ink or typewritten because they become part of the archives at the Copyright Office.

Ordinarily the term of a copyright starting on or after January 1, 1978, is the life of the author plus 50 years after his death. In the case of joint authors, 50 years after the last survivor's death. In the case of works for hire, 75 years from publication or 100 years from creation, whichever is shorter.

Works created before January 1, 1978, but not published or registered before January 1, 1978, are automatically given protection under the new law and in general under the same terms. One exception: at least 25 years of protection is guaranteed.

As stated previously, any or all of the exclusive rights in a copyright may be transferred. This is normally by contract (see previous example). The Copyright Office does not have forms for transfers.

Under certain conditions, the author who transfers his copyright may terminate the agreement if certain conditions are met—generally, by serving written notice on the transferee within 35 years after the agreement was made.

What about international protection?

The Copyright Office warns, "There is no such thing as an international copyright that will automatically protect an author's writing throughout the entire world." Protection depends on the laws of individual countries. The United States is a member of the Universal Copyright Convention which came into force in 1955. A list of countries that have copyright relations with the United States may be obtained from the Copyright Office. Ask for *Circular R38a*.

Application forms for registering a copyright may be filed by the author, a person or organization that has obtained ownership of all rights under the copyright initially belonging to the author, the owner of exclusive rights (he may transfer only part of his rights), and a duly authorized agent.

Application forms are supplied free by the Copyright Office. Forms may be ordered by telephoning 202/287-9100.

Forms include:

Form TX: for published and unpublished nondramatic literary works.
Form PA: for published and unpublished works of the performing arts.
Form RE: for claims to renew copyright in works copyrighted under the old law.

All material and communications sent to the Copyright Office should be addressed to the Register of Copyrights, Library of Congress, Washington, D.C. 20559. The application, deposit (copies), and fee should be mailed in the same package.

For a list of other material published by the Copyright Office, write for "Publications of the Copyright Office." Any requests for Copyright Office publications or special questions relating to copyright problems should be addressed to the Information and Publications Section, LM-455, Copyright Office, Library of Congress, Washington, D.C. 20559.

Closer Look at the Right of Fair Use Doctrine

The concept is of particular interest to writers who use published information in pieces they write. Two questions arise: When should the writer ask the copyright owner for permission to use copyrighted material? When—under the fair use doctrine—is this step unnecessary?

Permission is usually granted routinely. The author writes the owner of the copyright—typically the publisher of a periodical or book—and asks for permission, stating how the material will be used (as in an article or book), and specifically what will be used (such as the name of an article, date of publication, and portions of the text, or specific figures, or specific tables to be used). In addition, the request states the manner in which ownership of the copyright will be acknowledged—as, for example, a notice in a prominent place, such as "Reprinted with permission from *Metal Progress*."

The law on Fair Use (Section 107) reads:

". . . fair use of a copyrighted work, including such use by reproduction in copies . . . for purposes such as criticism, comment, news reporting, teaching (including copies for classroom use), scholarship, or research, is not an infringement of copyright.

"In determining whether the use . . . is a fair use the factors to be considered include—

"1. The purpose and character of the use, including whether [it is] of a commercial nature or is for nonprofit educational purposes.

"2. The nature of the copyrighted work.

"3. The amount and substantiality of the portion used in relation to the copyrighted work as a whole.

"4. The effect of the use upon the potential market for or value of the copyrighted work."

Further discussion is found in Chapter 8 of the *General Guide to the Copyright Act of 1976*. The doctrine of fair use, developed over the years by courts, was not included in copyright law prior to the 1976 act. However, the concept, the Copyright Office explains, "is not susceptible to exact definition." The need for amplification is suggested by this general rule: copying is allowed without permission from, or payment to, the copyright owner where the use is reasonable and not harmful to the rights of the copyright owner.

Section 107 of the Act, it is suggested, "is somewhat vague because it would be difficult to prescribe precise rules to cover all situations." The law, it is remembered, sets out four factors to consider in determining fair usage: Is it commercial? How much is copied? What is the nature of the copyrighted work? Will the use hurt the sale of the copyrighted work?

Courts, it is stated, may consider other factors. Use of the terms "including" and "such as" in the Act means that the examples are not intended to be restrictive. Courts, it is explained, apply "an equitable rule of reason."

Private arrangements have been worked out by special interest groups, including publishers of educational materials and educators and educational institutions using such materials in the classroom. No such relief has been fashioned for the technical writers who need to borrow bits and pieces from the copyrighted works of commercial publishers of books and periodicals and, on occasion, unpublished works of fellow technical writers.

In looking back to the body of the law, it seems clear that technical writing qualifies as "scholarship" and "research." It follows that the technical writer gains whatever protection the law and the courts allow. My experience as an editor of a periodical has been that the technical writer is invariably law-abiding. Whether he needs to or not, he routinely asks permission to use an excerpt from a text. He also promises to give full credit to the holder of the copyright.

Perhaps the practice is encouraged by the vagueness of the law. One may assume courts will be equitable. What one cannot predict is the charity or lack of same on the part of the owner of the copyright.

As a matter of curiosity, one may examine the treaty between publishers and educational institutions for telltale signs of how far copyright owners are willing to go in allowing fair usage.

In this instance, usage is limited to making copies of copyrighted materials for classroom use. Three tests are specified—brevity, spontaneity, and cumulative effect. In all instances, copies for classroom use "must include notice of copyright."

The brevity guideline relates to how music can be copied.

For poetry—not more than 250 words from a poem, which may be a complete poem.

APPLICATION FOR PERMISSION TO
USE MATERIAL UNDER COPYRIGHT
By American Society for Metals, Metals Park, Ohio

Applicant ______________________________Date ________________

Address ____________________________

____

____

Permission is requested to reproduce the following excerpts or illustrations from
METAL PROGRESS:

Selections from text (specify by date of issue, page number and paragraph):

____

____

____

Illustrations (specify by date of issue, page and figure number):

____

____

in a work to be entitled: ______________________________________

Author ________________________Publisher ____________________

Estimated date of publication ________Edition ____________________

or for the following purpose: _________________________________

It is agreed that credit will be given for excerpts and illustrations in the following
manner (indicate style, form and position of credit): __________________

____

____

Applicant agrees to furnish to Metal Progress, American Society for Metals, at no
charge, one copy of work in which material is used.

______________________________   ____________________

Signed   Date

Application approved by:

______________________________   ____________________

For American Society for Metals   Date

Example of an application for permission to use copyrighted material

For prose—a complete article, story, or essay of less than 2500 words, or an excerpt from any prose work of not more than 1000 words or 10% of the total, "whichever is less."

An exception for both poetry and prose: the limitations are relaxed "to permit the completion of an unfinished line of a poem or an unfinished prose paragraph."

For illustrations—one chart, graph, diagram, drawing, cartoon, or picture per book or per periodical issue.

Spontaneity relates to who initiated the copying for classroom usage. "The inspiration," it is explained, "must come from the teacher."

Schools can't turn to wholesale copying as an alternative to the purchase of texts. The rule also has a time element in this sense: timeliness of use in the classroom would make it "unreasonable to expect a timely report [from a publisher] to a request for permission."

Cumulative effect relates to the amount of usage of copies. Usage is limited to:

1. One course.

2. One short poem, article, story, or essay or two excerpts from the same author, "nor more than three from the same collective work or periodical volume during one class term."

3. Not more than "nine instances of multiple copying per class term."

Lastly, the limitations do not apply to "current news periodicals, newspapers, and current news sections of other periodicals."

The rules are more liberal for single copies for use by teachers in pursuit of "scholarly research, use in teaching, or preparation to teach a class."

Fair use permits copying:

1. One chapter from a book.

2. One article from a periodical or newspaper.

3. A short story, short essay, or short poem.

4. A chart, graph, diagram, drawing, cartoon, or picture from a book, periodical, or newspaper.

An example of an application for permission to use material under copyright is shown in the box on the preceding page.

An Anthology of Stylebooks

CHAPTER 26

Introduction to Part III

The term "style" implies a "way of doing something." To writers and editors and secretaries and typists, a stylebook answers questions such as "do you spell it 'insure' or 'ensure'?" In this sense, a stylebook gives us approved ways of doing things. Usually the approval (source of authority) is of the in-house variety—publisher, editor, company, the boss. You are expected to observe these rules, and innovation is usually frowned upon.

Compliance, I believe, would be improved if stylebooks were made easier to use and if, for those of inquiring mind, a better reason for compliance than "it's a condition of employment" were offered. I offer two: economics and professionalism.

An incident illustrates the former. I once worked for a magazine that required three designated editors to read pages for the final time as they were finished by the print shop downstairs. If Editor A decided to remove a comma from a line, he would put the corrected page in a tube and shoot it downstairs. The correction would be made, and the page would eventually return to Editor B, who might decide to put back the comma Editor A had excised. And so on.

After this sort of thing had gone too far, it was brought to the attention of the editor's office by the manager of the print shop who was being thrown off schedule by wholesale and often arbitrary changes. Each change, if nothing more than a comma in one line, cost a minimum of $2. I was called in and told to develop a stylebook that set forth approved ways of doing things, including when and when not to use commas. Henceforth things would be done the stylebook way and no other way.

The classic rationalization for a stylebook—economics—is illustrated by the example. A second motive is of equal importance in my opinion. Again an incident.

Years ago I was a member of a panel talking about stylebooks. One editor was with the U.S. Government Printing Office in Washington. He represented the quintessence in style. I was with a weekly business magazine and represented the middle ground of style.

The third panelist worked on an engineering magazine with a staff of three, including a secretary.

"When a question of style comes up," the engineering editor explained, "we have a staff meeting and flip a coin to decide how we will take care of a matter of style. The agreement is to handle the point of style the same way each time it comes up in the issue we are working on. The aim is to be consistent within a given issue."

Consistency, in my way of thinking, is part of being a professional. If you spell it "gage" on page 7 and "gauge" on page 97, you are not maintaining standards expected of a professional. I feel this self-discipline applies to all professionals who write, be they editors or engineers.

Once I have said that, and being absolutely sincere in what I said, I want to add what may appear to be an inconsistent qualification. I tend to treat stylebooks more lightly than others of my kind, and am impatient with people who read more than economics and professionalism into stylebooks. I do not want to sound emotional about this, but I feel such people do more harm than good. For this reason, I believe it is important to talk about what stylebooks aren't and what they will and won't do for the writer.

Style can be used as a haven for scoundrels who use this law of the land to whip recalcitrant editors or outside authors into shape. They tend to treat style as being part of the Constitution of the United States instead of an arbitrary, in-house rule of the stature—at least to unbiased observers—of a "keep off the grass" or a "no smoking" sign. There is a tendency among these zealots to think in terms of a "right way" and a "wrong way" of doing things. Both "gage" and "gauge" are correct, for example. In a style manual, we do no more than choose one of the acceptable spellings.

Sheriffs of style also give one the impression that knowing style is tantamount to knowing how to write. In fact, knowing style does not help any more than wearing approved golfing togs makes one a better golfer.

I once dealt with a writer who often turned in substandard copy and was always looking for the secret of success. In his quest for approval and excellence, he picked up a tip at a staff meeting. An enterprising editor had replaced numbers with large periods (called 1090 dots) in the enumeration of a series. He was congratulated by the powers that be and all heads nodded approval. In the next piece our errant scribe turned in, each paragraph, from beginning to end, was preceded by a 1090 dot. There was no apparent reason for the usage. I asked for an explanation and was told, "If it is so good for a series, why not anywhere?"

Success does not come that easily.

One more gripe. Those who take stylebooks too seriously, at least in my opinion, tend to make them too long and too omniscient. It is not possible or desirable to cover every conceivable question of style in a stylebook. Further, a stylebook should not go beyond the boundaries of in-house rules. Other matters are covered in standard references—and usually in a more competent manner.

The dictionary, for example, is often underutilized. A personal favorite is *Webster's New Collegiate Dictionary*, 2nd edition, G. & C. Merriam Co., Springfield, Mass., 1956. Following "zymurgy," the last entry, the writer in search of guidance will find sections on abbreviations; arbitrary signs and symbols; a pronouncing gazetteer (also a handy source for spellings); vocabulary of rhymes; rules for spelling, punctuation, compound words and hyphenation, use of capitals, and use of italic type; style for bibliographies; preparation of copy for the printer; printing terms; proofreader's marks; and a listing of colleges and universities.

I mention supplementary sources (a list is at the end of this section) because if such standard information is not duplicated in a stylebook, the result is a short stylebook. Brevity is a key to success in this instance. Long, exhaustive stylebooks are hard to use and scare their would-be beneficiaries; a short one stands a better chance of being observed than a long one.

Restraint, in the sense of favoring underregulation over overregulation, is also to be commended. Punctuation is a classic example. The rule of thumb is "if in doubt, punctuate." In my mellowing years, I tend to lean in the opposite direction, even though I may be violating another custom, "if there is any chance anyone will misunderstand, punctuate." Neither do I buy "insert a comma every time you have to stop to take a breath or pause in reading."

Another reason stylebooks are not universally loved or observed is explained by the nature of the subject matter. A stylebook is made up mostly of nits that must be memorized, and one must have the presence of mind to realize he is stepping on a matter of style before he can put himself into position to be a complier. When a matter of style can be reduced to a plausible concept, such as "if an abbreviation is a word, use a period," observance is likely to improve. Example: in. for inch.

What is a suitable scope for a stylebook?

In my opinion, a stylebook addresses topics like the following:

Capitalization—Do you capitalize spring, summer, autumn, fall?

Business names—Should "the" be included in a company name?

Personal titles—Do you spell out or abbreviate "governor" when it is used with only the last name?

Geographical terms—Is Chicago well enough known to let the name stand alone? Should Illinois be appended to the name? Is Illinois abbreviated in this instance?

Postal addresses—What is the zip code for Hawaii?

Metric practice—How do you convert inches to millimeters?

Abbreviations—What is the abbreviation for decibel?

Punctuation—Does style call for a period after the last word in a caption? Does a comma go between a last name and Jr.?

Plurals—What is the preferred plural for appendix?

Such topics are included in this stylebook. If it has any claim for distinction, it is in the manner its components were selected. Anthology is an accurate description. This stylebook is a collection of excerpts from a number of stylebooks—all are listed at the end of this section. This variety not only presents more than one viewpoint on occasion—a feature unique among stylebooks—but also includes items—such as a table for converting fractions of an inch to decimal equivalents and to millimeters—notable for their usefulness.

As stated previously, there are many outstanding references. They include the following:

Webster's Third New International Dictionary, G. & C. Merriam Co., Springfield, Mass.

Dictionary of Scientific and Technical Terms, McGraw-Hill Book Company, New York.

Thomas Register of American Manufacturers (annual), Thomas Publishing Co., New York.

Fraser's Canadian Trade Directory (annual), Fraser's Trade Directories, Div. of Maclean-Hunter Ltd., Toronto.

Glossary of Metallurgical Terms and Engineering Tables, American Society for Metals, Metals Park, Ohio.

A Manual of Style, 12th ed., revised, University of Chicago Press, Chicago, 1969; also, 13th ed., revised and expanded, 1982.

CBE Style Manual, 3rd ed., American Institute of Biological Sciences, Washington, D.C., 1972.

The Careful Writer, Theodore M. Bernstein, Atheneum, New York, 1966.

Words Into Type, 3rd ed., Prentice-Hall, Englewood Cliffs, N.J., 1974.

Bibliographic Guide for Editors and Authors, American Chemical Society, Washington, D.C., 1974.

Mathematics Into Type, American Mathematical Society, Providence, R.I., 1971.

Suggestions to Authors of the Reports of the United States Geological Survey, 5th ed., U.S. Government Printing Office, Washington, D.C., 1958.

Style Guide, U.S. Government Printing Office, Washington, D.C.

The Elements of Style, 2nd ed., William Strunk Jr. and E. B. White, The Macmillan Co., New York, 1972.

A Dictionary of Contemporary American Usage, Bergen Evans and Cornelia Evans, Random House, New York, 1957.

Sources for the 15 sections that follow are—in order of appearance:

Section 1—   *A Guide for Preparing Manuscripts*, National Academy of Sciences, Washington, D.C.

Section 2—   *Metric Conversion Tables*, Bulletin T-18A, Babcock & Wilcox Co., Beaver Falls, Pa.

Section 3—   *Style Manual*, American Iron and Steel Institute, Washington, D.C.

Section 4— Reprint, American Institute of Chemical Engineers, New York.

Section 5— Interoffice Memorandum, Climax Molybdenum Co. of Michigan, Ann Arbor, Mich.

Section 6— *Publication Style Guide* and Appendix F, North American Space Systems Div., Rockwell International.

Section 7— *Technosphere*, Fansteel Inc., North Chicago, Ill.

Section 8— Table Published by Teledyne Vasco, Latrobe, Pa.

Section 9— *A Guide to Authors for Preparation and Presentation of Technical Papers*, American Foundrymen's Society, Des Plaines, Ill.

Section 10— *Instructions and Suggestions for Authors*, American Welding Society, Inc., Miami, Fla.

Section 11— *Standard for Metric Practice*, American Society for Testing and Materials, Philadelphia.

Section 12— *Style Guide*, Northrop Corp., Hawthorne, Calif.,

Section 13— *Glossary of Metallurgical Terms and Engineering Tables*, American Society for Metals, Metals Park, Ohio.

Section 14— *Style Manual*, Periodical Publications, American Society for Metals, Metals Park, Ohio.

Section 15— *ASM Stylebook*, American Society for Metals, Metals Park, Ohio.

Section 1 ________________

Source: *A Guide for Preparing Manuscripts*, National Academy of Sciences, Washington, D.C. 1975. Available from Printing and Publishing Office, National Academy of Sciences, 2101 Constitution Ave., Washington, D.C. 20418.

Selections—in order of appearance:

>   Spelling and Compounding,
>   Including Hyphenation
>   Capitalization
>   Punctuation
>   Use of Italics
>   Abbreviations
>   Numbers
>   Tabular and Graphic Material
>   Line Drawings
>   Halftones
>   Figure Captions
>   Instructions to Typists Preparing
>   Manuscripts

Style

SPELLING AND COMPOUNDING OF WORDS

Webster's New Collegiate Dictionary (1973 edition) and *Webster's Third New International Dictionary* are our preferred guides to spelling. Some words, because of their newness or very special or limited use, may not be listed in any dictionary. In such cases, spelling must be determined by usage. In all cases, spelling must be consistent throughout a report.

Compounded words and expressions are formed by bringing two or more words together, either with or without a hyphen. Sometimes, previously separate words are combined to form a single new word, such as "lawgiver." Words can also be connected to convey units of thought or ideas, as in "go-between" or "long-term loan." Hyphens are used in such expressions if the previously separate words of the compound have not yet been converted into a single word or if an idea would not be as clearly or quickly conveyed by the component words in unconnected succession. The primary requirement in the use of hyphens, then, is clarity of meaning. A few preferences for use of hyphens are given below.

Hyphens are needed to avoid ambiguity in such adjectival combinations (unit modifiers) as "alternating-current motor," "6-foot lane." "3-

inch-diameter pipe," "large-scale project," "3- to 4-ton trucks." In combinations where the meaning is clear, however, and readability is not aided by use of hyphens, they may be omitted (for example, "atomic energy plant," "soil conservation measures," "national defense appropriations"). The hyphen is not used if the unit modifier contains an adverb ending in *ly*: "rapidly rising prices."

In spelled-out fractions, the numerator and denominator should be connected with a hyphen only if the number is used as an adjective (for example: "two-thirds rule"; *but* "two thirds of the governing body must concur"). Hyphens should also be used to avoid ambiguity in such expressions as "twenty-three thirty-seconds of an inch" or "thirty-one hundredths of an inch."

Rephrasing sometimes eliminates the necessity for hyphenation. For example, without a hyphen, the expression "the guided-missile development program" is ambiguous. On the other hand, the meaning is clear in "The program for development of guided missiles," and thus the hyphen is not necessary.

The dictionaries noted above should be consulted for further guidance concerning hyphenation of compounded words and expressions. The style of the Printing and Publishing Office differs from theirs in preferring hyphenation of some words containing double vowels, such as "re-entry," "anti-inflationary," and "intra-aortic" (but not "cooperate" or "coordinate").

CAPITALIZATION

The following statements and examples indicate our preferences regarding rules of capitalization. Although they are not intended to be accepted as final in every instance, they constitute useful guides.

1. Capitalize proper nouns and derivatives of proper nouns.

America	Peruvian
Italy	Californian
John Macadam	Italian

2. Do not capitalize words derived from proper nouns that have developed common, independent meanings.

pasteurize	macadam (crushed rock)
quixotic	watt (electrical unit)
mendelian	italicize

3. Capitalize common nouns or adjectives that form an essential part of a proper noun; but do not capitalize incomplete designations or the common noun used alone.

State of New York; the state
Mississippi River; the river
United States Government; the government
Department of the Interior; the department
Committee on Science and Public Policy; the committee

4. Capitalize descriptive terms used to denote definite regions, localities, or geographic features. Do not capitalize descriptive terms used to denote mere direction or position.

mountains in the West; *but* western mountains
the Midwest; *but* midwestern
the Occident; *but* occidental countries
Southern California; *but* southern France

5. In certain terms identified by proper names, capitalize the proper name, but not the remainder of the term.

Fourier transform	Boyle's law
Richter scale	Doppler effect
Planck's constant	

6. In titles or headings, capitalize the first word and all main words. Do not capitalize prepositions or articles. (*To*, when part of an infinitive, is not a preposition; nevertheless, like the University of Chicago, we prefer that it not be capitalized.) Do not capitalize scientific terms that by convention are not capitalized (for example, *in vitro*).

7. Capitalize registered trade names unless they have become generic terms no longer having significance as names of products of specific producers.

Orlon		nylon
Pyrex	*but*	cellophane
Anacin		aspirin

8. Capitalize an official, honorary, professional, or descriptive title standing before a name and forming part of it. Always capitalize President when referring to the President of the United States or to the office.

Professor James L. Brown; *but* the professor
President Lincoln
The President addressed the Congress
Many presidents of the organization have worked
  toward that goal

9. Do not capitalize words referring to governmental entities, unless used as parts of the names of specific entities.

federal government		U.S. Government
government agencies		Federal Aviation Agency
federal courts	*but*	United States Supreme Court
the state government		the State of Virginia

10. Capitalize the name of a phylum, division, class, order, family, or genus (but not a plural or derivative form). The genus, but *not* the specific epithet, is capitalized in the name of a species (for example, *Uta stansburiana*).

Generally, when more extensive or detailed guidance is required, problems of capitalization can be resolved by consulting *Webster's New Collegiate Dictionary* (1973 edition), *Webster's Third New International Dictionary*, or The University of Chicago *Manual of Style*.

PUNCTUATION

Many readily available handbooks of English offer complete discussions of the rules of punctuation. We suggest that most questions about these rules be resolved through use of such handbooks or a dictionary. For the treatment of some items, however, the Printing and Publishing Office follows a "house style," preferences selected from the many choices available. Examples of house style for punctuation follow.

Serial Commas   In a series of three or more items, such as "red, white, and blue," place a comma before the conjunction. The comma will clarify the number of items in the series and their relationship.

Ellipses   Use ellipsis marks to indicate any omission from quoted matter. If an ellipsis appears at the beginning or in the middle of a sentence, it is indicated by three evenly spaced points (. . .). If the ellipsis falls at the end of a sentence, the three evenly spaced points follow the period.

Quotation Marks   Place commas and periods before closing quotation marks. When question marks and exclamation marks are parts of the quoted matter, place them before closing quotation marks. When they punctuate entire sentences, place them after closing quotation marks. Other punctuation marks (semicolon, colon, dash) are placed after the closing quotation marks.

Apostrophes   The plurals of letters and numbers are formed by adding *'s* (dot the *i*'s, in the 1960's). The possessive forms of nouns are formed as follows:

1. Add *'s* to singular nouns and to proper names ending in *s* (acid's reaction, Charles's law) except where tradition or euphony dictates adding only an apostrophe (Archimedes' principle, Moses' law, for conscience' sake).

2. Add *'s* to plural nouns not ending in *s* (children's hospital, men's decision).

3. Add only an apostrophe to plural nouns ending in *s* (classes' requirements, drugs' effects).

ITALICS

Italics may be used to set apart or to emphasize words and phrases. Those most commonly italicized are:

1. The name of a genus when used in its capitalized, singular, noun form or when cited with a species or any of its subdivisions (such as *Staphylococcus aureus*, the genus *Clostridium*); but not the names of classifications higher than genus or the plural or derivative forms of genera (such as the family Bacillaceae, clostridia, staphylococcal)

2. The following Latin expressions when used in literary or legal references: *ibid.*, *circa*, *op, cit.*, *loc. cit.*, *et al.*, *infra*, *supra*; but not e.g., v., etc., or i.e.

3. Latin phrases containing the preposition *ad* (such as *ad hoc*, *ad interim*, *ad valorem*)

4. The names (or abbreviated names) and volume numbers of periodicals and the titles of books cited in the text of a manuscript

5. Some foreign words and expressions (such as *noblesse oblige*, *Weltanshauung*, *in vivo*, *in vitro*, *in situ*, *in vacuo*, *ipso facto*, *a priori*, *a posteriori*, *vis-à-vis*)

Italics used for emphasis should be kept to a minimum. If overused, they tend to lose their effect.

Some foreign words and phrases, such as percent, per diem, per capita,

en route, and versus, have been accepted into common English usage and
are not italicized.

If a typewriter is not equipped with italic characters, words and
phrases to be set in italics should be indicated in a manuscript by a single-
line underscore.

ABBREVIATIONS

When abbreviations are to be used in a report, the words or phrases to be
abbreviated are, in most cases, spelled out the first time they are used. If a
report is a collection of papers by individual authors, spell out the words
or phrases to be abbreviated the first time they appear in each paper and
place the abbreviation in parentheses immediately following the spelled-
out form. Thereafter the abbreviation can be used in place of the spelled-
out form.

Notable exceptions to this general rule are units of measurement. They
are spelled out only when they are associated with spelled-out numbers at
the beginning of a sentence, when they occur in the text unaccompanied
by numerals, or when they are not commonly used or seen. Abbreviations
for names of states or titles (such as Va., N.Y., Dr., Mr., and Capt.) and for
other common terms (such as Inc., Co., Corp., and p.) also can be used
without first presenting spelled-out forms.

In general, a scientific or technical abbreviation is written without a
period unless it spells a word (for example, "atm," the abbreviation for
atmosphere; "cal," the abbreviaton for calorie; but "in.," the abbreviation
for inch).

An acronym is a special kind of abbreviation. It is a word formed from
the first (or first few) letters of several words—for example, UNESCO,
formed from *U*nited *N*ations *E*ducational, *S*cientific, and *C*ultural *Or*-
ganization, and radar, formed from *r*adio *d*etecting *a*nd *r*anging. Acronyms, like abbreviations, should be defined when first introduced and
should be written without periods.

NUMBERS

Perhaps the most troublesome question in the handling of numbers in
written material is when to spell out the numbers and when to express
them as numerals. The diagram presented below answers the question
for most cases:

at beginning of sentence		
SPELL OUT		
not at beginning of sentence		
		not with unit of measurement
with unit of measurement	under 10	10 or higher
NUMERAL	SPELL OUT	NUMERAL

A few exceptions should be noted:

1. In most cases, spell out ordinal numbers.

   nineteenth century
   Eighty-ninth Congress
   third place

2. When several numbers appear in the same sentence, they must be treated uniformly. Either use numerals for all of them or spell them all out.

> two cats, nine mice, and twenty-three dogs
> 57 hospitals, 14 clinics, and 6 laboratories

3. Decimal numbers (for example, 2.4 mg) are always expressed as numerals; therefore, never place them at the beginning of a sentence.

When numbers containing decimal points are listed vertically, as in a table, the decimal points should be aligned. If numbers to be listed do not contain decimal points, they should be aligned on the final digit or, in special cases, they may be centered. With few exceptions, a zero is placed to the left of the decimal point in decimal numbers less than one; thus, .03 appears as 0.03. Always use a decimal point, not a comma, when expressing a decimal number, and place the point on the line (3.2), not as a raised period (3·2).

In most cases, commas are used in numbers that have four or more digits. However, commas are not used with numbers of four or more digits that apply to some units, such as measurements of wavelengths, units of heat, years, or journal and book page numbers.

Tabular and Graphic Material

TABLES

Tables are used to present short descriptions or numerical listings that are most clearly and effectively presented in tabular form. They should be self-explanatory and should supplement, not duplicate, information given in the text and illustrations. They are numbered with arabic numerals consecutively throughout a report, unless the report is made up of chapters or papers by individual authors, in which case they are numbered consecutively within each chapter or paper.

Type each table on a separate sheet of paper that contains no other manuscript copy, and provide a brief title that does not repeat material appearing in the table itself. No two tables in the same chapter or paper should have the same title.

All tables should be cited by number in the text. The data reported in the text should agree with the figures in the table. Authors should check tables carefully, especially those containing numerical data. Mistakes in simple arithmetic are among the most common errors in manuscripts.

If the same unit of measurement applies to all entries in a column, place the abbreviation for the unit or its spelled-out form in the column heading. This will eliminate the need to list the unit in each entry in the column.

All words (except prepositions) in the column headings (box heads) should be capitalized.

In tabular matter in which numbered footnotes can cause confusion, key footnotes with lowercase italic superscript letters. If this arrangement is not feasible, use a symbolic system.

The following sample table provides examples of the points discussed above.

TABLE 1　Cost of Mechanical Work as Related to Total Cost[a]

Year	Hospital		Total Cost ($)	Cost of Mechanical Work ($)			Percentage of Mechanical Work	
				HVAC	Plumbing	Total	With AC	Without AC[b]
1955	Teaching	A	37.70			6.60[c]	18.5	—
1957	Community	B	21.90	4.64 (21.2%)	2.20 (9.1%)	6.84	30.3	—
1957	State mental	C	23.26	2.12[b] (9.1%)	3.25 (14.0%)	5.37	—	23.1
1958	State mental	D	19.15	1.50[b] (7.9%)	1.25 (6.6%)	2.75	—	14.5
1962	Community	E	45.00			14.00[c]	32.0	—
1964	Nursing home	F	21.00	1.97[b] (9.4%)	2.25 (10.8%)	4.22	—	20.2
1965	Research	G	57.00	9.50 (16.6%)	5.44 (9.6%)	14.94	26.2	—
1965	Community	H	28.00	4.40 (15.7%)	2.01 (7.2%)	6.41	22.9	—
1966	Community	I	25.25	4.01 (15.9%)	1.45 (5.7%)	5.46	21.6	—
1967	Community	J	26.00			6.50[c]	25.0	—

[a]Costs are construction cost only and do not include planning, architectural, and engineering costs, site preparation or demolition work, and cost of movable hospital facilities. Areas are gross, not net. Costs are given in dollars per square foot.

[b]Not completely air-conditioned.

[c]Separation of total costs into HVAC and plumbing costs is not available.

Example of table that has been properly prepared for submission with manuscript. Footnotes are keyed with lowercase italic superscript letters, indicated on typewriter copy by raised, underlined lowercase letters.

ILLUSTRATIONS

Selection and preparation of illustrations are integral parts of manuscript prepartion. The printing process to be used, the kind of paper to be used, and the design of a publication are determined to a large degree by the nature of the artwork furnished for reproduction.

Except in very rare circumstances, illustrations should be used only where they are indispensable to the author as a means of conveying information or ideas. That is, they should not be used exclusively, or even primarily, to make publications "more attractive," nor should they be "added" to the text for other reasons. They should serve the author's purpose of conveying his message clearly and completely.

LINE DRAWINGS

Line drawings consist of black lines on a solid white or light background. Pen-and-ink drawings and diagrams are good examples of this kind of illustration. To ensure effective reproduction of these drawings, originating units should submit artists' *original* drawings whenever possible. The artwork should be large enough to allow for a reduction in final form to one half its size. Do not submit artwork that requires enlargement. Reduction tends to eliminate minor imperfections in the original art, whereas enlargement distorts and emphasizes minor imperfections.

Prepare drawings in India ink. If charts and diagrams are plotted on graph paper, use only paper with light-blue cross lines (the next paragraph explains the reason for this specification). Indicate all cross lines intended to be reproduced by making them sharp and black. If symbols used in figures are to be defined in a legend, it is best to key them by using standard symbols such as open and solid circles, squares, and triangles (○●, □ ■, △ ▲).

Use a light-blue pencil, never ink and never any other color, to indicate any corrections or changes in artwork or lettering or to make any other necessary comments for production or editorial personnel. Light-blue markings will not appear when copy is photographed, whereas ink and colors other than light blue will be picked up.

Lettering can be done either by mechanical typesetting or by the Leroy method. All lettering must be large enough to be legible when reduced to half size. In addition, sufficient space should be left between plotted graph lines; otherwise, they may tend to run together and blur when reduced.

Shading can be superimposed by the author on original art by use of Zip-a-Tone, benday, or other commercial tints. The Printing and Publishing Office can arrange to have the tints put on by a printer, who can usually obtain better results.

HALFTONES

A halftone is a reproduction of continuous-tone artwork or photographs, the image being formed by dots of various sizes. It is produced by photographing the original copy through a screen. Photographs in newspapers are familiar examples of halftones. On newsprint, the tiny dots characteristic of all halftones are clearly visible. A finer screen is used in preparing half tones for books than is used for newspapers; hence, there are more dots per square inch, and the resulting pictures are much clearer. Glossy photographs with good contrast are best adapted to the process used for reproducing halftones. Halftone prints of photographs ordinarily cannot be satisfactorily reproduced. Thus the original glossy photographs should always be submitted.

The slightest impression on the original copy will show up in reproduction. Therefore, avoid attaching paper clips or other objects that may leave impressions on the copy. Do not write on the back of the picture with a hard pencil or ballpoint pen. Any necessary identification of a photograph by figure number or any special instructions should be written on a piece of paper and pasted to the back of the picture.

It is sometimes necessary to crop a photograph, a process that involves cutting off unwanted parts at a straight line extending from one edge to another. Crop marks should be made in one of two ways: A light-blue grease pencil can be used on the margin of the photograph to show crop marks, or a tissue overlay can be used. In the latter process, the crop marks are shown on a transparent or translucent sheet that is placed over the photograph. This sheet should also be used to carry the figure identification or any special instructions or corrections.

Line drawings and halftones require captions (titles or explanatory material accompanying a picture). In contrast with legends, which are within the artwork and are integral parts of it, captions are outside the artwork.

FIGURE CAPTIONS

A figure caption is an explanatory comment or designation that accompanies an illustration. It appears outside the artwork (in contrast with a legend, which is within the artwork).

Each figure must be accompanied by a figure caption. Type captions double-spaced on a separate sheet of paper on which there is no other manuscript copy. Number both figures and accompanying captions with consecutive arabic numerals throughout an entire manuscript, unless the manuscript is a collection of papers by individual authors, in which case numbering begins anew with Figure 1 in each paper.

Every figure must be cited in the text. It is helpful for the author to indicate, in the margin of the text, where each figure is first mentioned. The word "figure" is always spelled out and capitalized in figure captions and in the text.

Instructions to Typists Preparing Manuscripts

GENERAL

All manuscripts must be typed on one side only of $8\frac{1}{2}'' \times 11''$ opaque white paper. All typewritten material, including footnotes, quotations, and references, must be double-spaced. Additional space should be left around mathematical equations or expressions. These specifications are designed to simplify the publishing process. Double-spacing provides room between lines for editorial marking and for instructions to the printer.

Each manuscript page should have liberal margins—about $1\frac{1}{2}''$ on the left and at the top, and $1\frac{1}{4}''$ on the right and at the bottom. Hyphenation of words that fall at the ends of lines should be avoided by running lines a little long or a little short. Each chapter or paper should begin on a new page. The original copy and one carbon or Xerox copy of the typescript should be submitted to the Printing and Publishing Office through the Publications Editor. At least one copy should be retained by the originating unit.

CORRECTIONS

If a correction must be made, cross out the error and type or write legibly, the correct version above it. The use of correction tape or correction fluid (such as Wite-Out or Snopake) to obliterate some errors is also recommended. Directly overtyping an incorrect letter with a correct letter will

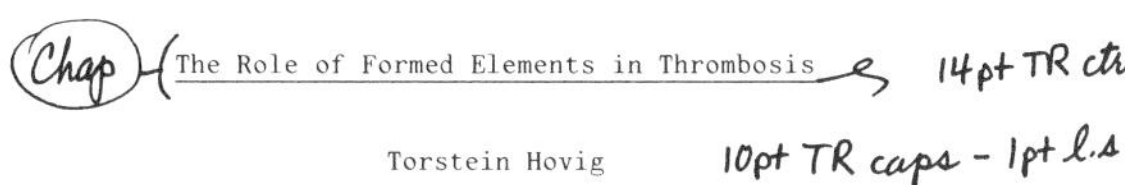

A typical manuscript page reduced by one half. Note editorial changes made by copy editor and instructions for typesetting.

usually result in illegibility. If a substantial part of the page requires correction, the page should be retyped. All handwritten remarks must be legible.

INSERTS

If inserts longer than a few words must be made, they should be typed on a separate sheet of 8½″ × 11″ paper and keyed to the text, and the sheet interleaved in the manuscript immediately after the manuscript page to which the insert is to be added. The following practices should be avoided: extensive handwritten revisions; writing in the margins or on the backs of pages; pasted-on or stapled-on bits of paper; use of tape where it covers text.

QUOTATIONS AND FOOTNOTES

Quotations should be double-spaced and set up in a way that is easy for editor and printer to follow: Place long quotations in separate paragraphs and indent the copy to set it off from the text (do not use quotation marks); run short quotations into the text and place quotation marks around them.

Footnotes should be double-spaced at the foot of the page, if there are only a few per page, and separated from the text by a short line, or they may be placed immediately following the paragraph to which they refer. In the latter case, a full-width solid line should be typed above and below the footnote to separate it from the text.

Section 2

Source: *Metric Conversion Tables*, Bulletin T-18A, Babcock & Wilcox Co., Tubular Products Div., Beaver Falls, Pa. 15010.

Selections—in order of appearance:
>   Metric Units and Abbreviations
>   Some Approximate Conversions

Metric System of Measurements

In the metric system of measurements, the principal unit for length is the meter; the principal unit for volume, the liter; and the principal unit for weight, the gram. The following prefixes are used for subdivisions and multiples: milli = 1/1000; centi = 1/100; deci = 1/10; deka = 10; hecto = 100; kilo = 1000. In abbreviations, the subdivisions are frequently used with a smaller letter and the multiples with a capital letter, although this practice is not universally followed everywhere the metric system is used. Not all the multiples and the subdivisions are used commercially. Those ordinarily used for length are kilometer, meter, centimeter, and millimeter; for area, square meter, square centimeter and square millimeter; for volume, cubic meter, cubic decimeter (liter), cubic centimeter, and cubic millimeter. The most commonly used weights are the kilogram and gram. The metric system was legalized in the United States by an Act of Congress in 1866.

Conversion from one subdivision or multiple of measurement to another in the metric system can be accomplished simply by moving the decimal point. Going from a larger to a smaller unit, the decimal is moved one place to the right for each multiple or subdivision. For example: 9.45 meters = 94.5 decimeters, 945 centimeters or 9450 millimeters.

Similarly, going from a smaller to a larger unit, the decimal is moved to the left. That is, 13.097 millimeters = 1.3097 centimeters, .13097 decimeters or .013097 meters.

The same technique can be applied to conversions of weight or volume. The following are examples:

>   4.609 kilograms = 46.09 hectograms,
>   460.9 dekagrams or 4609 grams.

>   47.09 milliliters = 4.709 centiliters,
>   .4709 deciliters or .04709 liters.

Metric Units and Abbreviations

	LENGTH		WEIGHT		VOLUME	
	Unit	Abbr.	Unit	Abbr.	Unit	Abbr.
1000	Kilometer	Km	Kilogram	Kg	Kiloliter	Kl
100	Hectometer	Hm	Hectogram	Hg	Hectoliter	Hl
10	Dekameter	Dam	Dekagram	Dag	Dekaliter	Dal
Base unit	Meter	m	Gram	g	Liter	l
1/10	Decimeter	dm	Decigram	dg	Deciliter	dl
1/100	Centimeter	cm	Centrigram	cg	Centiliter	cl
1/1000	Millimeter	mm	Milligram	mg	Milliliter	ml

Some Approximate Conversions

	If you know:	You can get:	If you multiply by*:
Length	inches	millimeters	25
	feet	centimeters	30
	yards	meters	0.9
	miles	kilometers	1.6
	millimeters	inches	0.04
	centimeters	inches	0.4
	meters	yards	1.1
	kilometers	miles	0.6
Area	square inches	square centimeters	6.5
	square feet	square meters	0.09
	square yards	square meters	0.8
	square miles	square kilometers	2.6
	acres	square hectometers (hectares)	0.4
	square centimeters	square inches	0.16
	square meters	square yards	1.2
	square kilometers	square miles	0.4
	square hectometers	acres	2.5

Some Approximate Conversions (continued)

Mass (and weight)	ounces	grams	28
	pounds	kilograms	0.45
	short tons	megagrams (metric tons)	0.9
	grams	ounces	0.035
	kilograms	pounds	2.2
	megagrams (metric tons)	short tons	1.1
	pounds per foot	kilograms per meter	1.5
Liquid volume	ounces	milliliters	30
	pints	liters	0.47
	quarts	liters	0.95
	gallons	liters	3.8
	milliliters	ounces	0.034
	liters	pints	2.1
	liters	quarts	1.06
	liters	gallons	0.26
Temperature	degrees Fahrenheit	degrees Celsius	5/9 (after subtracting 32)
	degrees Celsius	degrees Fahrenheit	9/5 (and then add 32)

*The conversion factors are all approximate, but close enough to be useful for all practical purposes. The greater the degree of precision desired, the less "quick" the conversions are likely to be.

Section 3 ——————————————————

Source: *Style Manual*, American Iron and Steel Institute, Washington, D.C.,
          1950.
Selection:
          Tables

Tables

TABULAR MATTER should be kept to a minimum since it is expensive
and unessential data are rarely studied. Avoid tables too large to be ac-
commodated within a printed page. Maximum width is four inches when
table is printed across the page and six and a half inches when it is
printed sideways.

Tables should be numbered consecutively throughout the paper with
arabic numerals.

Each table should bear a heading following the table number. The
headings should be as brief as possible.

Abbreviations should conform to the rules given in the section on abbreviations; but where space is limited, other than the listed words may
be abbreviated.

In a long column of numbers, all with a cipher preceding the decimal,
the cipher may be omitted from all of the sums except the first and last.

Center the figures in the column. If the number of digits is uneven, center on the longest line and align the figures at the right. Columns of figures containing decimals are aligned on the decimal point.

If the material used in the table is not original, its source should be
indicated below the table.

Avoid all ditto marks by either repeating the words or changing the
form of heading.

| JUNIOR BEAMS | | JUNIOR BEAMS | |
Nominal size	*Weight per lineal foot*	*Nominal size inches*	*Weight per lineal foot pounds*
12 × 3 in.	11.8 lb	12 × 3	11.8
11 × 2 in.	10.3 lb	11 × 2	10.3

Never use the symbols ' for feet, " for inches, @ for at, ° for degree.

If a table is continued from one page to another, repeat the table
number followed by the word *(continued)*. It is not necessary to repeat the
heading of the table. However, the column headings should be repeated.

Periods are omitted after headings and also after box heads.

The table is placed as near the first reference to it as possible without
needlessly breaking into the middle of a paragraph. Every care should be
taken to avoid the necessity of turning a page to refer to a figure or table.

Section 4 ______________________

Source: "Letter Symbols for Chemical Engineering," by E. Buck, Union Carbide Corp., South Charlestown, W. Va., from a reprint published by the American Institute of Chemical Engineers, Publications Dept., 345 E. 47th St., New York, N.Y. 10017.

Selections—in order of appearance:

> Symbols for Concentrations
> Symbols for Rate Concepts

Symbols for Concentrations

	Symbol	Unit or Definition
Absorption factor	A	.... $A = L/K^*V$
Concentration, mass or moles per unit volume	c	.... kg/m^3, $kmol/m^3$
Fraction,		
cumulative beyond a given size	ϕ	
by volume	x_v	
by weight	x_w	
Humidity,	H, Y_H	.... kg/kg dry air
at saturation	H_s, Y^*	.... kg/kg dry air
at wet-bulb temperature	H_w, Y_w	.... kg/kg dry air
at adiabatic saturation temperature	H_a, Y_a	.... kg/kg dry air
Mass concentration of particles	c_p	.... kg/m^3
Moisture content,		
total water to bone-dry stock	X_T	.... kg/kg dry stock
equilibrium water to bone-dry stock	X^*	.... kg/kg dry stock
free water to bone-dry stock	X	.... kg/kg dry stock
Mole or mass fraction,		
in heavy or extract phase	x	
in light or raffinate phase	y	
Mole or mass ratio,		
in heavy or extract phase	X	
in light or raffinate phase	Y	
Number concentration of particles	n_p	.... number/m^3
Phase equilibrium ratio	K^*	.... $K^* = y^*/x$
Relative distribution of two components, between two phases in equilibrium	α	.... $\alpha = K_i^*/K_j^*$
between successive stages	β	.... $\beta_n = (y_i/y_j)_n/(x_j/x_i)_{n+1}$
Relative humidity	H_R, R_H	
Slope of equilibrium curve	m	.... $m = dy^*/dx$
Stripping factor	S	.... $S = K^*V/L$

Symbols for Rate Concepts

	Symbol	Unit or Definition
Quantity per unit time, in general	q	
Angular velocity	ω	
Feed rate	F	kg/s, kmol/s
Frequency	f, N_f	
Friction velocity	u^*	$u^* = (\tau_w\rho)^{1/2}$, m/s
Heat transfer rate	q	J/s
Heavy or extract phase rate	L	kg/s, kmol/s
Heavy or extract product rate	B	kg/s, kmol/s
Light or raffinate phase rate	V	kg/s, kmol/s
Light or raffinate product rate	D	kg/s, kmol/s
Mass rate of flow	w	kg/s, kg/h
Molal rate of transfer	N	kmol/s
Power	P	W
Revolutions per unit time	n	
Velocity, in general	u	m/s
instantaneous, local		
longitudinal (x) component of	u	m/s
lateral (y) component of	v	m/s
normal (z) component of	w	m/s
Volumetric rate of flow	q	m^3/s, m^3/h
Quantity per unit time, unit area		
Emissive power, total	W	W/m^2
Mass velocity, average	G	$G = w/S$, kg/(s · m²)
vapor or light phase	$G, \overline{G}$	kg/(s · m²)
liquid or heavy phase	$L, \overline{L}$	kg/(s · m²)
Radiation, intensity of	I	W/m^2
Velocity,		
nominal, basis total cross section of		
packed vessel	v_s	m/s
Volumetric average	$V, \overline{V}$	m^3/(s · m²), m/s
Quantity per unit time, unit volume		
Quantity reacted per unit time, reactor		
volume	N_R	kmol/(s · m³)
Space velocity, volumetric	Λ	m^3/(s · m³)
Quantity per unit time, unit area, unit		
driving force, in general	k	
Eddy diffusivity	δ_E	m^2/s
Eddy viscosity	ν_E	m^2/s
Eddy thermal diffusivity	α_E	m^2/s
Heat transfer coefficient		
individual	h	W/(m² · K)
overall	U	W/(m² · K)

	Symbol	Unit or Definition
Mass transfer coefficient		kmol/(s · m²) (driving force)
Individual	k	To define driving force,
gas film	k_G	use subscript:
liquid film	k_L	c for kmol/m³
Overall	K	p for bar
gas film basis	K_G	x for mole fraction
liquid film basis	K_L	
Stefan-Boltzmann constant	σ	5.6703×10^{-8} W/(m² · K⁴)

Section 5 ______________________________

Source: Interoffice Memorandum, Climax Molybdenum Co. of Michigan,
           Ann Arbor, Mich.
Selection:
           References and Footnotes

References are used to acknowledge borrowed material and to inform the reader of the source of statements and quotations. Footnotes, on the other hand, are used to present explanatory remarks not appropriate to the text. They should be used sparingly.

Position—All references appear on a separate page at the end of the text—before the figures. (If only two or three references are used, they are placed at the bottom of the page on which they are first cited.) Footnotes appear at the bottom of the page on which they are cited; they are separated from the text by a bar extending two inches from the left-hand margin.

Numbering—References are numbered (with Arabic numbers) consecutively throughout the text. The reference number is placed as a superscript (without parentheses or other markings) in the text. All numbers are placed outside commas and periods and inside colons and semicolons.

Symbols rather than Arabic numbers are used to show the sequence of footnotes on each page. The following set of symbols should be used in the indicated sequence: *, †, §. All symbols are placed outside commas and periods and inside colons and semicolons.

In a table, lower cased letters (a, b, c) are used to number references and footnotes.

Style—The style of footnotes is flexible owing to the nature of footnotes themselves; if possible, however, the footnote should be a complete sentence.

Many styles are available for references. However, the style illustrated in the following examples has been adopted at this Laboratory:

1. J. C. Thomas, "Study of Cast Irons," Trans. ASM 98 (1945), 247-289.
2. J. C. Thomas and A. F. Root, *Study of Cast Iron*, John Wiley and Sons, Inc., New York, 1950, pp. 58-59.
3. J. C. Thomas, A. F. Root, and S. Norton, "Study of Cast Irons," in *Survey of Ferrous Alloys*, edited by A. E. Axe, McGraw-Hill Book Company, Inc., New York, 1948, pp. 58-59.
4. J. C. Thomas, "Investigation of a Pearlitizing Method," Climax Report L-212-64, November 15, 1966.

Here are some things to notice. The name of the author(s) is given first in the natural order (the first name or initials and then the surname). The names of two authors are joined by "and"; however, when three or more authors appear, "and" preceded by a comma is placed before the last author's name.

Section 6 ___________

Source: *Publication Style Guide*, North American Space Systems Div., Rock-
           well International.
           Appendix F, *Publication Style Guide,*, "Commonly Used Abbrevia-
           tions and Acronymns."

Selections—in order of appearance:

           Examples, Abbreviations,
               Acronyms
           Editorial Standards   Greek Alphabet
           Abbreviations   Prefixes to SI Units
           Acronyms   Common Metric Equivalents
           Conventional Symbols   and Conversions
           Arabic Numbers   Approximate Equivalents
           Fractions   Conversions Accurate to
           Decimals   Parts Per Million
           Tables and Listings

A	acceleration	A/D	analog to digital
A	alpha	A/F	airframe
A	analog	A/G	air to ground
A	analog signal	A/L	approach and landing
A	anode	A/N	alphanumeric
A-G	air to ground	A/P	airport
A-V	audiovisual	A/P	autopilot
A&A	advertise and award	A/R	as required
A&E	architects and engineers	A/S	arm/safe
A&E	architectural and engineer-ing	A/S	auxiliary stage
		AA	accelerated assemblies
A&L	approach and landing	AA	accelerometer assembly
A&PS	Administration and Program Support (MSFC Directorate)	AA	air to air
		AA	American Airlines
		AA	antiaircraft
A&R	assembly and recycle	AA	associate administrator
A&RC	application and resource control	AA	Associate of Arts degree
		AA/AL	airplane avionics/autoland
A&T	assembly and test	AA/AL	automatic approach/autoland
A/A	air to air		
A/A	airplane avionics	AA/SF	Associate Administrator for Space Flight
A/A	analog to analog		
A/A	angular accelerometer	AA/TDA	AA for tracking and data acquisition
A/B	airborne		
A/C	air conditioning	AADR	applications approval design review
A/C	aircraft		
A/C	associate contractor	AADS	ascent air data system

AAE	abort advisory equipment	ACA	Associate Contractor Administration
AAE	aerospace ancillary equipment	ACA	attitude controller assembly
AAE	aerospace auxiliary equipment	ACB	air-cushion barge
AAE	American Association of Engineers	ACC	automatic control console
AAFE	advanced applications flight equipment	ACCE	augmented continuous control evaluator
AAIR	advanced atmospheric sounder and imaging radiometer	ACCEL	acceleration
		ACCEL	accelerometer
AAO	Astronaut Activities Office (JSC)	ACCN	audit central control network
AAS	abort advisory system	ACCU	audio central control unit
AAS	advanced antenna system	ACD	accuracy control document
AASM	Association of American Steel Manufacturers	ACE	acceptance checkout equipment
AB	adapter booster	ACE	attitude control electronics
AB	afterburner	ACE	automatic checkout equipment
AB	air bearing	ACEL	Air Crew Equipment Laboratory (U.S. Navy)
AB	airborne	ACES	acceptance checkout and evaluation system
AB	anchor bolt	ACES	acceptance control equipment section
ABC	automatic brightness control	ACES	automatic checkout equipment sequencer
ABCF	as-built configuration file	ACF	American Car and Foundry
ABCL	as-built configuration list	ACI	age-controlled item
ABCR	as-built configuration record	ACI	allocated configuration identification
ABD	Airborne Ballistics Division (NASA)	ACIL	automatic controlled instrument landing
ABE	air-breathing engine	ACIP	aerodynamic coefficient identification package
ABES	ABE system	ACL	allowable container load
ABETS	airborne beacon test set	ACL	ascent closed loop
ABM	advanced bill of materials	ACLC	adaptive communication live controller
ABMA	Army Ballistic Missile Agency	ACM	acquisition control module
ABPS	air-breathing propulsion system	ACM	Association of Computing Machinery
ABS	air-break switch	ACME	attitude control and maneuvering electronics
ABT	abort	ACN	acceptance change notice
ABT	air-bearing table	ACN	Ascension Island (STDN)
AC	aerodynamic center	ACO	acceptance checkout
AC	aircraft		
AC	audio center		
AC	auxiliary computer		

3. EDITORIAL STANDARDS

To promote clarity and to ensure consistency of style within and among the numerous types of . . . documents, standard editorial practices have been established. At times, it may be necessary to modify these practices or to establish a new approach; but any such departure should always be governed by sound editorial judgment, and the objective should always be clear communication.

The writer should be able to resolve many questions pertaining to compound words, abbreviations, preferred spellings, etc., by observing the rules given in this section. For words not covered by this style guide, consult the *United States Government Printing Office Style Manual* or *Webster's Third New International Dictionary of the English Language, Unabridged.*

USE OF ABBREVIATIONS, ACRONYMS, AND CONVENTIONAL SYMBOLS

Abbreviated forms of words are sometimes used to conserve space or to avoid cumbersome repetition. They are used extensively in tables, illustrations, and briefing visuals but sparingly in text. Whenever an abbreviation is used, its meaning should be clear to the reader. The following general rules apply to the use of abbreviated forms.

Abbreviations

1. Most abbreviations should be defined the first time they appear in a publication. The usual method is to place the abbreviation in parentheses immediately following the word or phrase it represents: modulated continuous wave (mcw).
2. Certain abbreviations may be used without definition: those that are in common use outside the technical fields (TNT and ID, for example) and those that are widely employed in technical work (ac, dc, rms, lb/in.^{2}, and Hz, for example). Always leave a space between the number and the abbreviated unit of measure unless it is temperature:

   28 V, 14 Hz, 40 W, but 75°F, 6°C

3. When the need for a large number of abbreviations makes it impractical to define each term in text, a separate nomenclature should be included.
4. When an abbreviation has two or more acceptable forms (e.g., ft/s or fps for feet per second), one form should be selected and used consistently.
5. The singular form is used for most abbreviations (e.g., "in." represents both "inch" and "inches"; "lb" represents both "pound" and "pounds"). When it is necessary to pluralize an abbreviation, add an apostrophe and a lowercase "s" (e.g., CRT's for cathode ray tubes).
6. Periods are omitted in abbreviations unless the form spells a common word, as "in." for "inch" or "press." for "pressure," or unless the abbreviation is normally punctuated in nontechnical English, as "Co.," "vs.," "e.g.," "i.e.," and "etc."
7. The words "figure," "reference," and "equation" (as in "Figure 1," "Reference 2," and "Equation 3") are not abbreviated in . . . documents. The word "page" is not abbreviated in text. It may be abbreviated in bibliographical entries, and a period is used, i.e., p. (page), pp. (pages).
8. The names of months and states are not abbreviated in text. To conserve space, they may be abbreviated in tables, illustrations, footnotes, bibliographical entries, and briefing charts. "May," "June," and "July" are spelled out. The preferred abbreviation for the other months uses their first three letters, except "September," which uses the first four letters. However, if abbreviations of all 12 months appear together (e.g., on a time line or bar chart), use only their first 3 letters.
9. Briefing visuals generally employ more abbreviated forms than reports to maintain a simple, uncluttered appearance.

Acronyms

Many abbreviations are composed of the initial letters of a series of words. Some of these abbreviations form words and are referred to as "acronyms." In this style guide, the term "acronym" is applied only to abbreviated forms that are normally pronounced as words rather than as a series of letters (e.g., "NASA" and "SAC" are considered acronyms; "DOD"—usually pronounced D...O...D—is considered an abbreviation). The following general rules apply to the use of acronyms in . . . documents.

1. Most acronyms should be defined the first time they appear in a document. The usual method is to place the acronym in parentheses immediately following the word or phrase it represents: distant early warning (DEW).
2. Acronyms, such as radar and laser, that have passed into common usage need not be defined.

Conventional Symbols

Conventional symbols such as "%" for "percent" are sometimes used in the tables and illustrations of documents to conserve space. Such signs should not be used in text. The following rules apply to some of the more common symbols.

1. The percent sign (%) should not be used in text; it may be used in tables, illustrations, and briefing charts.
2. The use of the inch (") and foot (') signs should be avoided.
3. The sign "#" is not acceptable for either "pound" or "number."
4. The slash sign for "per" is used only with abbreviated units of measure: 10 ft/s.
5. An "x" for "by" is used only in abbreviated units of measure: 8 x 10 ft.
6. The word "to" or "and," rather than a hyphen, is used in text to indicate a spread of values expressed by numerals: e.g., between 15 and 20 pounds and from 10 to 15 feet. Hyphens may be used in tabular material and references to conserve space: 10-15 feet, pp. 26-30.
7. Use of the dollar sign ($) is acceptable in text and preferred in tables, illustrations, and briefings.
8. Avoid the use of Greek symbols in text; they present layout problems. For instance, substitute "three sigma" for "3σ" in text; but use the Greek in tables, illustrations, and briefings.

NUMBER TERMS

Arabic Numbers

Normally, an Arabic number should not be used at the beginning of a sentence. But when a number is very large, as in "2,138 flights were conducted," or when it would be awkward to rewrite the sentence, it is acceptable to begin the sentence with a numeral. The rules that follow define usage for Arabic numerals in the text of . . . documents. Numerals are always used in tables, illustrations, and briefing visuals.

1. Use numerals in text for explicit numbers greater than ten; numbers from one through ten should be spelled out.
2. Where several numbers, some greater and some less than ten, appear in a sentence, use numerals exclusively: 11 capacitors, 8 diodes, and 27 feet of wire. However, a numeral used with a unit of measurement does not affect other numbers in the sentence (see Rule 4): "Each of the five men earned $200 a week."
3. Spell out numbers in succession where confusion may occur: "On August 12, eleven transformers burned out."
4. Use numerals for numbers associated with units of measure: 100 pounds, 640 knots, and 3 feet. A numeral used with a unit of measure does not

affect other numbers in the sentence: "The smallest of the four cylinders was 3 feet in diameter."

5. Spell out numbers of less than 100 that precede a compound modifier containing a numeral: twelve 6-foot cables, ninety 3/4-inch boards, 120 8-inch boards.
6. Place positive and negative tolerances of different values in parentheses and separate them with a comma: 28 (+2, –0) volts, 10.0 (+0.5 –0) inches. Do not leave a space between the tolerance sign and the number.
7. Do not follow a spelled-out number by a numeral in parentheses: 20 missiles, not twenty (20) missiles.
8. Avoid the use of "K" and "M" to express thousands and millions (e.g., 10K for 10,000) in text. When they must be used to conserve space (in a proposal, illustration, briefing, etc.), do not leave a space between the number and the letter.
9. Hyphenate military and space government contract numbers and . . . specification numbers in the following manner:

NASA	NAS9-14000
Air Force	AF 33(907)-5432
	F04701-74-C-0527
Army	DA-04-495
Navy	NOw 65-0375D
	NONr 4669(00)
NASO	MJ072-0001

10. Write dates in traditional civilian style (July 12, 1980) in documents prepared for nonmilitary recipients. Use military style (12 July 1980) in documents issued to military recipients.

Fractions

The use of fractions in . . . publications should conform to the following rules.

1. Express fractions in Arabic numbers when used with units of measure: 1/2 foot, 2/3 mile.
2. Use either a singular or a plural verb with a fraction depending on whether the fraction is regarded as a unit or as a certain number of a group:
   Three fourths of the surface of the earth is sea.
   Three fourths of the people are illiterate.
3. Use one hyphen in a fraction, but not two: one-twentieth, one twenty-fourth. (Note: As in the preceding examples, a hyphen is not used if the fraction is followed by a preposition.)
4. When a fraction is used with a number greater than one, separate the fraction and the whole number with a hyphen: 8-1/2, 10-1/4.
5. Use the singular form of the noun for a unit of measurement when the total quantity is less than one: 3/4 gram, 1-3/4 grams; 1/2 inch, 1-1/2 inches.

Decimals

The use of decimals in . . . publications should conform to the following rules.

1. Place a zero before the decimal point in a number less than one to prevent misreading: 0.58, 0.34.

2. Do not place a decimal point or a decimal point plus zeros after a whole number (e.g., 50. or 50.00) unless the decimal point by itself has significance or unless the decimal point and zeros are used to indicate that the whole number is an exact measurement.
3. Use the singular form of the noun for a unit of measurement when the total quantity is less than one: 0.75 gram, 1.75 grams; 0.05 inch, 1.05 inches.

TABLES AND LISTINGS

A table is a concise and orderly form for the presentation of data. The appearance of tables should be as simple as possible to facilitate analysis of the data. Tabulations in text take the following forms:

1. Simple listings without headings (one or two columns)
2. Listings with columnar headings
3. Ruled listings with columnar headings

Tabular listings without titles are placed immediately after the paragraphs or sentences introducing them and may be open-ruled. Tables with titles are numbered, usually placed on the same page with their text reference or on the page immediately following, and box-ruled.

The following general rules should be observed in preparing tables for final typing. Their application is illustrated in Figure 10.

1. Center the table number and title over the table. The word "Table," the table number, and a period precede the title:

   Table 1.   Comparison of Type Weight and Type Size

2. Use column headings to avoid repetition. A heading is assumed to refer to all entries in the column below it. Initial letters of major words in the headings are capitalized. Units of measure should be lowercase and in parentheses.

Table 1.   Data on Helium-Storage Spheres

Material	Wall Thickness (in.)	Pressure Test (psig)	Yield Strength (lb/in.^{2})	Ultimate Tensile Strength (lb/in.^{2})
6Al-4V titanium	0.185	4,000	150,000	160,000
6Al-4V titanium	0.213	4,000	150,000	160,000
410 stainless steel	0.094	1,400	147,000	170,000
6Al-4V titanium	0.214	4,850	150,000	160,000

3. Use standard abbreviations and symbols to conserve space.
4. Leave a space when an entry in a column is omitted. But a dash or hyphen may be used when the omission occurs in a table that has a high density of data.
5. Divide a table into major parts, when necessary, by using uppercase, centered subtitles (Figure 10). If column headings apply to all subdivisions of the table, they appear above the initial subtitle and are not repeated. If column headings differ for the various subdivisions, they appear below each subtitle.

6. Use footnotes when it is necessary to clarify data entries, column headings, or table titles.
7. Capitalize only the first word in column entries.

Table 3.   Results of Helium Leakage Tests

Fitting Size	Fitting Type	Number Tested	Fitting Leaks at Max Torque or Turns		Amount of Leakage at Max Torque or Turns (scim)	
			Number	Percent	Range	Average
ROOM TEMPERATURE (SPECIMEN IMMERSED IN WATER)						
−4	Flared	48	0	0		
	MS flareless	48	1	2	Trace	Trace
	Super flareless	96	1	1	1.9	1.9
−6	Flared	48	2	4	Trace	Trace
	MS flareless	34	9	26	Trace	Trace
	Super flareless	96	1	1	0.8	0.8
−8	Flared	168	67	40	Trace to 19.6	0.4
	MS flareless	168	14	8	Trace to 19.8	1.7
	Super flareless	120	1	1	0.8	0.8
−12	Flared	114	97	85	Trace to 60	15.6
	Super flareless	120	0			
−16	Flared	114	84	74	Trace to 101	13.8
	Super flareless	112	5	5	0.8 to 2.8	1.0
−320°F (SPECIMEN IMMERSED IN LIQUID NITROGEN)*						
−4	Flared	60	2	3		
	MS flareless	66	1	1		
	Super flareless	96	6	6		
−6	Flared	24	0			
	MS flareless	28	0			
	Super flareless	96	4	4		
−8	Flared	120	0			
	MS flareless	168	0			
	Super flareless	120	5	4		
−12	Flared	96	88	92		
	Super flareless	120	0			
−16	Flared	114	80	70		
	MS flareless	24	0			
	Super flareless	112	15	13		

*Leakage was detected visually but could not be measured.

Figure 10.   Sample Table Arrangement

APPENDIX B.   GREEK ALPHABET

Greek Letter	Standard		Handwritten	
	Cap	LC	Cap	LC
Alpha	A	a	A	α
Beta	B	β	B	β
Gamma	Γ	γ	Γ	γ
Delta	Δ	δ	Δ	δ
Epsilon	E	ϵ	E	ϵ, ε
Zeta	Z	ζ	Z	ζ
Eta	H	η	H	η
Theta	Θ	θ	Θ	θ, ϑ
Iota	I	ι	I	ι
Kappa	K	κ	K	κ
Lambda	Λ	λ	Λ	λ
Mu	M	μ	M	μ
Nu	N	ν	N	ν
Xi	Ξ	ξ	Ξ	ξ
Omicron	O	o	O	o
Pi	Π	π	Π	π
Rho	P	ρ	P	ρ
Sigma	Σ	σ	Σ	σ
Tau	T	τ	T	τ
Upsilon	Υ	υ	Υ	ν
Phi	Φ	ϕ	Φ	ϕ, φ
Chi	X	χ	X	χ
Psi	Ψ	ψ	Ψ	ψ
Omega	Ω	ω	Ω	ω

PREFIXES TO SI UNITS

Multiples or Submultiples	Prefix	Symbol
$1\ 000\ 000\ 000\ 000 = 10^{12}$	tera	T
$1\ 000\ 000\ 000 = 10^{9}$	giga	G
$1\ 000\ 000 = 10^{6}$	mega	M
$1\ 000 = 10^{3}$	kilo	k
$100 = 10^{2}$	hecto	h
$10 = 10$	deka	da
$0.1 = 10^{-1}$	deci	d
$0.01 = 10^{-2}$	centi	c
$0.001 = 10^{-3}$	milli	m
$0.000\ 001 = 10^{-6}$	micro	μ
$0.000\ 000\ 001 = 10^{-9}$	nano	n
$0.000\ 000\ 000\ 001 = 10^{-12}$	pico	p
$0.000\ 000\ 000\ 000\ 001 = 10^{-15}$	femto	f
$0.000\ 000\ 000\ 000\ 000\ 001 = 10^{-18}$	atto	a

APPENDIX D.   COMMON METRIC EQUIVALENTS
AND CONVERSIONS

Approximate Equivalents		*Conversions Accurate to Parts per Million*	
1 inch	= 25 millimeters	inches $\times$ 25.4*	= millimeters
1 foot	= 0.3 meter	feet $\times$ 0.3048*	= meters
1 yard	= 0.9 meter	yards $\times$ 0.944*	= meters
1 mile	= 1.6 kilometers	miles $\times$ 1.609	= kilometers
1 nautical mile	= 1.852 kilometers		
1 inch^2	= 6.5 centimeters^2	inches^2 $\times$ 6.4516*	= centimeters^2
1 foot^2	= 0.09 meter^2	feet^2 $\times$ 0.0929030	= meters^2
1 yard^2	= 0.8 meter^2	yards^2 $\times$ 0.836127	= meters^2
1 acre	= 0.4 hectare†	acres $-$ 0.404686	= hectares
1 inch^3	= 16 centimeters^3	inches^3 $\times$ 16.3871	= centimeters^3
1 foot^3	= 0.03 meter^3	feet^3 $\times$ 0.0283168	= meters^3
1 yard^3	= 0.8 meter^3	yards^3 $\times$ 0.764555	= meters^3
1 quart (liq)	= 1 liter†	quarts (liq) $\times$ 0.946353	= liters
1 gallon	= 0.004 meter^3	gallons $\times$ 0.00378541	= meters^3
1 ounce	= 28 grams	ounces $-$ 28.3495	= grams
1 pound	= 0.45 kilogram	pounds $\times$ 0.45392	= kilograms
1 horsepower	= 0.75 kilowatt	horsepower $\times$ 0.745700	= kilowatts
1 millimeter	= 0.04 inch	millimeters $\times$ 0.0393701	= inches
1 meter	= 3.3 feet	meters $\times$ 3.28084	= feet
1 meter	= 1.1 yards	meters $\times$ 1.09361	= yards
1 kilometer	= 0.6 mile	kilometers $\times$ 0.621371	= miles
1 centimeter^2	= 0.16 inch^2	centimeters^2 $\times$ 0.155000	= inches^2
1 meter^2	= 11 feet^2	meters^2 $\times$ 10.7639	= feet^2
1 meter^2	= 1.2 yards^2	meters^2 $\times$ 1.19599	= yards^2
1 hectare†	= 2.5 acres	hectares $\times$ 2.47105	= acres
1 centimeter^3	= 0.06 inch^3	centimeters^3 $\times$ 0.0610237	= inches^3
1 meter^3	= 35 feet^3	meters^3 $\times$ 35.3147	= feet^3
1 meter^3	= 1.3 yards^3	meters^3 $\times$ 1.30795	= yards^3
1 liter†	= 1 quart (liq)	liters $\times$ 1.05669	= quarts (liq)
1 meter^3	= 250 gallons	meters^3 $\times$ 264.172	= gallons
1 gram	= 0.035 ounce	grams $\times$ 0.0352470	= ounces
1 kilogram	= 2.2 pounds	kilograms $\times$ 2.20462	= pounds
1 kilowatt	= 1.3 horsepower	kilowatts $\times$ 1.34102	= horsepower

*exact

†common term not used in SI

Section 7

Source: *Technosphere*, Fansteel Inc., Number One Tantalum Place, North Chicago, Ill. 60064—from information published by the Subcommittee on Energy, U.S. House of Representatives.

Selection:

Measuring Energy

The following tables of equivalents contain those figures commonly used to compare different types of energy sources and their various measurements.

Btu—a British thermal unit—the amount of heat required to raise one pound of water one degree Fahrenheit; equivalent to 1055 joules or about 252 gram calories. A **therm** is usually 100,000 Btu but is sometimes used to refer to other units.

Calorie—the amount of heat required to raise one gram of water one degree centigrade: abbreviated cal.: equivalent to about .003968 Btu. More common is the kilogram calorie, also called a **kilocalorie** and abbreviated Cal. or Kcal: equivalent to about 3.97 Btu. (One Kcal is equivalent to one food calorie.)

Btu Values of Energy Sources

(These are conventional or average values, not precise equivalents.)

Coal (per 2,000 lb. ton):

Anthracite	$= 25.4 \times 10^6$ Btu
Bituminous	$= 26.2 \times 10^6$
Sub-bituminous	$= 19.0 \times 10^6$
Lignite	$= 13.4 \times 10^6$

Average heating value of coal used to generate electricity in 1969 was 27.7×10^6 Btu.

Natural Gas (per cubic foot):

Dry	$= 1.031$ Btu
Wet	$= 1.103$
Liquid (avg.)	$= 4.100$

Electricity—1 kwh = 3.413 Btu

Petroleum (per barrel):

Crude oil	$= 5.60 \times 10^6$ Btu
Residual fuel oil	$= 6.29 \times 10^6$
Distillate fuel oil	$= 5.83 \times 10^6$
Gasoline (including av gas)	$= 5.25 \times 10^6$
Jet fuel (kerosene)	$= 5.67 \times 10^6$
Jet fuel (naphtha)	$= 5.36 \times 10^6$
Kerosene	$= 5.67 \times 10^6$

Nuclear
 1 gram of fissioned U-235—74,000,000 Btu

The Btu and cal., being small amounts of energy, are usually expressed as follows when large numbers are involved.

1×10^3 Btu = 1,000
1×10^6 Btu = 1,000,000
1×10^9 Btu = 1,000,000,000
1×10^{12} Btu = 1 trillion
1×10^{15} Btu = 1 quadrillion
1×10^{18} Btu = 1 quintillion or 1 Q unit
One Q unit = 38.46 billion tons of coal
         = 172.4 billion barrels of oil
         = 968.9 trillion cubic ft. of natural gas

Other Conversion Factors

Electricity—1 kwh = 0.88 lbs. of coal
 = 0.076 gallons of oil
 = 10.4 cu. ft. of natural gas

Natural Gas—1 tcf = 39.3×10^6 tons of coal
 (trillion cubic feet) = 184×10^6 barrels of oil

Coal—1 mtce = 4.48×10^6 barrels of oil
 (million tons of coal equivalent) = 67 tons of oil
 = 25.19×10^{12} cu. ft. of natural gas

Oil—1 million tons = 4×10^9 kwh of electricity
 (6.65×10^6 barrels)   (when used to generate power)
 = 12×10^9 kwh unconverted
 = 1.5×10^6 tons of coal
 = 41.2×10^9 cu. ft. of natural gas

Approximate Conversion Factors for Oils

To convert	Barrels to Metric tons	Metric tons to barrels Multiply by—	Barrels/days to tons/year	Tons/year to barrels/day
Crude oil[1]	0.136	7.33	49.8	0.0201
Gasoline	.118	8.45	43.2	.0232
Kerosene	.128	7.80	46.8	.0214
Diesel fuel	.133	7.50	48.7	.0205
Fuel oil	.149	6.70	54.5	.0184

[1]Based on world average gravity (excluding natural gas liquids).

Section 8

Source: Table, Teledyne Vasco, Latrobe, Pa. 15650.

Selection—Decimal Equivalents, Metric Conversion

FRACTION	DECIMAL	MM	FRACTION	DECIMAL	MM
1/64	.015625	.3969	33/64	.515625	13.0969
1/32	.03125	.7938	17/32	.53125	13.4938
3/64	.046875	1.1906	35/64	.546875	13.8906
1/16	.0625	1.5875	9/16	.5625	14.2875
5/64	.078125	1.9844	37/64	.578125	14.6844
3/32	.09375	2.3812	19/32	.59375	15.0812
7/64	.109375	2.7781	39/64	.609375	15.4781
1/8	.125	3.175	5/8	.625	15.875
9/64	.140625	3.5719	41/64	.640625	16.2719
5/32	.15625	3.9688	21/32	.65625	16.6688
11/64	.171875	4.3656	43/64	.671875	17.0656
3/16	.1875	4.7625	11/16	.6875	17.4625
13/64	.203125	5.1594	45/64	.703125	17.8594
7/32	.21875	5.5562	23/32	.71875	18.2562
15/64	.234375	5.9531	47/64	.734375	18.6531
1/4	.25	6.35	3/4	.75	19.05
17/64	.265625	6.7469	49/64	.765625	19.4469
9/32	.28125	7.1438	25/32	.78125	19.8438
19/64	.296875	7.5406	51/64	.796875	20.2406
5/16	.3125	7.9375	13/16	.8125	20.6375
21/64	.328125	8.3344	53/64	.828125	21.0344
11/32	.34375	8.7312	27/32	.84375	21.4312
23/64	.359375	9.1281	55/64	.859375	21.8281
3/8	.375	9.525	7/8	.875	22.225
25/64	.390625	9.9219	57/64	.890625	22.6219
13/32	.40625	10.3188	29/32	.90625	23.0188
27/64	.421875	10.7156	59/64	.921875	23.4156
7/16	.4375	11.1125	15/16	.9375	23.8125
29/64	.453125	11.5094	61/64	.953125	24.2094
15/32	.46875	11.9062	31/32	.96875	24.6062
31/64	.484375	12.3031	63/64	.984375	25.0031
1/2	.5	12.7	1	1.0	25.4

MM	INCHES	MM	INCHES	MM	INCHES	MM	INCHES	MM	INCHES
1	.0394	21	.8268	41	1.6142	61	2.4016	81	3.1890
2	.0787	22	.8661	42	1.6535	62	2.4410	82	3.2284
3	.1181	23	.9055	43	1.6929	63	2.4803	83	3.2677
4	.1575	24	.9449	44	1.7323	64	2.5197	84	3.3071
5	.1969	25	.9843	45	1.7717	65	2.5591	85	3.3465
6	.2362	26	1.0236	46	1.8110	66	2.5984	86	3.3858
7	.2756	27	1.0630	47	1.8504	67	2.6378	87	3.4252
8	.3150	28	1.1024	48	1.8898	68	2.6772	88	3.4646
9	.3543	29	1.1417	49	1.9291	69	2.7165	89	3.5039
10	.3937	30	1.1811	50	1.9685	70	2.7559	90	3.5433
11	.4331	31	1.2205	51	2.0079	71	2.7953	91	3.5827
12	.4724	32	1.2598	52	2.0472	72	2.8347	92	3.6221
13	.5118	33	1.2992	53	2.0866	73	2.8740	93	3.6614
14	.5512	34	1.3386	54	2.1260	74	2.9134	94	3.7008
15	.5906	35	1.3780	55	2.1654	75	2.9528	95	3.7402
16	.6299	36	1.4173	56	2.2047	76	2.9921	96	3.7795
17	.6693	37	1.4567	57	2.2441	77	3.0315	97	3.8189
18	.7087	38	1.4961	58	2.2835	78	3.0709	98	3.8583
19	.7480	39	1.5354	59	2.3228	79	3.1102	99	3.8976
20	.7872	40	1.5748	60	2.3622	80	3.1496	100	3.9370

Section 9

Source: *Guide to Authors for Preparation and Presentation of Technical Papers*, American Foundrymen's Society, Golf and Wolf Rds., Des Plaines, Ill. 60016.

Selections—in order of appearance:
Visual Aids
Supplementary Reading

Visual Aids

Well selected and properly prepared projection slides are a valuable aid in presentation of the paper. It should be remembered, however, that they are to be viewed on the screen for only a short time and are to be accompanied by an oral explanation by the speaker. It is therefore neither necessary nor desirable that the slides be complete within themselves. The audience loses interest in the presentation if they are required to digest detailed slides or more information than can be assimilated in the brief showing time allowed.

Projection slides prepared from typewritten copy rarely are readable when projected on the screen. If a typewritten copy must be used, it is recommended that an electric typewriter (carbon ribbon) be used. (Pica type capital letters are the smallest size acceptable.) To be legible, letters and symbols must have a certain minimum size on the screen. This size depends on the distance from the screen to the most distant viewer in the audience. For typical viewing, the maximum viewing distance is approximately *6 times the width of the screen image*. As a guide for typical 6W viewing, if a printed table is 4-1/2 in. wide it should be viewed from a distance of 27 in. If it is readable, it is suitable for copying and projection. The same applies for a wall chart or graph 5 ft wide. If it is legible at a distance of 30 ft, it is acceptable as a projected image for 6W viewing.

For clarity, lettering on slides should be a minimum of 1/8 in. The minimum line size should be the width of a No. 00 pen. Other lettering aids are available, e.g. instant transfer letters, gummed paper die-cut letters, lettering guides and stencil sets. The use of color in art, as well as background, improves the communication impact on the audience, but cannot be used for publishing.

A template 8-1/2×5-3/4 in. may be used as the typical guide for 35-mm (2 × 2-in.) slides. When designing projection slides for the presentation of your paper, observing the following professional advice will assure good results:

- Use one idea per slide.
- Do not use complex graphs and mathematical tables.

Figures 3 and 4 may be used as guides, and it is suggested that the sources mentioned in "Supplementary Reading" be consulted.

IMPORTANT—All projection slides must be 35 mm (2 × 2-in.) in size.

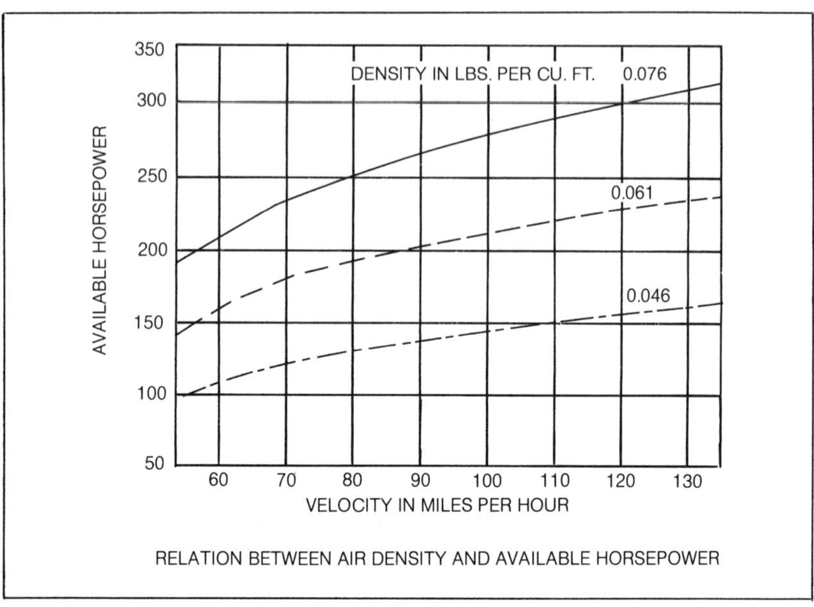

Fig. 3. Chart properly prepared for standard slide.

COMPARISON OF FILTER PERFORMANCE CHARACTERISTICS FOR OVERLOAD TESTS

FILTER	PER CENT	DUST MIXTURE	FILTER PERFORMANCE BASED ON 0.4 IN. H_2O FINAL RESISTANCE		
			EFFICIENCY, PER CENT	LIFE, HOURS	DUST HOLDING CAP, GRAMS
B-1	80 20	POCAHONTAS ASH CARBON DUST	86.9	7.0	243.0
B-1	80 20	POCAHONTAS ASH LAMPBLACK	83.8	42	140.7
C	80 20	POCAHONTAS ASH CARBON DUST	47.4	21.5	408.0
D	50 20 20 10	POCAHONTAS ASH ILLINOIS FLY ASH LAMPBLACK FULLERS EARTH	94.7	7.2	272.8
D	80 20	POCAHONTAS ASH CARBON DUST	96.0	8.6	330.4

Fig. 4. Table suitable for projection.

Supplementary Reading

Designing Good Slides, Public Health Service, Publication No. 2196. Write to Supt. of Documents, U.S. Government Printing Office, Washington, D.C. 20402 (Stock No. 1723-0050).

Producing Slides and Filmstrips (publication S-8), Eastman Kodak Company, Rochester, N.Y. 14650.

Artwork Size Standards for Projected Visuals, Eastman Kodak Company, Rochester, N.Y. 14650.

Section 10

Source: *Instructions and Suggestions for Authors*, American Welding Society,
          Inc., 2501 W. 7th St., Miami, Fla. 33125.
Selections—in order of appearance:

Technical Terms, Abbreviations, and Symbols
Welding Terms and Definitions
Metric Practice

Technical Terms, Abbreviations and Symbols

Proper use of terms, abbreviations and symbols which are part of the language of technical papers is an important consideration in processing a manuscript for publication.

Because welding cuts across many different scientific disciplines, various standards of technical usage can be applied. However, in the interest of simplicity, the *Welding Journal* adheres to the standards of the American Welding Society. For usage not covered by the above standards, authors are encouraged to apply the recognized standards of the discipline involved.

AWS A3.0-76, *Welding Terms and Definitions*, includes approved process abbreviations and should be used where applicable. Papers from authors outside the U.S. will be edited with care to preserve original meanings. A list of the more commonly used terms showing preferred and nonpreferred forms is given below.

AWS Preferred	*Nonpreferred*
arc welding	electric arc welding; electric welding
backing	back up (except in flash or upset welding)
base metal (material)	parent metal
braze welding	bronze welding
brazing filler metal	brazing alloy, silver solder
covered electrode	coated electrode, stick electrode
diffusion brazing, diffusion welding	diffusion bonding
electrode	wire
filler metal	filler alloy
gas metal arc welding (GMAW)	metal inert gas (MIG); CO_2 welding
gas tungsten arc welding (GTAW)	tungsten inert gas (TIG)
groove radius	root radius
incomplete fusion	lack of fusion
joint buildup sequence	joint welding procedure
oxygen cutter*	burner,* gas cutter*
oxygen cutting	flame cutting; burning
porosity	blowhole; gas pocket
residual stress	locked-up stress; shrinkage stress
root face	land; nose
root opening	root gap; gap
shielded metal-arc welding	manual arc welding; stick electrode welding
solder	soft solder

surfacing	buildup; overlaying
welder,* welding operator*	weldor; operator
welding machine	welder
welding torch or cutting torch	blowpipe (except in brazing and soldering); gas torch
work connection	welding ground

*Refers to the individual, not to equipment or machines

Metric Practice

The policy of the American Welding Society is to facilitate the transition from U.S. customary units of measurement to the International System of Units (SI). Because SI units are based on the metric system and most U.S. customary units are not, a gradual transition is underway. To encompass both ends of the transition period, it is suggested that authors present measurements first in the units by which the measurements were actually made and then accompany them in the following manner:

1. U.S. customary (nonmetric) units should be accompanied, in parentheses, by SI conversions.
2. SI units should be accompanied, in parentheses, by U.S. customary units. This request is made in the interest of the readership of the *Welding Journal*, and is especially for readers who are learning to use the SI units.
3. Where a large number of measurements and their conversions will appear together in a text, the author should omit the conversions and, instead, present a table of conversions for each of the units used. The point here is to avoid impairing the readability of the article (and thus the *Welding Journal*).

These suggestions apply also to drawings, graphs and tables.

Section 11___________________________

Source: *Standard for Metric Practice*, E 380-79, American Society for Testing and Materials, 1916 Race St., Philadelphia, Pa. 19103.

Selections—in order of appearance:

    Metric Units
    Multiples and Submultiples
    Metric Conversions—Alphabetical List of Units

Metric Units

A short list of the International System of Units is given next; metric decimal prefixes follow. Conversion factors to common engineering units and a more comprehensive listing can be found in the *ASTM Metric Practice Guide* (E 380-72). A copy of the Guide is available on request from ASTM.

Quantity	*Unit*	*Symbol*
	Elemental Units	
Length	metre	m
Mass	kilogram	kg
Time	second	s
Electric current	ampere	A
Temperature	degree Kelvin	deg K
Luminous intensity	candela	cd
	Supplementary Units	
Plane angle	radian	rad
Solid angle	steradian	sr
	Derived Units	
Area	square metre	m^2
Volume	cubic metre	m^3
Frequency	hertz	$Hz\ (s^{-1})$
Density	kilogram per cubic metre	kg/m^3
Velocity	metre per second	m/s
Angular velocity	radian per second	rad/s
Acceleration	metre per second squared	m/s^2
Angular acceleration	radian per second squared	rad/s^2
Force	newton	$N\ (kg \cdot m/s^2)$
Pressure	newton per sq metre	N/m^2
Kinematic viscosity	sq metre per second	m^2/s
Dynamic viscosity	newton-second per sq metre	$N \cdot s/m^2$
Work, energy, quantity of heat	joule	$J\ (N \cdot m)$
Power	watt	$W\ (J/s)$

Derived Units (continued)

Electric charge	coulomb	C (A · s)
Voltage, potential difference electromotive force	volt	V (W/A)
Electric field strength	volt per metre	V/m
Electric resistance	ohm	Ω (V/A)
Electric capacitance	farad	F (A · s/V)
Magnetic flux	weber	Wb (V · s)
Inductance	henry	H (V · s/A)
Magnetic flux density	tesla	T (Wb/m^2)
Magnetic field strength	ampere per metre	A/m
Magnetomotive force	ampere	A
Luminous flux	lumen	lm (cd · sr)
Luminance	candela per sq metre	cd/m^2
Illumination	lux	lx (lm/m^2)

Multiples and Submultiples

The International Committee on Weights and Measures has adopted the following prefixes for denoting multiples and submultiples:

Multiple	*Prefix*	*Abbreviation*
10^{12}	tera	T
10^9	giga	G
10^6	mega	M
10^3	kilo	k
10^2	hecto	h
10	deka	da
10^{-1}	deci	d
10^{-2}	centi	c
10^{-3}	milli	m
10^{-6}	micro	μ
10^{-9}	nano	n
10^{-12}	pico	p
10^{-15}	femto	f
10^{-18}	atto	a

Metric Conversions—Alphabetical List of Units

Factors with an asterisk (*) are exact

(Symbols of SI units given in parentheses)

To convert from	to	Multiply by
abampere	ampere (A)	1.000 000*E+01
abcoulomb	coulomb (C)	1.000 000*E+01
abfarad	farad (F)	1.000 000*E+09
abhenry	henry (H)	1.000 000*E−09
abmho	siemens (S)	1.000 000*E+09
abohm	ohm (Ω)	1.000 000*E−09
abvolt	volt (V)	1.000 000*E−08
acre foot[14]	cubic metre (m^3)	1.233 489 E+03
acre[14]	square metre (m^2)	4.046 873 E+03
ampere hour	coulomb (C)	3.600 000*E+03
angstrom	metre (m)	1.000 000*E−10
are	square metre (m^2)	1.000 000*E+02
astronomical unit	metre (m)	1.495 979 E+11
atmosphere (standard)	pascal (Pa)	1.013 250*E+05
atmosphere (technical = 1 kgf/cm^2)	pascal (Pa)	9.806 650*E+04
bar	pascal (Pa)	1.000 000*E+05
barn	square metre (m^2)	1.000 000*E−28
barrel (for petroleum, 42 gal)	cubic metre (m^3)	1.589 873 E−01
board foot	cubic metre (m^3)	2.359 737 E−03
British thermal unit (International Table)[15]	joule (J)	1.055 056 E+03
British thermal unit (mean)	joule (J)	1.055 87　E+03
British thermal unit (thermochemical)	joule (J)	1.054 350 E+03
British thermal unit (39° F)	joule (J)	1.059 67　E+03
British thermal unit (59° F)	joule (J)	1.054 80　E+03

To convert from	to	Multiply by
British thermal unit (60° F)	joule (J)	1.054 68 E+03
Btu (International Table) · ft/(h · ft² · ° F) (k, thermal conductivity)	watt per metre kelvin [W/(m · K)]	1.730 735 E+00
Btu (thermochemical · ft/(h· ft² · ° F) (k, thermal conductivity)	watt per metre kelvin [W/(m · K)]	1.729 577 E+00
Btu (International Table) · in/(h · ft² · ° F) (k, thermal conductivity)	watt per metre kelvin [W/(m · K)]	1.442 279 E−01
Btu (thermochemical · in/(h · ft² · ° F) (k, thermal conductivity)	watt per metre kelvin [W/(m · K)]	1.441 314 E−01
Btu (International Table) · in/(s · ft² · ° F) (k, thermal conductivity)	watt per metre kelvin [W/(m · K)]	5.192 204 E+02
Btu (thermochemical) · in/(s · ft² · ° F) (k, thermal conductivity)	watt per metre kelvin [W/(m · K)]	5.188 732 E+02
Btu (International Table)/h	watt (W)	2.930 711 E−01
Btu (International Table)/s	watt (W)	1.055 056 E+03
Btu (thermochemical)/h	watt (W)	2.928 751 E−01
Btu (thermochemical)/min	watt (W)	1.757 250 E+01
Btu (thermochemical)/s	watt (W)	1.054 350 E+03
Btu (International Table)/ft²	joule per square metre (J/m²)	1.135 653 E+04
Btu (thermochemical)/ft²	joule per square metre (J/m²)	1.134 893 E+04
Btu (thermochemical)/ft² · h)	watt per square metre (W/m²)	3.152 481 E+00
Btu (thermochemical)/(ft² · min)	watt per square metre (W/m²)	1.891 489 E+02
Btu (thermochemical)/(ft² · s)	watt per square metre (W/m²)	1.134 893 E+04
Btu (thermochemical)/(in² · s)	watt per square metre (W/m²)	1.634 246 E+06

[14] The U.S. Metric Law of 1866 gave the relationship, 1 metre equals 39.37 inches. Since 1893 the U.S. yard has been derived from the metre. In 1959 a refinement was made in the definition of the yard to bring the U.S. yard and the yard used in other countries into agreement. The U.S. yard was changed from 3600/3937 m to 0.9144 m to 0.9144 m exactly. The new length is shorter by exactly two parts in a milion.

At the same time it was decided that any data in feet derived from and published as a result of geodetic surveys within the U.S. would remain with the old standard (1 ft = 1200/3937 m) until further decision. This foot is named the U.S. survey foot.

All conversion factors for units of land measure in these tables referenced to this footnote are based on the U.S. survey foot and the following relationships: 1 fathom = 6 feet; 1 rod (pole or perch) = 16½ feet; 1 chain = 66 feet; 1 mile (U.S. statute) = 5280 feet.

[15] This value was adopted in 1956. Some of the older International Tables use the value 1.055 04 E+03. The exact conversion factor is 1.055 055 852 62* E+03.

To convert from	to	Multiply by
Btu (International Table)/(h · ft^2 · ° F) (C, thermal conductance)[16]	watt per square metre kelvin [W/(m^2 · K)]	5.678 263 E+00
Btu (thermochemical)/(h · ft^2 · ° F) (C, thermal conductance)[16]	watt per square metre kelvin [W/(m^2 · K)]	5.674 466 E+00
Btu (International Table)/(s · ft^2 · ° F)	watt per square metre kelvin [W/(m^2 · K)]	2.044 175 E+04
Btu (thermochemical)/(s · ft^2 · ° F)	watt per square metre kelvin [W/(m^2 · K)]	2.042 808 E+04
Btu (International Table)/lb	joule per kilogram (J/kg)	2.326 000*E+03
Btu (thermochemical)/lb	joule per kilogram (J/kg)	2.324 444 E+03
Btu (International Table)/(lb · ° F) (c, heat capacity)	joule per kilogram kelvin [J/(kg · K)]	4.186 800*E+03
Btu (thermochemical)/(lb. · F) (c, heat capacity	joule per kilogram kelvin [J/(kg · K)]	4.184 000*E+03
Btu (International Table)/ft^3	joule per cubic metre (J/m^3)	3.725 895 E+04
Btu (thermochemical)/ft^3	joule per cubic metre (J/m^3)	3.723 402 E+04
bushel (U.S.)	cubic metre (m^3)	3.523 907 E–02
calorie (International Table)	joule (J)	4.186 800*E+00
calorie (mean)	joule (J)	4.190 02 E+00
calorie (thermochemical)	joule (J)	4.184 000*E+00
calorie (15 ° C)	joule (J)	4.185 80 E+00
calorie (20 ° C)	joule (J)	4.181 90 E+00
calorie (kilogram, International Table)	joule (J)	4.186 800*E+03
calorie (kilogram, mean)	joule (J)	4.190 02 E+03
calorie (kilogram, thermochemical)	joule (J)	4.184 000*E+03
cal (thermochemical)/cm^2	joule per square metre (J/m^2)	4.184 000*E+04
cal (International Table)/g	joule per kilogram (J/kg)	4.186 800*E+03
cal (thermochemical)/g	joule per kilogram (J/kg)	4.184 000*E+03
cal (International Table)/(g · ° C)	joule per kilogram kelvin [J/(kg · K)]	4.186 800*E+03
cal (thermochemical)/(g · ° C)	joule per kilogram kelvin [J/(kg · K)]	4.184 000*E+03
cal (thermochemical)/min.	watt (W)	6.973 333 E–02

cal (thermochemical)/s	watt (W)	4.184 000*E+00
cal (thermochemical)/(cm^2 · min)	watt per square metre (W/m^2)	6.973 333 E+02
cal (thermochemical)/(cm^2 · s)	watt per square metre (W/m^2)	4.184 000*E+04
cal (thermochemical)/(cm · s · ° C)	watt per metre kelvin [W/(m · K)]	4.184 000*E+02
cd/in^2	candela per square metre (cd/m^2)	1.550 003 E+03
carat (metric)	kilogram (kg)	2.000 000*E−04
centimetre of mercury (0° C)	pascal (Pa)	1.333 22 E+03
centimetre of water (4° C)	pascal (Pa)	9.806 38 E+01
centipoise	pascal second (Pa · s)	1.000 000*E−03
centistokes	square metre per second (m^2/s)	1.000 000*E−06
chain[14]	metre (m)	2.011 684 E+01
circular mil	square metre (m^2)	5.067 075 E−10
clo	kelvin square metre per watt (K · m^2/W)	2.003 712 E−01
cup	cubic metre (m^3)	2.365 882 E−04
curie	becquerel (Bq)	3.700 000*E+10
darcy[17]	square metre (m^2)	9.869 233 E−13
day	second (s)	8.640 000*E+04
day (sidereal)	second (s)	8.616 409 E+04
degree (angle)	radian (rad)	1.745 329 E−02
degree Celsius	kelvin (K)	$T_K = t°_C + 273.15$
degree centigrade	[see 3.4.2]	
degree Fahrenheit	degree Celsius	$t°_C = (t°_F − 32)/1.8$
degree Fahrenheit	kelvin (K)	$T_K = (t°_F + 459.67)/1.8$
degree Rankine	kelvin (K)	$T_K = T°_R 1.8$
° F · h · ft^2/Btu (International Table) (R, thermal resistance)[18]	kelvin square metre per watt (K · m^2/W)	1.761 102 E−01
° F · h · ft^2/Btu (thermochemical) (R, thermal resistance)[18]	kelvin square metre per watt (K · m^2/W)	1.762 280 E−01

[16] In ISO 31 this quantity is called *coefficient of heat transfer.*
[17] The darcy is a unit for measuring permeability of porous solids.
[18] In ISO 31 this quantity is called *thermal insulance* and the quantity *thermal resistance* has the unit K/W.

To convert from	to	Multiply by
°F · h · ft^2/(Btu (International Table) · in) (thermal resistivity)	kelvin metre per watt (K · m/W)	6.933 471 E+00
°F · h · ft^2/(Btu (thermochemical) · in) (thermal resistivity)	kelvin metre per watt (K · m/W)	6.938 113 E+00
denier	kilogram per metre (kg/m)	1.111 111 E–07
dyne	newton (N)	1.000 000*E–05
dyne · cm	newton metre (N · m)	1.000 000*E–07
dyne/cm^2	pascal (Pa)	1.000 000*E–01
electronvolt	joule (J)	1.602 19 E–19
EMU of capacitance	farad (F)	1.000 000*E+09
EMU of current	ampere (A)	1.000 000*E+01
EMU of electric potential	volt (V)	1.000 000*E–08
EMU of inductance	henry (H)	1.000 000*E–09
EMU of resistance	ohm (Ω)	1.000 000*E–09
ESU of capacitance	farad (F)	1.112 650 E–12
ESU of current	ampere (A)	3.335 6 E–10
ESU of electric potential	volt (V)	2.997 9 E+02
ESU of inductance	henry (H)	8.987 554 E+11
ESU of resistance	ohm (Ω)	8.987 554 E+11
erg	joule (J)	1.000 000*E–07
erg/(cm^2 · s)	watt per square metre (W/m^2)	1.000 000*E–03
erg/s	watt (W)	1.000 000*E–07
faraday (based on carbon-12)	coulomb (C)	9.648 70 E+04
faraday (chemical)	coulomb (C)	9.649 57 E+04
faraday (physical)	coulomb (C)	9.652 19 E+04
fathom[14]	metre (m)	1.828 804 E+00
fermi (femtometre)	metre (m)	1.000 000*E–15
fluid ounce (U.S.)	cubic metre (m^3)	2.957 353 E–05
foot	metre (m)	3.048 000*E–01
foot (U.S. survey)[14]	metre (m)	3.048 006 E–01
foot of water (39.2° F)	pascal (Pa)	2.988 98 E+03

ft^2	square metre (m^2)	9.290 304*E–02
ft^2/h (thermal diffusivity)	square metre per second (m^2/s)	2.580 640*E–05
ft^2/s	square metre per second (m^2/s)	9.290 304*E–02
ft^3 (volume; section modulus)	cubic metre (m^3)	2.831 685 E–02
ft^3/min	cubic metre per second (m^3/s)	4.719 474 E–04
ft^3/s	cubic metre per second (m^3/s)	2.831 685 E-02
ft^4 (second moment of area)[19]	metre to the fourth power (m^4)	8.630 975 E–03
ft/h	metre per second (m/s)	8.466 667 E–05
ft/min	metre per second (m/s)	5.080 000*E–03
ft/s	metre per second (m/s)	3.048 000*E–01
ft/s^2	metre per second squared (m/s^2)	3.048 000*E–01
footcandle	lux (lx)	1.076 391 E+01
footlambert	candela per square metre (cd/m^2)	3.426 259 E+00
ft · lbf	joule (J)	1.355 818 E+00
ft · lbf/h	watt (W)	3.766 161 E–04
ft · lbf/min	watt (W)	2.259 697 E–02
ft · lbf/s	watt (W)	1.355 818 E+00
ft-poundal	joule (J)	4.214 011 E–02
free fall, standard (*g*)	metre per second squared (m/s^2)	9.806 650*E+00
gal	metre per second squared (m/s^2)	1.000 000*E–02
gallon (Canadian liquid)	cubic metre (m^3)	4.546 090 E–03
gallon (U.K. liquid)	cubic metre (m^3)	4.546 092 E–03
gallon (U.S. dry)	cubic metre (m^3)	4.404 884 E–03
gallon (U.S. liquid)	cubic metre (m^3)	3.785 412 E–03
gallon (U.S. liquid) per day	cubic metre per second (m^3/s)	4.381 264 E–08
gallon (U.S. liquid) per minute	cubic metre per second (m^3/s)	6.309 020 E–05
gallon (U.S. liquid) per hp · h (SFC, spe- cific fuel consumption)	cubic metre per joule (m^3/J)	1.410 089 E–09
gamma	tesla (T)	1.000 000*E–09
gauss	tesla (T)	1.000 000*E–04
gilbert	ampere (A)	7.957 747 E–01
gill (U.K.)	cubic metre (m^3)	1.420 654 E–04

[19] This is sometimes called the moment of section or area moment of inertia of a plane section about a specified axis.

To convert from	to	Multiply by
gill (U.S.)	cubic metre (m^3)	1.182 941 E−04
grad	degree (angular)	9.000 000*E−01
grad	radian (rad)	1.570 796 E−02
grain	kilogram (kg)	6.479 891*E−05
grain/gal (U.S. liquid)	kilogram per cubic metre (kg/m^3)	1.711 806 E−02
gram	kilogram (kg)	1.000 000*E−03
g/cm^3	kilogram per cubic metre (kg/m^3)	1.000 000*E+03
gf/cm^2	pascal (Pa)	9.806 650*E+01
hectare	square metre (m^2)	1.000 000*E+04
horsepower (550 ft · lbf/s)	watt (W)	7.456 999 E+02
horsepower (boiler)	watt (W)	9.809 50 E+03
horsepower (electric)	watt (W)	7.460 000*E+02
horsepower (metric)	watt (W)	7.354 99 E+02
horsepower (water)	watt (W)	7.460 43 E+02
horsepower (U.K.)	watt (W)	7.457 0 E+02
hour	second(s)	3.600 000*E+03
hour (sidereal)	second(s)	3.590 170 E+03
hundredweight (long)	kilogram (kg)	5.080 235 E+01
hundredweight (short)	kilogram (kg)	4.535 924 E+01
inch	metre (m)	2.540 000*E−02
inch of mercury (32° F)	pascal (Pa)	3.386 38 E+03
inch of mercury (60° F)	pascal (Pa)	3.376 85 E+03
inch of water (39.2° F)	pascal (Pa)	2.490 82 E+02
inch of water (60° F)	pascal (Pa)	2.488 4 E+02
in^2	square metre (m^2)	6.451 600*E−04
in^3 (volume; section modulus)[20]	cubic metre (m^3)	1.638 706 E−05
in^3/min	cubic metre per second (m^3/s)	2.731 177 E−07
in^4 (second moment of area)[19]	metre to the fourth power (m^4)	4.162 314 E−07
in/s	metre per second (m/s)	2.540 000*E−02
in/s^2	metre per second squared (m/s^2)	2.540 000*R−02
kayser	1 per metre (1/m)	1.000 000*E+02

To convert from	to	multiply by
kelvin	degree Celsius	$t^\circ{}_\mathrm{C} = T_\mathrm{K}-273.15$
kilocalorie (International Table)	joule (J)	4.186 800*E+03
kilocalorie (mean)	joule (J)	4.190 02 E+03
kilocalorie (thermochemical)	joule (J)	4.184 000*E+03
kilocalorie (thermochemical)/min	watt (W)	6.973 333 E+01
kilocalorie (thermochemical)/s	watt (W)	4.184 000*E+03
kilogram-force (kgf)	newton (N)	9.806 650*E+00
kgf · m	newton metre (N · m)	9.806 650*E+00
kgf · s^2/m (mass)	kilogram (kg)	9.806 650*E+00
kgf/cm^2	pascal (Pa)	9.806 650*E+04
kgf/m^2	pascal (Pa)	9.806 650*E+00
kgf/mm^2	pascal (Pa)	9.806 650*E+06
km/h	metre per second (m/s)	2.777 778 E−01
kilopond (1 kp = 1 kgf)	newton (N)	9.806 650*E+00
kW · h	joule (J)	3.600 000*E+06
kip (1000 lbf)	newton (N)	4.448 222 E+03
kip/in^2 (ksi)	pascal (Pa)	6.894 757 E+06
knot (international)	metre per second (m/s)	5.144 444 E−01
lambert	candela per square metre (cd/m^2)	$1/\pi$ *E+04
lambert	candela per square metre (cd/m^2)	3.183 099 E+03
langley	joule per square metre (J/m^2)	4.184 000*E+04
light year	metre (m)	9.460 55 E+15
litre[21]	cubic metre (m^3)	1,000 000*E−03
maxwell	weber (Wb)	1,000 000*E−08
mho	siemens (S)	1,000 000*E+00
microinch	metre (m)	2.540 000*E−08
micron	metre (m)	1,000 000*E−06
mil	metre (m)	2.540 000*E−05
mile (international)	metre (m)	1.609 344*E+03
mile (U.S. statute)[14]	metre (m)	1.609 347 E+03

[20] The exact conversion factor is 1.638 706 4*E−05.

[21] In 1964 the General Conference on Weights and Measures reestablished the name litre as a special name for the cubic decimetre. Between 1901 and 1964 the litre was slightly larger (1.000 028 dm^3); in the use of high-accuracy volume data of that time interval, this fact must be kept in mind.

To convert from	to	Multiply by
mile (international nautical)	metre (m)	1.852 000*E+03
mile (U.S. nautical)	metre (m)	1.852 000*E+03
mi^2 (international)	square metre (m^2)	2.589 988 E+06
mi^2 (U.S. statute)[14]	square metre (m^2)	2.589 998 E+06
mi/h (international)	metre per second (m/s)	4.470 400*E−01
mi/h (international)	kilometre per hour (km/h)	1.609 344*E+00
mi/min (international)	metre per second (m/s)	2.682 240*E+01
mi/s (international)	metre per second (m/s)	1.609 344*E+03
millibar	pascal (Pa)	1.000 000*E+02
millimetre of mercury (0° C)	pascal (Pa)	1.333 22 E+02
minute (angle)	radian (rad)	2.908 882 E−04
minute	second (s)	6.000 000*E+01
minute (sidereal)	second (s)	5.983 617 E+01
oersted	ampere per metre (A/m)	7.957 747 E+01
ohm centimetre	ohm meter ($\Omega \cdot m$)	1.000 000*E−02
ohm circular-mil per foot	ohm metre ($\Omega \cdot m$)	1.662 426 E−09
ounce (avoirdupois)	kilogram (kg)	2.834 952 E−02
ounce (troy or apothecary)	kilogram (kg)	3.110 348 E−02
ounce (U.K. fluid)	cubic metre (m^3)	2.841 307 E−05
ounce (U.S. fluid)	cubic metre (m^3)	2.957 353 E−05
ounce-force	newton (N)	2.780 139 E−01
ozf · in	newton metre (N · m)	7.061 552 E−03
oz (avoirdupois)/gal (U.K. liquid)	kilogram per cubic metre (kg/m^3)	6.236 021 E+00
oz (avoirdupois)/gal (U.S. liquid)	kilogram per cubic metre (kg/m^3)	7.489 152 E+00
oz (avoirdupois)/in^3	kilogram per cubic metre (kg/m^3)	1.729 994 E+03
oz (avoirdupois)/ft^2	kilogram per square metre (kg/m^2)	3.051 517 E−01
oz (avoirdupois)/yd^2	kilogram per square metre (kg/m^2)	3.390 575 E−02
parsec	metre (m)	3.085 678 E+16
peck (U.S.)	cubic metre (m^3)	8.809 768 E−03
pennyweight	kilogram (kg)	1.555 174 E−03

perm (0° C)	kilogram per pascal second square metre [kg/(Pa · s · m^2)]	5.721 35 E-11
perm (23° C)	kilogram per pascal second square metre [kg/(Pa · s · m^2)]	5.745 25 E-11
perm · in (0° C)	kilogram per pascal second metre [kg/(Pa · s · m)]	1.453 22 E-12
perm · in (23° C)	kilogram per pascal second metre [kg/(Pa · s · m)]	1.459 29 E-12
phot	lumen per square metre (lm/m^2)	1.000 000*E+04
pica (printer's)	metre (m)	4.217 518 E-03
pint (U.S. dry)	cubic metre (m^3)	5.506 105 E-04
pint (U.S. liquid)	cubic metre (m^3)	4.731 765 E-04
point (printer's)	metre (m)	3.514 598*E-04
poise (absolute viscosity)	pascal second (Pa · s)	1.000 000*E-01
pound (lb avoirdupois)[22]	kilogram (kg)	4.535 924 E-01
pound (troy or apothecary)	kilogram (kg)	3.732 417 E-01
lb · ft^2 (moment of inertia)	kilogram square metre (kg · m^2)	4.214 011 E-02
lb · in^2 (moment of inertia)	kilogram square metre (kg · m^2)	2.926 397 E-04
lb/ft · h	pascal second (Pa · s)	4.133 789 E-04
lb/ft · s	pascal second (Pa · s)	1.488 164 E+00
lb/ft^2	kilogram per square metre (kg/m^2)	4.882 428 E+00
lb/ft^3	kilogram per cubic metre (kg/m^3)	1.601 846 E+01
lb/gal (U.K. liquid)	kilogram per cubic metre (kg/m^3)	9.977 633 E+01
lb/gal (U.S. liquid)	kilogram per cubic metre (kg/m^3)	1.198 264 E+02
lb/h	kilogram per second (kg/s)	1.259 979 E-04
lb/hp · h (SFC, specific fuel consumption)	kilogram per joule (kg/J)	1.689 659 E-07
lb/in^3	kilogram per cubic metre (kg/m^3)	2.767 990 E+04
lb/min	kilogram per second (kg/s)	7.559 873 E-03
lb/s	kilogram per second (kg/s)	4.535 924 E-01
lb/yd^3	kilogram per cubic metre (kg/m^3)	5.932 764 E-01
poundal	newton (N)	1.382 550 E-01
poundal/ft^2	pascal (Pa)	1.488 164 E+00

[22] The exact conversion factor is 4.535 923 7*E-01.

To convert from	to	Multiply by
poundal $\cdot$ s/ft^2	pascal second (Pa $\cdot$ s)	1.488 164 E+00
pound-force (lbf)[23]	newton (N)	4.448 222 E+00
lbf $\cdot$ ft	newton metre (N $\cdot$ m)	1.355 818 E+00
lbf $\cdot$ ft/in	newton metre per metre (N $\cdot$ m/m)	5.337 866 E+01
lbf $\cdot$ in	newton metre (N $\cdot$ m)	1.129 848 E−01
lbf $\cdot$ in/in	newton metre per metre (N $\cdot$ m/m)	4.448 222 E+00
lbf $\cdot$ s/ft^2	pascal second (Pa $\cdot$ s)	4.788 026 E+01
lbf $\cdot$ s/in^2	pascal second (Pa $\cdot$ s)	6.894 757 E+03
lbf/ft	newton per metre (N/m)	1.459 390 E+01
lbf/ft^2	pascal (Pa)	4.788 026 E+01
lbf/in	newton per metre (N/m)	1.751 268 E+02
lbf/in^2 (psi)	pascal (Pa)	6.894 757 E+03
lbf/lb (thrust/weight [mass] ratio)	newton per kilogram (N/kg)	9.806 650 E+00
quart (U.S. dry)	cubic metre (m^3)	1.101 221 E−03
quart (U.S. liquid)	cubic metre (m^3)	9.463 529 E−04
rad (absorbed dose)	gray (Gy)	1.000 000*E−02
rem (dose equivalent)	sievert (Sv)	1.000 000*E−02
rhe	1 per pascal second [1/(Pa $\cdot$ s)]	1.000 000*E+01
rod[14]	metre (m)	5.029 210 E+00
roentgen	coulomb per kilogram (C/kg)	2.58 E−04
second (angle)	radian (rad)	4.848 137 E−06
second (sidereal)	second (s)	9.972 696 E−01
shake	second (s)	1.000 000*E−08
slug	kilogram (kg)	1.459 390 E+01
slug/ft $\cdot$ s	pascal second (Pa $\cdot$ s)	4.788 026 E+01
slug/ft^3	kilogram per cubic metre (kg/m^3)	5.153 788 E+02
statampere	ampere (A)	3.335 640 E−10
statcoulomb	coulomb (C)	3.335 640 E−10
statfarad	farad (F)	1.112 650 E−12
stathenry	henry (H)	8.987 554 E+11
statmho	siemens (S)	1.112 650 E−12
statohm	ohm (Ω)	8.987 554 E+11

statvolt	volt (V)	2.997 925 E+02
stere	cubic metre (m³)	1.000 000*E+00
stilb	candela per square metre (cd/m²)	1.000 000*E+04
stokes (kinematic viscosity)	square metre per second (m²/s)	1.000 000*E−04
tablespoon	cubic metre (m³)	1.478 676 E−05
teaspoon	cubic metre (m³)	4.928 922 E−06
tex	kilogram per metre (kg/m)	1.000 000*E−06
therm	joule (J)	1.055 056 E+08
ton (assay)	kilogram (kg)	2,916 667 E−02
ton (long, 2240 lb)	kilogram (kg)	1.016 047 E+03
ton (metric)	kilogram (kg)	1.000 000*E+03
ton (nuclear equivalent of TNT)	joule (J)	4.184 E+09[24]
ton (refrigeration)	watt (W)	3.516 800 E+03
ton (register)	cubic metre (m³)	2.831 685 E+00
ton (short, 2000 lb)	kilogram (kg)	9.071 847 E+02
ton (long)/yd³	kilogram per cubic metre (kg/m³)	1.328 939 E+03
ton (short)/yd³	kilogram per cubic metre (kg/m³)	1.186 553 E+03
ton (short)/h	kilogram per second (kg/s)	2.519 958 E−01
ton-force (2000 lbf)	newton (N)	8.896 444 E+03
tonne	kilogram (kg)	1.000 000*E+03
torr (mmHg, 0° C)	pascal (Pa)	1.333 22 E+02
unit pole	weber (Wb)	1.256 637 E−07
W · h	joule (J)	3.600 000*E+03
W · s	joule (J)	1.000 000*E+00
W/cm²	watt per square metre (W/m²)	1.000 000*E+04
W/in²	watt per square metre (W/m²)	1.550 003 E+03
yard	metre (m)	9.144 000*E−01
yd²	square metre (m²)	8.361 274 E−01
yd³	cubic metre (m³)	7.645 549 E−01
yd³/min	cubic metre per second (m³/s)	1.274 258 E−02
year (365 days)	second (s)	3.153 600*E+07
year (sidereal)	second (s)	3.155 815 E+07
year (tropical)	second (s)	3.155 693 E+07

[23] The exact conversion factor is 4.448 221 615 260 5*E+00.
[24] Defined (not measured) value.

Section 12______________________________

Source: *Style Guide*, Materials and Process Dept., Northrop Corp., Aircraft Group, Aircraft Div., 83901 W. Broadway, Hawthorne, Calif. 90250, April, 1979.

Selections—in order of appearance:

Choice of Words	Choice of Mood
Misused Words	Decimal Equivalents of Fractions
Words to Avoid	Preferred Spelling
Choice of Phrases	

____

2.2   *Choice of Words*

2.2.1   *General*—Words should be selected that effectively communicate, that contribute to the desired meaning, and that are active in mood and positive in approach. However, certain words and word forms are used incorrectly (misused words) or should not be used (words to avoid). Incorrect use of words reflects upon the writer and may distract the reader.

2.2.2   *Misused Words*

2.2.2.1   The words "addenda," "criteria," and "data" are plural and as subjects are followed by plural verbs.

2.2.2.2   The word "as" is often misused:
a. When used in place of "since," "because," or "for"—each one is preferred to "as"
b. When used in place of "that" or "whether"
c. In correlative "as . . . as." "So . . . as" is preferred generally.
d. "As per" is not acceptable for "according to."
e. "As to" is doubtful substitution for "of," "on," "about," and "concerning."

2.2.2.3   "Agree to" meaning "to give assent" and "agree with" meaning "to be in accord with" are not interchangeable.

2.2.2.4   "To compose" meaning "to put together, to make, to fashion" and "to comprise" meaning "to include, to contain, to consist of" are not interchangeable.
Examples:
a. The committee is composed of one representative from each organization.
b. The committee is composed of 10 representatives.
c. The committee comprises 10 representatives.

2.2.2.5 The choice between "shall," "will," "should,'" and "may" sometimes seems difficult.
a. "Shall" is used whenever a specification expresses a provision that is binding or mandatory.
b. "Should" and "may" are used when necessary to express a declaration of purpose on the part of Northrop.
c. "Will" may be required to express simple futurity; for example, "Power for the motors will be supplied by the ground equipment."

2.2.2.6 "Affect" and "effect" are problem words; in technical work, as a general rule, "affect" will be used as a verb, not as a noun; "effect" will be used as either a verb or a noun.

Affect—verb: To influence, to make an impression on, to produce an effect upon. Also, to have an affection for; to put on a pretense of.

    —noun: Conscious subjective aspect of an emotion considered apart from bodily changes.

Effect—verb: To bring about, accomplish; cause (produce, perform, make)

    —noun: A result, consequence, or outcome; something produced by an agent or cause.

Examples:
a. Bach's music *affects* me deeply.
b. He *affects* a British accent and manner.
c. Overproduction *affects* prices.
d. Congress *effected* a change in the law.
e. The doctor *effected* a marvelous cure.
f. A change was *effected* that *affected* all areas.
g. We shall long feel the *effects* of World War II.
h. He did this chiefly for *effect*.
i. We shall do all we can to put your plan into *effect*.

NOTE: When the use of "affect" or "effect" becomes involved and distracting, rewriting with other words should be considered.

2.2.2.7 "That" and "which" are not necessarily interchangeable.

"That" leads into a clause that is defining (restrictive) and is not set off by commas.

"Which" leads into a clause that is nondefining (nonrestrictive) and is set off by commas.

"Which" can be used restrictively without commas, but in almost every restrictive instance, "that" could be and should be used for "which".

Examples:

Restrictive— Metal details that have been verifilmed shall be washed with . . .

Nonrestrictive—This material, which has been used up to the 300 ksi strength level, will completely air harden through a . . .

2.2.2.8 A "catalyst" is a substance that accelerates or retards a reaction; a catalyst is not consumed by the reaction and a very small amount will catalyze a large mixture of reactants in chemical synthesis. In plastics, "catalyst" has long been a generic term for any additive that causes solidification of a liquid plastic resin. A number of other terms can be used to describe such an additive, some of which are listed below.

    a. Accelerator or promoter (most often added to polyester resin)

    b. Converter (applies to a polyamide added to an epoxy resin)

    c. Hardener or curing agent (applies to additive for epoxy resins)

2.2.2.9   Words describing the flammability of a material are often confused:

    a. *Fireproof*—will not burn.

    b. *Fire resistant*—may burn when subjected to an igniting source.

    c. *Flame resistant*—not susceptible to combustion to point of propagating a flame after ignition source is removed.

    d. *Flammable*—susceptible to igniting readily or exploding.

    e. *Nonflammable*—will not ignite or explode when subjected to igniting source.

    f. *Self-extinguishing*—will not support combustion after removal of igniting source.

Do not use inflammable, noninflammable, or uninflammable.

2.2.2.10 "Using" is sometimes a poor choice of a word, particularly with time or temperature.

    a. Awkward: Adhesive shall be applied using a spatula . . .
       Preferred: Adhesive shall be applied with a spatula . . .

    b. Awkward: Cure adhesive in an autoclave using 30 to 40 psi.
       Preferred: Cure adhesive in an autoclave at 30 to 40 psi.

2.2.2.11 Acceptability of a material or process must be described with words not open to interpretation. Use words that conform to accepted usage in industry or are clearly defined. Such words as "defect," "imperfection," and "discontinuity" are often incorrectly made interchangeable. There cannot be an "acceptable defect" since a "defect" means the material or process is unacceptable.

Defect—A shortcoming or flaw that makes the process or material unacceptable.

Imperfection—A fault or blemish that might prevent the process or material from meeting requirements but may still be judged acceptable.

Discontinuity—Not continuous, interruption or termination of an ordered sequence, discernible inhomogeneity, nonuniform structure or composition; may still be judged acceptable.

2.2.2.12 Another set of misused words is the set, "departure," "deviation," and "variation." "Deviation" and "variation" reflect degrees of "departure" from requirements.

Departure—Differences from agreed requirements; changes to the accepted practices; does not convey magnitude of difference. Acceptability must be determined.

Variation—Minor acceptable departure from strict specification con-
(based on   formance which does not:
Air Force   a. Adversely affect safety, performance, or durability.
definition)   b. Materially affect weight.
        c. Adversely affect interchangeability of parts or assemblies.
        d. Depart from the basic objectives of the specification.

         Deviation—Acceptable departure from the requirement of the specification which does not adversely affect safety but is more extensive than a variation. Air Force (Wright Field) approval is required for all deviations.

2.2.2.13 The word "insure" is often incorrectly used in place of "ensure" or "assure."

         Insure— Insure against loss; implies guarantee with monetary penalty; implies to give, take, or procure a policy or contract to indemnify or guarantee against loss, hurt, or damage.

         Ensure—Make certain the occurrence of; implies a making certain; implies guarantee without monetary penalty.

         Assure—(transitive: requires object)—Make certain of a fact; implies making sure by removing all doubt and suspense.

2.2.2.14 Care must be exercised in use of "or" and "and"; for example, in corrosion prevention, the following condition exists:

    a. Aluminum— Treat with coating in accordance with FP-28 *or* coat with NAI 1269 or TT-P-1757.

    b. Magnesium—Treat with coating in accordance with FP-28 *and* coat with NAI-1269 or TT-P-1757.

2.2.3 *Words to Avoid*

    a. "And/or." "X and y," "x or y," or "x or y or both" is preferred.

    b. "Per." "Per" should not be used to mean "in accordance with" but should be used to describe rates such as "feet per second."

    c. "It" or "its." Also, avoid any other indefinite pronoun (such as this, their, and they) that vaguely could refer to a number of antecedents.

2.3 *Choice of Phrases*

2.3.1 The writer who used several words where only one or two are necessary is trying to impress the reader. The skillful writer avoids this practice; he uses simple, direct phrases.

2.3.2 One of the most overworked and best examples of a wordy phrase is the use of ". . . in order to . . ." when in every case ". . . to . . ." says the same thing.

2.3.3 Another example is "We would like to talk with you . . .

    in connection with = about . . ."
    with regard to   = about . . ."
    with reference to = about . . . "
    on the subject of = about . . ."
    in relation to   = about . . ."

2.3.4 Other examples follow.

    a. At the present time   = now
    b. Prior to the start of   = before
    c. In the near future   = soon
    d. In the event that   = if
    e. In order that   = so
    f. The reason is because = the reason is, because (not both)
    g. Due to the fact that   = since or because
    h. For the reason that   = since or because
    i. Subsequent to   = after

2.3.5   Avoid the repetitive prepositional phrase "of the . . . of the . . . of the." An example would be:
    . . . of the level of the knowledge of the reader = reader's knowledge level.

2.4   *Choice of Mood*—The indicative (emphatic) mood is preferred over the imperative mood. The imperative is best used only for direct instructions, as in the following example.

3.8.3   Facings to be bonded shall be primed with liquid adhesive as follows: indicative

a. Adjust viscosity of adhesive by adding thinner,

b. Prime the surface to a dry film thickness of 0.003 inch maximum,

c. Air dry for 30 minutes minimum.   imperative

In any case, the indicative (or emphatic) and the imperative moods should not be used in the same instructional element or sentence.

TABLE 2.   Decimal Equivalents of Fractions

Fractions	Decimals to 3 Places	Fractions	Decimals to 3 Places
1/64	0.016	33/64	0.516
1/32	0.031	17/32	0.531
3/64	0.047	35/64	0.547
1/16	0.062	9/16	0.562
5/64	0.078	37/64	0.578
3/32	0.094	19/32	0.594
7/64	0.109	39/64	0.609
1/8	0.125	5/8	0.625
9/64	0.141	41/64	0.641
5/32	0.156	21/32	0.656
11/64	0.172	43/64	0.672
3/16	0.188	11/16	0.688
13/64	0.203	45/64	0.703
7/32	0.219	23/32	0.719
15/64	0.234	47/64	0.734
1/4	0.250	3/4	0.750
17/64	0.266	49/64	0.766
9/32	0.281	25/32	0.781
19/64	0.297	51/64	0.797
5/16	0.312	13/16	0.812
21/64	0.328	53/64	0.828
11/32	0.344	27/32	0.844
23/64	0.359	55/64	0.859
3/8	0.375	7/8	0.875
25/64	0.391	57/64	0.891
13/32	0.406	29/32	0.906
27/64	0.422	59/64	0.922
7/16	0.438	15/16	0.938
29/64	0.453	61/64	0.953
15/32	0.469	31/32	0.969
31/64	0.484	63/64	0.984
1/2	0.500	1	1.000

Preferred Spelling List
This list contains words commonly misspelled, words with more than one correct way for spelling, and compound words where there is a question of hyphenation. Where a choice is possible, the choice has been made and should be observed. Quite often compound words may be open in the verb form but hyphenated when modifying another word. The philosophy in a decision to hyphenate (or not) should be that, where meaning is clear and readability is not helped, a hyphen is unnecessary. In the following list, when use of the word as verb, noun, or modifier is a factor, a code is used to indicate the open or hyphenated compound words; in the code, adj is adjective, n is noun, v is verb, um is unit modifier, and pl is plural.

a priori (2 words)
acid-fast
acidproof (um)
acknowledgment (no "e" after "g")
admiralty metal (n)
aging (not ageing)
airborne
air-cool (v)
air-dry (adj)
airframe
air lock (n)
airtight (adj)
alkalies (not alkalis)
antecedent
antennas
antiaircraft
antibody
anti-icing (adj)
antioxidant
appendixes (pl)
artificially

backup (n, um)
-base, as a suffix is hyphenated; is
   not "-based"; example: paraffin-
   base chemical, not paraffin-based
   chemical
bimetallic
bimolecular
bistable
blowout (n)
boattail
bondline
breadboard
breakdown (n)
breakthrough (n)
Brinell
build in (v)
built-in (um)
buildup (n)
built-up (um)

burnup (n, um)
buses
bypass
byproduct

call out (v)
callout (n, adj)
canceling
cannot—one word
castings
centerline
checkout (n, adj)
Chem-mill (upper case: proprietary
   name; "chemical mill" is a process.)
cleanup (n, um)
clean up (v)
coexist
concurrence
contaminants (n)
contaminate (v)
conveyor (mechanical conveying)
coordinate
crisscross
crosshatch
cross-reference (n, v)
cross section (n)
cross-section (um)
cross-sectional (adj)
cutback (n, adj)
cutoff (n, adj)
cutout (n, adj)

degassed
deionize
dimout (n, um)
disk
double space (n, v)
double-space (um)
drapeability

Du Pont, not E.I. DuPont de Nem-
ours Co., Inc.
dust-free
dusttight

electro-optical
embed (not imbed)
equiaxial
explosionproof

feedwater
fiberglass or Fiberglas (trade name)
filleting
foam-in-place (um)
focusing
foreword
forgings
formulas (pl)
fueled

gage (preferred variation of gauge)
gases
germ-free
germproof
go/no-go
gray (not grey)

half-life
hand-built (um)
heat treat (v)
heat treated (adj)
heat treatment (n)
heat treating (n)
hookup (n, um)
hot plate (n)

inconsistent
in-flight (um)
in-process (um)
insofar
insure—see 2.2.2.13 of Style Guide
(Use assure, ensure.)
indexes (pl)
interconnecting

jelled

kerosene
knife-edge
know-how
kraft

labeled
lapping
lay off (v)
layoff (n, adj)
lay out (v)
layout (n, adj)
lay up (v)
lay-up (n, um)
line up (v)
lineup (n, um)
lockwire (n, v)
lock-wire (v)

machinable
make up (v)
makeup (n, adj)
Materials and Process Department
Materials and Process Directive
Material Specification
melt-thru
metal-to-metal
microetch
misalignment
moistureproof
multiple-purpose (um)
multipurpose

(Non is not open styled or hyphenated but close styled.)
nonabsorbent
nonabrasive
nonadhesive
nonalcoholic
nonaqueous
nonaromatic
nonbearing
noncorrosive
nonconductor
nonelectrolyte
non-Euclidean
non-flame-sprayed coating
non-heat treated
nonionic
nonluminous
nonmagnetic
nonmetal
nonmetallic
nonoil
nonoperating
nonorganic
nonporous
non-self-aligning
non-self-governing

nonscientific
nonstandard
nonrestrictive
non-tumor-bearing
nontechnical
nontoxic
nonwhite
nonuse
nut plate

O-ring
offcenter (um)
off-the-shelf (um)
oilproof
overall
overheat
overlapping
overtemper
oxyacetylene
oxyhydrogen

paraffin-base
permissible
pick up (v)
pickup (n)
post-age (postage is another word)
postcure
postheat
posttreat
postweld
preclean
preheat
preimpregnate
premixed
preproduction
preventive (not preventative)
procuring activity (not procuring agency)
-proof (as adj suffix is closed up)
propagate

® (first occurence, use note)
    NOTE: Registered in U.S. Patent Office
reenter
reentry
re-etch
refinish
re-form
re-heat treat
repairable (vs. reparable)
repeen
reposition

reprime
respray
re-solution heat treat
restress
re-stress relieved
retest
reuse
rustproof

sawed (not sawn)
selvage
semiannual
semiaustenitic
semiautomatic
semifinished
set up (v)
setup (n)
shut down (v)
shutdown (n, um)
sign off (v)
signoff (n, um)
silverplate (v)
silver-plated
Solder Sleeve (proprietary)
specialty (another word: speciality)
spot check (n, v)
spot-checked (um)
spot face (v)
spotface (n, um)
spotweld (v)
spot-welded (um)
squeezeout (n, um)
start up (v)
startup (n, um)
subzero
subtier
sulfur (not sulphur)
superalloy
susceptible

through (not thru; "thru" may be used in tables where space is limited)
torquing (not torqueing)
touch up (v)
touchup (n, um)
toweling

ultrahigh
usable (not useable)
up-to-date (um)

vendor—use supplier
vibro-etch

watchglass
water glass
waterline
waterproof
wavelength
wavemeter
weldbond
welder (person or machine)—preferred

weldor (person)
windowpane
Wrap Around (proprietary)

X ray (n)
X-ray (v, adj)

Section 13______________

Source: *Glossary of Metallurgical Terms and Engineering Tables*, American
Society for Metals, Metals Park, Ohio 44073, 1979.

Selections—in order of appearance:

Metric Conversion Factors
Temperature Conversions
Stress or Pressure Conversions
Stress Intensity Conversions
Energy Conversions
Approximate Equivalent
    Hardness Numbers and Ten-
    sile Strength for Vickers
    Hardness Numbers for Steel
Approximate Equivalent
    Hardness Numbers and Ten-
    sile Strengths for Brinell
    Hardness Numbers for Steel
Approximate Equivalent
    Hardness Numbers and Ten-
    sile Strengths for Rockwell C
    Hardness Numbers for Steel
Approximate Equivalent
    Hardness Numbers and Ten-
    sile Strengths for Rockwell
    B Hardness Numbers for
    Steel

Metric Conversion Factors

To convert from	To	Multiply by	To convert from	To	Multiply by
angstrom	m	1.0000×10^{-10}(a)	hp(e)	W	7.4570×10^2
atm	Pa	1.0133×10^5	hp(f)	W	7.4600×10^2
Btu(b)	J	1.054×10^3	in.	m	2.5400×10^{-2}
Btu(b)/ft²·h	W/m²	3.1525	in.²	m²	6.4516×10^{-4}
Btu(b)/ft²·h·°F	W/m²·K	5.6745	in.³	m³	1.6387×10^{-5}
Btu(b)·ft/h·ft²·°F	W/m·K	1.7296	in. of Hg(g)	Pa	3.3864×10^3
Btu(b)/ft²·s	W/m²	1.135×10^4	in. of water(c)	Pa	2.4908×10^2
Btu(b)·in./ft²·h·°F	W/m·K	1.4413×10^{-1}	K	°C	$t_{°C} = t_K - 273.15$
Btu(b)·in./s·ft²·°F	W/m·K	5.1887×10^2	kgf	N	9.80665(a)
Btu(b)/lbm·°F	J/kg·K	4.1840×10^3	kgf/mm²	Pa	9.80665×10^6(a)
cal(b)	J	4.1840 (a)	ksi	MPa	6.8948
cal(b)/cm·s·°C	W/m·K	4.1840×10^2(a)	ksi	Pa	6.8948×10^6
cal(b)/g	J/kg	4.1840×10^3(a)	ksi√in.	MPa√m	1.089
cal(b)/g·°C	J/kg·K	4.1840×10^3(a)	lb(h)	kg	4.5359×10^{-1}
circ mil	m²	5.0671×10^{-10}	lb/in.³	kg/m³	2.7680×10^4
°C	K	$t_K = t_{°C} + 273.15$	lbf	N	4.4482
degree	rad	1.7453×10^{-2}	lbf·in.	N·m	1.1298×10^{-1}
dyne/cm²	Pa	1.0000×10^{-1}(a)	lbf·ft	N·m	1.3558
°F	°C	$t_{°C} = (t_{°F} - 32)/1.8$	MPa√m	MNm$^{-3/2}$	1.0000(a)
°F	K	$t_K = (t_{°F} + 459.67)/1.8$	μin.	m	2.5400×10^8(a)
ft	m	3.0480×10^{-1}	mil	m	2.5400×10^{-5}(a)
ft²	m²	9.2903×10^{-2}	N/m²	Pa	1.0000(a)
ft³	m³	2.8317×10^{-2}	oersted	A/m	79.578
ft of water(c)	Pa	2.9890×10^3	oz/ft²	kg/m²	3.0515×10^{-1}
ft²/h (thermal diffusivity)	m²/s	2.58064×10^{-5}(a)	psi	Pa	6.8948×10^3
ft·lbf	J	1.3558	°R	K	$t_K = t_{°R}/1.8$
ft·lbf/s	W	1.3558	ton(j)	kg	9.0718×10^2
ft/s	m/s	3.0480×10^{-1}	ton(k)	kg	1.0160×10^3
gauss	T	1.0000×10^{-4}(a)	ton/in.²	Pa	1.3786×10^4
gallon(d)	m³	3.7854×10^{-3}	tonne	kg	1.0000×10^3(a)
g/cm³	kg/m³	1.0000×10^3(a)	torr	Pa	1.3332×10^2
g/cm³	Mg/m³	1.0000(a)	Ω/circ mil·ft	Ω·m	1.6624×10^{-9}

(a) Exactly. (b) Thermochemical. (c) At 4 °C (39.2 °F). (d) U.S. liquid. (e) Mechanical (1 hp = 550 ft·lbf/s). (f) Electrical. (g) At 0 °C (32 °F). (g) Avoirdupois. (j) Short; equal to 2000 lbm. (k) Long; 2240 lbm.

Temperature Conversions

The general arrangement of this table was devised by Sauveur and Boylston more than 40 years ago. The middle column of figures (in bold-faced type) contains the reading (°F or °C) to be converted. If converting from degrees Fahrenheit to degrees Centigrade, read the Centigrade equivalent in the column Headed "°C". If converting from Centigrade to Fahrenheit, read the Fahrenheit equivalent in the column headed "°F". °C = 5/9 (°F −32).

°F		°C	°F		°C	°F		°C	°F		°C	°F		°C
···	**−458**	−272.22	···	**−358**	−216.67	−432.4	**−258**	−161.11	−252.4	**−158**	−105.56	−72.4	**−58**	−50.00
···	**−456**	−271.11	···	**−356**	−215.56	−428.8	**−256**	−160.00	−248.8	**−156**	−104.44	−68.8	**−56**	−48.89
···	**−454**	−270.00	···	**−354**	−214.44	−425.2	**−254**	−158.89	−245.2	**−154**	−103.33	−65.2	**−54**	−47.78
···	**−452**	−268.89	···	**−352**	−213.33	−421.6	**−252**	−157.78	−241.6	**−152**	−102.22	−61.6	**−52**	−46.67
···	**−450**	−267.78	···	**−350**	−212.22	−418.0	**−250**	−156.67	−238.0	**−150**	−101.11	−58.0	**−50**	−45.56
···	**−448**	−266.67	···	**−348**	−211.11	−414.4	**−248**	−155.56	−234.4	**−148**	−100.00	−54.4	**−48**	−44.44
···	**−446**	−265.56	···	**−346**	−210.00	−410.8	**−246**	−154.44	−230.8	**−146**	−98.89	−50.8	**−46**	−43.33
···	**−444**	−264.44	···	**−344**	−208.89	−407.2	**−244**	−153.33	−227.2	**−144**	−97.78	−47.2	**−44**	−42.22
···	**−442**	−263.33	···	**−342**	−207.78	−403.6	**−242**	−152.22	−223.6	**−142**	−96.67	−43.6	**−42**	−41.11
···	**−440**	−262.22	···	**−340**	−206.67	−400.0	**−240**	−151.11	−220.0	**−140**	−95.56	−40.0	**−40**	−40.00
···	**−438**	−261.11	···	**−338**	−205.56	−396.4	**−238**	−150.00	−216.4	**−138**	−94.44	−36.4	**−38**	−38.89
···	**−436**	−260.00	···	**−336**	−204.44	−392.8	**−236**	−148.89	−212.8	**−136**	−93.33	−32.8	**−36**	−37.78
···	**−434**	−258.89	···	**−334**	−203.33	−389.2	**−234**	−147.78	−209.2	**−134**	−92.22	−29.2	**−34**	−36.67
···	**−432**	−257.78	···	**−332**	−202.22	−385.6	**−232**	−146.67	−205.6	**−132**	−91.11	−25.6	**−32**	−35.56
···	**−430**	−256.67	···	**−330**	−201.11	−382.0	**−230**	−145.56	−202.0	**−130**	−90.00	−22.0	**−30**	−34.44
···	**−428**	−255.56	···	**−328**	−200.00	−378.4	**−228**	−144.44	−198.4	**−128**	−88.89	−18.4	**−28**	−33.33
···	**−426**	−254.44	···	**−326**	−198.89	−374.8	**−226**	−143.33	−194.8	**−126**	−87.78	−14.8	**−26**	−32.22
···	**−424**	−253.33	···	**−324**	−197.78	−371.2	**−224**	−142.22	−191.2	**−124**	−86.67	−11.2	**−24**	−31.11
···	**−422**	−252.22	···	**−322**	−196.67	−367.6	**−222**	−141.11	−187.6	**−122**	−85.56	−7.6	**−22**	−30.00
···	**−420**	−251.11	···	**−320**	−195.56	−364.0	**−220**	−140.00	−184.0	**−120**	−84.44	−4.0	**−20**	−28.89
···	**−418**	−250.00	···	**−318**	−194.44	−360.4	**−218**	−138.89	−180.4	**−118**	−83.33	−0.4	**−18**	−27.78
···	**−416**	−248.89	···	**−316**	−193.33	−356.8	**−216**	−137.78	−176.8	**−116**	−82.22	+3.2	**−16**	−26.67
···	**−414**	−247.78	···	**−314**	−192.22	−353.2	**−214**	−136.67	−173.2	**−114**	−81.11	+6.8	**−14**	−25.56
···	**−412**	−246.67	···	**−312**	−191.11	−349.6	**−212**	−135.56	−169.6	**−112**	−80.00	+10.4	**−12**	−24.44
···	**−410**	−245.56	···	**−310**	−190.00	−346.0	**−210**	−134.44	−166.0	**−110**	−78.89	+14.0	**−10**	−23.33
···	**−408**	−244.44	···	**−308**	−188.89	−342.4	**−208**	−133.33	−162.4	**−108**	−77.78	+17.6	**−8**	−22.22
···	**−406**	−243.33	···	**−306**	−187.78	−338.8	**−206**	−132.22	−158.8	**−106**	−76.67	+21.2	**−6**	−21.11
···	**−404**	−242.22	···	**−304**	−186.67	−335.2	**−204**	−131.11	−155.2	**−104**	−75.56	+24.8	**−4**	−20.00
···	**−402**	−241.11	···	**−302**	−185.56	−331.6	**−202**	−130.00	−151.6	**−102**	−74.44	+28.4	**−2**	−18.89
···	**−400**	−240.00	···	**−300**	−184.44	−328.0	**−200**	−128.89	−148.0	**−100**	−73.33	+32.0	**±0**	−17.78
···	**−398**	−238.89	···	**−298**	−183.33	−324.4	**−198**	−127.78	−144.4	**−98**	−72.22	+35.6	**+2**	−16.67
···	**−396**	−237.78	···	**−296**	−182.22	−320.8	**−196**	−126.67	−140.8	**−96**	−71.11	+39.2	**+4**	−15.56
···	**−394**	−236.67	···	**−294**	−181.11	−317.2	**−194**	−125.56	−137.2	**−94**	−70.00	+42.8	**+6**	−14.44
···	**−392**	−235.56	···	**−292**	−180.00	−313.6	**−192**	−124.44	−133.6	**−92**	−68.89	+46.4	**+8**	−13.33
···	**−390**	−234.44	···	**−290**	−178.89	−310.0	**−190**	−123.33	−130.0	**−90**	−67.78	+50.0	**+10**	−12.22
···	**−388**	−233.33	···	**−288**	−177.78	−306.4	**−188**	−122.22	−126.4	**−88**	−66.67	+53.6	**+12**	−11.11
···	**−386**	−232.22	···	**−286**	−176.67	−302.8	**−186**	−121.11	−122.8	**−86**	−65.56	+57.2	**+14**	−10.00
···	**−384**	−231.11	···	**−284**	−175.56	−299.2	**−184**	−120.00	−119.2	**−84**	−64.44	+60.8	**+16**	−8.89
···	**−382**	−230.00	···	**−282**	−174.44	−295.6	**−182**	−118.89	−115.6	**−82**	−63.33	+64.4	**+18**	−7.78
···	**−380**	−228.89	···	**−280**	−173.33	−292.0	**−180**	−117.78	−112.0	**−80**	−62.22	+68.0	**+20**	−6.67
···	**−378**	−227.78	···	**−278**	−172.22	−288.4	**−178**	−116.67	−108.4	**−78**	−61.11	+71.6	**+22**	−5.56
···	**−376**	−226.67	···	**−276**	−171.11	−284.8	**−176**	−115.56	−104.8	**−76**	−60.00	+75.2	**+24**	−4.44
···	**−374**	−225.56	···	**−274**	−170.00	−281.2	**−174**	−114.44	−101.2	**−74**	−58.89	+78.8	**+26**	−3.33
···	**−372**	−224.44	−457.6	**−272**	−168.89	−277.6	**−172**	−113.33	−97.6	**−72**	−57.78	+82.4	**+28**	−2.22
···	**−370**	−223.33	−454.0	**−270**	−167.78	−274.0	**−170**	−112.22	−94.0	**−70**	−56.67	+86.0	**+30**	−1.11
···	**−368**	−222.22	−450.4	**−268**	−166.67	−270.4	**−168**	−111.11	−90.4	**−68**	−55.56	+89.6	**+32**	±0.00
···	**−366**	−221.11	−446.8	**−266**	−165.56	−266.8	**−166**	−110.00	−86.8	**−66**	−54.44	+93.2	**+34**	+1.11
···	**−364**	−220.00	−443.2	**−264**	−164.44	−263.2	**−164**	−108.89	−83.2	**−64**	−53.33	+96.8	**+36**	+2.22
···	**−362**	−218.89	−439.6	**−262**	−163.33	−259.6	**−162**	−107.78	−79.6	**−62**	−52.22	+100.4	**+38**	+3.33
···	**−360**	−217.78	−436.0	**−260**	−162.22	−256.0	**−160**	−106.67	−76.0	**−60**	−51.11	+104.0	**+40**	+4.44

Temperature Conversions (continued)

°F		°C	°F		°C	°F		°C	°F		°C	°F		°C
107.6	42	5.56	305.6	152	66.67	503.6	262	127.78	701.6	372	188.89	899.6	482	250.00
111.2	44	6.67	309.2	154	67.78	507.2	264	128.89	705.2	374	190.00	903.2	484	251.11
114.8	46	7.78	312.8	156	68.89	510.8	266	130.00	708.8	376	191.11	906.8	486	252.22
118.4	48	8.89	316.4	158	70.00	514.4	268	131.11	712.4	378	192.22	910.4	488	253.33
122.0	50	10.00	320.0	160	71.11	518.0	270	132.22	716.0	380	193.33	914.0	490	254.44
125.6	52	11.11	323.6	162	72.22	521.6	272	133.33	719.6	382	194.44	917.6	492	255.56
129.2	54	12.12	327.2	164	73.33	525.2	274	134.44	723.2	384	195.56	921.2	494	256.67
132.8	56	13.33	330.8	166	74.44	528.8	276	135.56	726.8	386	196.67	924.8	496	257.78
136.4	58	14.44	334.4	168	75.56	532.4	278	136.67	730.4	388	197.78	928.4	498	258.89
140.0	60	15.56	338.0	170	76.67	536.0	280	137.78	734.0	390	198.89	932.0	500	260.00
143.6	62	16.67	341.6	172	77.78	539.6	282	138.89	737.6	392	200.00	935.6	502	261.11
147.2	64	17.78	345.2	174	78.89	543.2	284	140.00	741.2	394	201.11	939.2	504	262.22
150.8	66	18.89	348.8	176	80.00	546.8	286	141.11	744.8	396	202.22	942.8	506	263.33
154.4	68	20.00	352.4	178	81.11	550.4	288	142.22	748.4	398	203.33	946.4	508	264.44
158.0	70	21.11	356.0	180	82.22	554.0	290	143.33	752.0	400	204.44	950.0	510	265.56
161.6	72	22.22	359.6	182	83.33	557.6	292	144.44	755.6	402	205.56	953.6	512	266.67
165.2	74	23.33	363.2	184	84.44	561.2	294	145.56	759.2	404	206.67	957.2	514	267.78
168.8	76	24.44	366.8	186	85.56	564.8	296	146.67	762.8	406	207.78	960.8	516	268.89
172.4	78	25.56	370.4	188	86.67	568.4	298	147.78	766.4	408	208.89	964.4	518	270.00
176.0	80	26.67	374.0	190	87.78	572.0	300	148.89	770.0	410	210.00	968.0	520	271.11
179.6	82	27.78	377.6	192	88.89	575.6	302	150.00	773.6	412	211.11	971.6	522	272.22
183.2	84	28.89	381.2	194	90.00	579.2	304	151.11	777.2	414	212.22	975.2	524	273.33
186.8	86	30.00	384.8	196	91.11	582.8	306	152.22	780.8	416	213.33	978.8	526	274.44
190.4	88	31.11	388.4	198	92.22	586.4	308	153.33	784.4	418	214.44	982.4	528	275.56
194.0	90	32.22	392.0	200	93.33	590.0	310	154.44	788.0	420	215.56	986.0	530	276.67
197.6	92	33.33	395.6	202	94.44	593.6	312	155.56	791.6	422	216.67	989.6	532	277.78
201.2	94	34.44	399.2	204	95.56	597.2	314	156.67	795.2	424	217.78	993.2	534	278.89
204.8	96	35.56	402.8	206	96.67	600.8	316	157.78	798.8	426	218.89	996.8	536	280.00
208.4	98	36.67	406.4	208	97.78	604.4	318	158.89	802.4	428	220.00	1000.4	538	281.11
212.0	100	37.78	410.0	210	98.89	608.0	320	160.00	806.0	430	221.11	1004.0	540	282.22
215.6	102	38.89	413.6	212	100.00	611.6	322	161.11	809.6	432	222.22	1007.6	542	283.33
219.2	104	40.00	417.2	214	101.11	615.2	324	162.22	813.2	434	223.33	1011.2	544	284.44
222.8	106	41.11	420.8	216	102.22	618.8	326	163.33	816.8	436	224.44	1014.8	546	285.56
226.4	108	42.22	424.4	218	103.33	622.4	328	164.44	820.4	438	225.56	1018.4	548	286.67
230.0	110	43.33	428.0	220	104.44	626.0	330	165.56	824.0	440	226.67	1022.0	550	287.78
233.6	112	44.44	431.6	222	105.56	629.6	332	166.67	827.6	442	227.78	1040.0	560	293.33
237.2	114	45.56	435.2	224	106.67	633.2	334	167.78	831.2	444	228.89	1058.0	570	298.89
240.8	116	46.67	438.8	226	107.78	636.8	336	168.89	834.8	446	230.00	1076.0	580	304.44
244.4	118	47.78	442.4	228	108.89	640.4	338	170.00	838.4	448	231.11	1094.0	590	310.00
248.0	120	48.89	446.0	230	110.00	644.0	340	171.11	842.0	450	232.22	1112.0	600	315.56
251.6	122	50.00	449.6	232	111.11	647.6	342	172.22	845.6	452	233.33	1130.0	610	321.11
255.2	124	51.11	453.2	234	112.22	651.2	344	173.33	849.2	454	234.44	1148.0	620	326.67
258.8	126	52.22	456.8	236	113.33	654.8	346	174.44	852.8	456	235.56	1166.0	630	332.22
262.4	128	53.33	460.4	238	114.44	658.4	348	175.56	856.4	458	236.67	1184.0	640	337.78
266.0	130	54.44	464.0	240	115.56	662.0	350	176.67	860.0	460	237.78	1202.0	650	343.33
269.6	132	55.56	467.6	242	116.67	665.6	352	177.78	863.6	462	238.89	1220.0	660	348.89
273.2	134	56.67	471.2	244	117.78	669.2	354	178.89	867.2	464	240.00	1238.0	670	354.44
276.8	136	57.78	474.8	246	118.89	672.8	356	180.00	870.8	466	241.11	1256.0	680	360.00
280.4	138	58.89	478.4	248	120.00	676.4	358	181.11	874.4	468	242.22	1274.0	690	365.56
284.0	140	60.00	482.0	250	121.11	680.0	360	182.22	878.0	470	243.33	1292.0	700	371.11
287.6	142	61.11	485.6	252	122.22	683.6	362	183.33	881.6	472	244.44	1310.0	710	376.67
291.2	144	62.22	489.2	254	123.33	687.2	364	184.44	885.2	474	245.56	1328.0	720	382.22
294.8	146	63.33	492.8	256	124.44	690.8	366	185.56	888.8	476	246.67	1346.0	730	387.78
298.4	148	64.44	496.4	258	125.56	694.4	368	186.67	892.4	478	247.78	1364.0	740	393.33
302.0	150	65.56	500.0	260	126.67	698.0	370	187.78	896.0	480	248.89	1382.0	750	398.89

Temperature Conversions (continued)

°F		°C
1400.0	760	404.44
1418.0	770	410.00
1436.0	780	415.56
1454.0	790	421.11
1472.0	800	426.67
1490.0	810	432.22
1508.0	820	437.76
1526.0	830	443.33
1544.0	840	448.89
1562.0	850	454.44
1580.0	860	460.00
1598.0	870	465.56
1616.0	880	471.11
1634.0	890	476.67
1652.0	900	482.22
1670.0	910	487.78
1688.0	920	493.33
1706.0	930	498.89
1724.0	940	504.44
1742.0	950	510.00
1760.0	960	515.56
1778.0	970	521.11
1796.0	980	526.67
1814.0	990	532.22
1832.0	1000	537.78
1850.0	1010	543.33
1868.0	1020	548.89
1886.0	1030	554.44
1904.0	1040	560.00
1922.0	1050	565.56
1940.0	1060	571.11
1958.0	1070	576.67
1976.0	1080	582.22
1994.0	1090	587.78
2012.0	1100	593.33
2030.0	1110	598.89
2048.0	1120	604.44
2066.0	1130	610.00
2084.0	1140	615.56
2102.0	1150	621.11
2120.0	1160	626.67
2138.0	1170	632.22
2156.0	1180	637.78
2174.0	1190	643.33
2192.0	1200	648.89
2210.0	1210	654.44
2228.0	1220	660.00
2246.0	1230	665.56
2264.0	1240	671.11
2282.0	1250	676.67
2300.0	1260	682.22
2318.0	1270	687.78
2336.0	1280	693.33
2354.0	1290	698.89
2372.0	1300	704.44

°F		°C
2390.0	1310	710.00
2408.0	1320	715.56
2426.0	1330	721.11
2440.0	1340	726.67
2462.0	1350	732.22
2480.0	1360	737.78
2498.0	1370	743.33
2516.0	1380	748.89
2534.0	1390	754.44
2552.0	1400	760.00
2570.0	1410	765.56
2588.0	1420	771.11
2606.0	1430	776.67
2624.0	1440	782.22
2642.0	1450	787.78
2660.0	1460	793.33
2678.0	1470	798.89
2696.0	1480	804.44
2714.0	1490	810.00
2732.0	1500	815.56
2750.0	1510	821.11
2768.0	1520	826.67
2786.0	1530	832.22
2804.0	1540	837.78
2822.0	1550	843.33
2840.0	1560	848.89
2858.0	1570	854.44
2876.0	1580	860.00
2894.0	1590	865.56
2912.0	1600	871.11
2930.0	1610	876.67
2948.0	1620	882.22
2966.0	1630	887.78
2984.0	1640	893.33
3002.0	1650	898.89
3020.0	1660	904.44
3038.0	1670	910.00
3056.0	1680	915.56
3074.0	1690	921.11
3092.0	1700	926.67
3110.0	1710	932.22
3128.0	1720	937.78
3146.0	1730	943.33
3164.0	1740	948.89
3182.0	1750	954.44
3200.0	1760	960.00
3218.0	1770	965.56
3236.0	1780	971.11
3254.0	1790	976.67
3272.0	1800	982.22
3290.0	1810	987.78
3308.0	1820	993.33
3326.0	1830	998.89
3344.0	1840	1004.4
3362.0	1850	1010.0

°F		°C
3380.0	1860	1015.6
3398.0	1870	1021.1
3416.0	1880	1026.7
3434.0	1890	1032.2
3452.0	1900	1037.8
3470.0	1910	1043.3
3488.0	1920	1048.9
3506.0	1930	1054.4
3524.0	1940	1060.0
3542.0	1950	1065.6
3560.0	1960	1071.1
3578.0	1970	1076.7
3596.0	1980	1082.2
3614.0	1990	1087.8
3632.0	2000	1093.3
3650.0	2010	1098.9
3668.0	2020	1104.4
3686.0	2030	1110.0
3704.0	2040	1115.6
3722.0	2050	1121.1
3740.0	2060	1126.7
3758.0	2070	1132.2
3776.0	2080	1137.8
3794.0	2090	1143.3
3812.0	2100	1148.9
3830.0	2110	1154.4
3848.0	2120	1160.0
3866.0	2130	1165.6
3884.0	2140	1171.1
3902.0	2150	1176.7
3920.0	2160	1182.2
3938.0	2170	1187.8
3956.0	2180	1193.3
3974.0	2190	1198.9
3992.0	2200	1204.4
4010.0	2210	1210.0
4028.0	2220	1215.6
4046.0	2230	1221.1
4064.0	2240	1226.7
4082.0	2250	1232.2
4100.0	2260	1237.8
4118.0	2270	1243.3
4136.0	2280	1248.9
4154.0	2290	1254.4
4172.0	2300	1260.0
4190.0	2310	1265.6
4208.0	2320	1271.1
4226.0	2330	1276.7
4244.0	2340	1282.2
4262.0	2350	1287.8
4280.0	2360	1293.3
4298.0	2370	1298.9
4316.0	2380	1304.4
4334.0	2390	1310.0
4352.0	2400	1315.6

°F		°C
4370.0	2410	1321.1
4388.0	2420	1326.7
4406.0	2430	1332.2
4424.0	2440	1337.8
4442.0	2450	1343.3
4460.0	2460	1348.9
4478.0	2470	1354.4
4496.0	2480	1360.0
4514.0	2490	1365.6
4532.0	2500	1371.1
4550.0	2510	1376.7
4568.0	2520	1382.2
4586.0	2530	1387.8
4604.0	2540	1393.3
4622.0	2550	1398.9
4640.0	2560	1404.4
4658.0	2570	1410.0
4676.0	2580	1415.6
4694.0	2590	1421.1
4712.0	2600	1426.7
4730.0	2610	1432.2
4748.0	2620	1437.8
4766.0	2630	1443.3
4784.0	2640	1448.9
4802.0	2650	1454.4
4820.0	2660	1460.0
4838.0	2670	1465.6
4856.0	2680	1471.1
4874.0	2690	1476.7
4892.0	2700	1482.2
4910.0	2710	1487.8
4928.0	2720	1493.3
4946.0	2730	1498.9
4964.0	2740	1504.4
4982.0	2750	1510.0
5000.0	2760	1515.6
5018.0	2770	1521.1
5036.0	2780	1526.7
5054.0	2790	1532.2
5072.0	2800	1537.8
5090.0	2810	1543.3
5108.0	2820	1548.9
5126.0	2830	1554.4
5144.0	2840	1560.0
5162.0	2850	1565.6
5180.0	2860	1571.1
5198.0	2870	1576.7
5216.0	2880	1582.2
5234.0	2890	1587.8
5252.0	2900	1593.3
5270.0	2910	1598.9
5288.0	2920	1604.4
5306.0	2930	1610.0
5324.0	2940	1615.6
5342.0	2950	1621.1
5360.0	2960	1626.7
5378.0	2970	1632.2
5396.0	2980	1637.8
5414.0	2990	1643.3
5432.0	3000	1648.9

°F		°C
5450.0	3010	1654.4
5468.0	3020	1660.0
5486.0	3030	1665.6
5504.0	3040	1671.1
5522.0	3050	1676.7
5540.0	3060	1682.2
5558.0	3070	1687.8
5576.0	3080	1693.3
5594.0	3090	1698.9
5612.0	3100	1704.4
5702.0	3150	1732.2
5792.0	3200	1760.0
5882.0	3250	1787.7
5972.0	3300	1815.5
6062.0	3350	1843.3
6152.0	3400	1871.1
6242.0	3450	1898.8
6332.0	3500	1926.6
6422.0	3550	1954.4
6512.0	3600	1982.2
6602.0	3650	2010.0
6692.0	3700	2037.7
6782.0	3750	2065.5
6872.0	3800	2093.3
6962.0	3850	2121.1
7052.0	3900	2148.8
7142.0	3950	2176.6
7232.0	4000	2204.4
7322.0	4050	2232.2
7412.0	4100	2260.0
7502.0	4150	2287.7
7592.0	4200	2315.5
7682.0	4250	2343.3
7772.0	4300	2371.1
7862.0	4350	2398.8
7952.0	4400	2426.6
8042.0	4450	2454.4
8132.0	4500	2482.2
8222.0	4550	2510.0
8312.0	4600	2537.7
8402.0	4650	2565.5
8492.0	4700	2593.3
8582.0	4750	2621.1
8672.0	4800	2648.8
8762.0	4850	2676.6
8852.0	4900	2704.4
8942.0	4950	2732.2
9032.0	5000	2760.0
9122.0	5050	2787.7
9212.0	5100	2815.5
9302.0	5150	2843.3
9392.0	5200	2871.1
9482.0	5250	2898.8
9572.0	5300	2926.6
9662.0	5350	2954.4
9752.0	5400	2982.2
9842.0	5450	3010.0
9932.0	5500	3037.7
10 022.0	5550	3065.5
10 112.0	5600	3093.3

Stress or Pressure Conversions

The middle column of figures (in bold-faced type) contains the reading (in MPa or ksi) to be converted. If converting from ksi to MPa, read the MPa equivalent in the column headed "MPa". If converting from MPa to ksi, read the ksi equivalent in the column headed "ksi". 1 ksi = 6.894757 MPa. 1 psi = 6.894757 kPa.

ksi		MPa	ksi		MPa	ksi		MPa	ksi		MPa
0.14504	**1**	6.895	8.2672	**57**	393.00	33.359	**230**	1585.8	114.58	**790**	...
0.29008	**2**	13.790	8.4122	**58**	399.90	34.809	**240**	1654.7	116.03	**800**	...
0.43511	**3**	20.684	8.5572	**59**	406.79	36.259	**250**	1723.7	117.48	**810**	...
0.58015	**4**	27.579	8.7023	**60**	413.69	37.710	**260**	1792.6	118.93	**820**	...
0.72519	**5**	34.474	8.8473	**61**	420.58	39.160	**270**	1861.6	120.38	**830**	...
0.87023	**6**	41.369	8.9923	**62**	427.47	40.611	**280**	1930.5	121.83	**840**	...
1.0153	**7**	48.263	9.1374	**63**	434.37	42.061	**290**	1999.5	123.28	**850**	...
1.1603	**8**	55.158	9.2824	**64**	441.26	43.511	**300**	2068.4	124.73	**860**	...
1.3053	**9**	62.053	9.4275	**65**	448.16	44.962	**310**	2137.4	126.18	**870**	...
1.4504	**10**	68.948	9.5725	**66**	455.05	46.412	**320**	2206.3	127.63	**880**	...
1.5954	**11**	75.842	9.7175	**67**	461.95	47.862	**330**	2275.3	129.08	**890**	...
1.7405	**12**	82.737	9.8626	**68**	468.84	49.313	**340**	2344.2	130.53	**900**	...
1.8855	**13**	89.632	10.008	**69**	475.74	50.763	**350**	2413.2	131.98	**910**	...
2.0305	**14**	96.527	10.153	**70**	482.63	52.214	**360**	2482.1	133.43	**920**	...
2.1756	**15**	103.42	10.298	**71**	489.53	53.664	**370**	2551.1	134.89	**930**	...
2.3206	**16**	110.32	10.443	**72**	496.42	55.114	**380**	2620.0	136.34	**940**	...
2.4656	**17**	117.21	10.588	**73**	503.32	56.565	**390**	2689.0	137.79	**950**	...
2.6107	**18**	124.11	10.733	**74**	510.21	58.015	**400**	2757.9	139.24	**960**	...
2.7557	**19**	131.00	10.878	**75**	517.11	59.465	**410**	2826.9	140.69	**970**	...
2.9008	**20**	137.90	11.023	**76**	524.00	60.916	**420**	2895.8	142.14	**980**	...
3.0458	**21**	144.79	11.168	**77**	530.90	62.366	**430**	2964.7	143.59	**990**	...
3.1908	**22**	151.68	11.313	**78**	537.79	63.817	**440**	3033.7	145.04	**1000**	...
3.3359	**23**	158.58	11.458	**79**	544.69	65.267	**450**	3102.6	147.94	**1020**	...
3.4809	**24**	165.47	11.603	**80**	551.58	66.717	**460**	3171.6	150.84	**1040**	...
3.6259	**25**	172.37	11.748	**81**	558.48	66.168	**470**	3240.5	153.74	**1060**	...
3.7710	**26**	179.26	11.893	**82**	565.37	69.618	**480**	3309.5	156.64	**1080**	...
3.9160	**27**	186.16	12.038	**83**	572.26	71.068	**490**	3378.4	159.54	**1100**	...
4.0611	**28**	193.05	12.183	**84**	579.16	72.519	**500**	3447.4	162.44	**1120**	...
4.2061	**29**	199.95	12.328	**85**	586.05	73.969	**510**	...	165.34	**1140**	...
4.3511	**30**	206.84	12.473	**86**	592.95	75.420	**520**	...	168.24	**1160**	...
4.4962	**31**	213.74	12.618	**87**	599.84	76.870	**530**	...	171.14	**1180**	...
4.6412	**32**	220.63	12.763	**88**	606.74	78.320	**540**	...	174.05	**1200**	...
4.7862	**33**	227.53	12.909	**89**	613.63	79.771	**550**	...	176.95	**1220**	...
4.9313	**34**	234.42	13.053	**90**	620.53	81.221	**560**	...	179.85	**1240**	...
5.0763	**35**	241.32	13.198	**91**	627.42	82.672	**570**	...	182.75	**1260**	...
5.2214	**36**	248.21	13.343	**92**	634.32	84.122	**580**	...	185.65	**1280**	...
5.3664	**37**	255.11	13.489	**93**	641.21	85.572	**590**	...	188.55	**1300**	...
5.5114	**38**	262.00	13.634	**94**	648.11	87.023	**600**	...	191.45	**1320**	...
5.6565	**39**	268.90	13.779	**95**	655.00	88.473	**610**	...	194.35	**1340**	...
5.8015	**40**	275.79	13.924	**96**	661.90	89.923	**620**	...	197.25	**1360**	...
5.9465	**41**	282.69	14.069	**97**	668.79	91.374	**630**	...	200.15	**1380**	...
6.0916	**42**	289.58	14.214	**98**	675.69	92.824	**640**	...	203.05	**1400**	...
6.2366	**43**	296.47	14.359	**99**	682.58	94.275	**650**	...	205.95	**1420**	...
6.3817	**44**	303.37	14.504	**100**	689.48	95.725	**660**	...	208.85	**1440**	...
6.5267	**45**	310.26	15.954	**110**	758.42	97.175	**670**	...	211.76	**1460**	...
6.6717	**46**	317.16	17.405	**120**	827.37	98.626	**680**	...	214.66	**1480**	...
6.8168	**47**	324.05	18.855	**130**	896.32	100.08	**690**	...	217.56	**1500**	...
6.9618	**48**	330.95	20.305	**140**	965.27	101.53	**700**	...	220.46	**1520**	...
7.1068	**49**	337.84	21.756	**150**	1034.2	102.98	**710**	...	223.36	**1540**	...
7.2519	**50**	344.74	23.206	**160**	1103.2	104.43	**720**	...	226.26	**1560**	...
7.3969	**51**	351.63	24.656	**170**	1172.1	105.88	**730**	...	229.16	**1580**	...
7.5420	**52**	358.53	26.107	**180**	1241.1	107.33	**740**	...	232.06	**1600**	...
7.6870	**53**	365.42	27.557	**190**	1310.0	108.78	**750**	...	234.96	**1620**	...
7.8320	**54**	372.32	29.008	**200**	1379.0	110.23	**760**	...	237.86	**1640**	...
7.9771	**55**	379.21	30.458	**210**	1447.9	111.68	**770**	...	240.76	**1660**	...
8.1221	**56**	386.11	31.908	**220**	1516.8	113.13	**780**	...	243.66	**1680**	...

Stress or Pressure Conversions (continued)

ksi		MPa	ksi		MPa	ksi		MPa	ksi		MPa
246.56	1700	· · ·	278.47	1920	· · ·	310.38	2140	· · ·	342.29	2360	· · ·
249.46	1720	· · ·	281.37	1940	· · ·	313.28	2160	· · ·	345.19	2380	· · ·
252.37	1740	· · ·	284.27	1960	· · ·	316.18	2180	· · ·	348.09	2400	· · ·
255.27	1760	· · ·	287.17	1980	· · ·	319.08	2200	· · ·	350.99	2420	· · ·
258.17	1780	· · ·	290.08	2000	· · ·	321.98	2220	· · ·	353.89	2440	· · ·
261.07	1800	· · ·	292.98	2020	· · ·	324.88	2240	· · ·	356.79	2460	· · ·
263.97	1820	· · ·	295.88	2040	· · ·	327.79	2260	· · ·	359.69	2480	· · ·
266.87	1840	· · ·	298.78	2060	· · ·	330.69	2280	· · ·	362.59	2500	· · ·
269.77	1860	· · ·	301.68	2080	· · ·	333.59	2300	· · ·			
272.67	1880	· · ·	304.58	2100	· · ·	336.49	2320	· · ·			
275.57	1900	· · ·	307.48	2120	· · ·	339.39	2340	· · ·			

Stress Intensity Conversions

The middle column of figures (in bold-faced type) contains the reading (in MPa$\sqrt{\text{m}}$ or ksi$\sqrt{\text{in.}}$) to be converted. If converting from ksi$\sqrt{\text{in.}}$ to MPa$\sqrt{\text{m}}$, read the MPa$\sqrt{\text{m}}$ equivalent in the column headed "MPa$\sqrt{\text{m}}$". If converting from MPa$\sqrt{\text{m}}$ to ksi$\sqrt{\text{in.}}$, read the ksi$\sqrt{\text{in.}}$ equivalent in the column headed "ksi$\sqrt{\text{in.}}$." 1 ksi$\sqrt{\text{in.}}$ = 1.098845 MPa$\sqrt{\text{m}}$.

ksi, $\sqrt{\text{in.}}$		MPa, $\sqrt{\text{m}}$	ksi, $\sqrt{\text{in.}}$		MPa, $\sqrt{\text{m}}$	ksi, $\sqrt{\text{in.}}$		MPa, $\sqrt{\text{m}}$	ksi, $\sqrt{\text{in.}}$		MPa, $\sqrt{\text{m}}$	ksi, $\sqrt{\text{in.}}$		MPa, $\sqrt{\text{m}}$
0.91005	**1**	1.0988	31.852	**35**	38.458	62.793	**69**	75.817	93.735	**103**	113.18	124.68	**137**	150.54
1.8201	**2**	2.1976	32.762	**36**	39.557	63.703	**70**	76.916	94.645	**104**	114.28	125.59	**138**	151.63
2.7301	**3**	3.2964	33.672	**37**	40.656	64.613	**71**	78.015	95.555	**105**	115.37	126.50	**139**	152.73
3.6402	**4**	4.3952	34.582	**38**	41.754	65.523	**72**	79.114	96.465	**106**	116.47	127.41	**140**	153.83
4.5502	**5**	5.4940	35.492	**39**	42.853	66.433	**73**	80.212	97.375	**107**	117.57	128.32	**141**	154.93
5.4603	**6**	6.5928	36.402	**40**	43.952	67.343	**74**	81.311	98.285	**108**	118.67	129.23	**142**	156.03
6.3703	**7**	7.6916	37.312	**41**	45.051	68.253	**75**	82.410	99.195	**109**	119.77	130.14	**143**	157.13
7.2804	**8**	8.7904	38.222	**42**	46.150	69.164	**76**	83.509	100.11	**110**	120.87	131.05	**144**	158.23
8.1904	**9**	9.8892	39.132	**43**	47.248	70.074	**77**	84.608	101.02	**111**	121.97	131.96	**145**	159.33
9.1005	**10**	10.988	40.042	**44**	48.347	70.984	**78**	85.706	101.93	**112**	123.07	132.87	**146**	160.42
10.011	**11**	12.087	40.952	**45**	49.446	71.893	**79**	86.805	102.84	**113**	124.16	133.78	**147**	161.52
10.921	**12**	13.186	41.862	**46**	50.545	72.804	**80**	87.904	103.75	**114**	125.26	134.69	**148**	162.62
11.831	**13**	14.284	42.772	**47**	51.644	73.714	**81**	89.003	104.66	**115**	126.36	135.60	**149**	163.72
12.741	**14**	15.383	43.682	**48**	52.742	74.624	**82**	90.102	105.57	**116**	127.46	136.51	**150**	164.82
13.651	**15**	16.482	44.592	**49**	53.841	75.534	**83**	91.200	106.48	**117**	128.56	137.42	**151**	165.92
14.561	**16**	17.581	45.502	**50**	54.940	76.444	**84**	92.300	107.39	**118**	129.66	138.33	**152**	167.02
15.471	**17**	18.680	46.412	**51**	56.039	77.354	**85**	93.398	108.30	**119**	130.76	139.24	**153**	168.12
16.381	**18**	19.778	47.322	**52**	57.138	78.264	**86**	94.497	109.21	**120**	131.86	140.15	**154**	169.22
17.291	**19**	20.877	48.232	**53**	58.236	79.174	**87**	95.596	110.12	**121**	132.95	141.06	**155**	170.31
18.201	**20**	21.976	49.143	**54**	59.335	80.084	**88**	96.694	111.03	**122**	134.05	141.97	**156**	171.41
19.111	**21**	23.075	50.053	**55**	60.434	80.994	**89**	97.793	111.94	**123**	135.15	142.88	**157**	172.51
20.021	**22**	24.174	50.963	**56**	61.533	81.904	**90**	98.892	112.85	**124**	136.25	143.79	**158**	173.61
20.931	**23**	25.272	51.873	**57**	62.632	82.814	**91**	99.991	113.76	**125**	137.35	144.70	**159**	174.71
21.841	**24**	26.371	52.783	**58**	63.730	83.724	**92**	101.09	114.67	**126**	138.45	145.61	**160**	175.81
22.751	**25**	27.470	53.693	**59**	64.829	84.634	**93**	102.19	115.58	**127**	139.55	146.52	**161**	176.91
23.661	**26**	28.569	54.603	**60**	65.928	85.544	**94**	103.29	116.49	**128**	140.65	147.43	**162**	178.01
24.571	**27**	29.668	55.513	**61**	67.027	86.454	**95**	104.39	117.40	**129**	141.75	148.34	**163**	179.10
25.481	**28**	30.766	56.423	**62**	68.126	87.364	**96**	105.48	118.31	**130**	142.84	149.25	**164**	180.20
26.391	**29**	31.865	57.333	**63**	69.224	88.275	**97**	106.58	119.22	**131**	143.94	150.16	**165**	181.30
27.301	**30**	32.964	58.243	**64**	70.323	89.185	**98**	107.68	120.13	**132**	145.04	151.07	**166**	182.40
28.211	**31**	34.063	59.153	**65**	71.422	90.095	**99**	108.78	121.04	**133**	146.14	151.98	**167**	183.50
29.121	**32**	35.162	60.063	**66**	72.521	91.005	**100**	109.88	121.95	**134**	147.24	152.89	**168**	184.60
30.032	**33**	36.260	60.973	**67**	73.620	91.915	**101**	110.98	122.86	**135**	148.34	153.80	**169**	185.70
30.942	**34**	37.359	61.883	**68**	74.718	92.825	**102**	112.08	123.77	**136**	149.44	154.71	**170**	186.80

Stress Intensity Conversions (continued)

ksi√in.		MPa√m	ksi√in.		MPa√m	ksi√in.		MPa√m	ksi√in.		MPa√m	ksi√in.		MPa√m
155.62	**171**	187.90	161.08	**177**	194.49	166.54	**183**	201.08	172.00	**189**	207.67	177.46	**195**	214.27
156.53	**172**	189.00	161.99	**178**	195.59	167.45	**184**	202.18	172.91	**190**	208.77	178.37	**196**	215.36
157.44	**173**	190.10	162.90	**179**	196.69	168.36	**185**	203.28	173.82	**191**	209.87	179.28	**197**	216.46
158.35	**174**	191.19	163.81	**180**	197.78	169.27	**186**	204.38	174.73	**192**	210.97	180.19	**198**	217.56
159.26	**175**	192.29	164.72	**181**	198.88	170.18	**187**	205.48	175.64	**193**	212.07	181.10	**199**	218.66
160.17	**176**	193.39	165.63	**182**	199.98	171.09	**188**	206.57	176.55	**194**	213.17	182.01	**200**	219.76

Energy Conversions

The middle column of figures (in bold-faced type) contains the reading (in J or ft·lb) to be converted. If converting from ft·lb to J, read the J equivalent in the column headed "J". If converting from J to ft·lb, read the equivalent in the column headed "ft·lb". 1 ft·lb = 1.355818 J.

ft·lb		J	ft·lb		J	ft·lb		J	ft·lb		J
0.7376	**1**	1.3558	28.7649	**39**	52.8769	56.7923	**77**	104.3980	129.0734	**175**	237.2681
1.4751	**2**	2.7116	29.5025	**40**	54.2327	57.5298	**78**	105.7538	132.7612	**180**	244.0472
2.2127	**3**	4.0675	30.2400	**41**	55.5885	58.2674	**79**	107.1096	136.4490	**185**	250.8263
2.9502	**4**	5.4233	30.9776	**42**	56.9444	59.0050	**80**	108.4654	140.1368	**190**	257.6054
3.6878	**5**	6.7791	31.7152	**43**	58.3002	59.7425	**81**	109.8212	143.8246	**195**	264.3845
4.4254	**6**	8.1349	32.4527	**44**	59.6560	60.4801	**82**	111.1771	147.5124	**200**	271.1636
5.1629	**7**	9.4907	33.1903	**45**	61.0118	61.2177	**83**	112.5329	154.8880	**210**	284.7218
5.9005	**8**	10.8465	33.9279	**46**	62.3676	61.9552	**84**	113.8887	162.2637	**220**	298.2799
6.6381	**9**	12.2024	34.6654	**47**	63.7234	62.6928	**85**	115.2445	169.6393	**230**	311.8381
7.3756	**10**	13.5582	35.4030	**48**	65.0793	63.4303	**86**	116.6003	177.0149	**240**	325.3963
8.1132	**11**	14.9140	36.1405	**49**	66.4351	64.1679	**87**	117.9562	184.3905	**250**	338.9545
8.8507	**12**	16.2698	36.8781	**50**	67.7909	64.9055	**88**	119.3120	191.7661	**260**	352.5126
9.5883	**13**	17.6256	37.6157	**51**	69.1467	65.6430	**89**	120.6678	199.1418	**270**	366.0708
10.3259	**14**	18.9815	38.3532	**52**	70.5025	66.3806	**90**	122.0236	206.5174	**280**	379.6290
11.0634	**15**	20.3373	39.0908	**53**	71.8583	67.1182	**91**	123.3794	213.8930	**290**	393.1872
11.8010	**16**	21.6931	39.8284	**54**	73.2142	67.8557	**92**	124.7452	221.2686	**300**	406.7454
12.5386	**17**	23.0489	40.5659	**55**	74.5700	68.5933	**93**	126.0911	228.6442	**310**	420.3036
13.2761	**18**	24.4047	41.3035	**56**	75.9258	69.3308	**94**	127.4469	236.0199	**320**	433.8617
14.0137	**19**	25.7605	42.0410	**57**	77.2816	70.0684	**95**	128.8027	243.3955	**330**	447.4199
14.7512	**20**	27.1164	42.7786	**58**	78.6374	70.8060	**96**	130.1585	250.7711	**340**	460.9781
15.4888	**21**	28.4722	43.5162	**59**	79.9933	71.5435	**97**	131.5143	258.1467	**350**	474.5363
16.2264	**22**	29.8280	44.2537	**60**	81.3491	72.2811	**98**	132.8702	265.5224	**360**	488.0944
16.9639	**23**	31.1838	44.9913	**61**	82.7049	73.0186	**99**	134.2260	272.8980	**370**	501.6526
17.7015	**24**	32.5396	45.7288	**62**	84.0607	73.7562	**100**	135.5818	280.2736	**380**	515.2108
18.4390	**25**	33.8954	46.4664	**63**	85.4165	77.4440	**105**	142.3609	287.6492	**390**	528.7690
19.1766	**26**	35.2513	47.2040	**64**	86.7723	81.1318	**110**	149.1400	295.0248	**400**	542.3272
19.9142	**27**	36.6071	47.9415	**65**	88.1282	84.8196	**115**	155.9191	302.4005	**410**	555.8854
20.6517	**28**	37.9629	48.6791	**66**	89.4840	88.5075	**120**	162.6982	309.7761	**420**	569.4435
21.3893	**29**	39.3187	49.4167	**67**	90.8398	92.1953	**125**	169.4772	317.1517	**430**	583.0017
22.1269	**30**	40.6745	50.1542	**68**	92.1956	95.8831	**130**	176.2563	324.5273	**440**	596.5599
22.8644	**31**	42.0304	50.8918	**69**	93.5514	99.5709	**135**	183.0354	331.9029	**450**	610.1181
23.6020	**32**	43.3862	51.6293	**70**	94.9073	103.2587	**140**	189.8145	339.2786	**460**	623.6762
24.3395	**33**	44.7420	52.3669	**71**	96.2631	106.9465	**145**	196.5936	346.6542	**470**	637.2344
25.0771	**34**	46.0978	53.1045	**72**	97.6189	110.6343	**150**	203.3727	354.0298	**480**	650.7926
25.8147	**35**	47.4536	53.8420	**73**	98.9747	114.3221	**155**	210.1518	361.4054	**490**	664.3508
26.5522	**36**	48.8094	54.5796	**74**	100.3305	118.0099	**160**	216.9308	368.7811	**500**	677.9090
27.2898	**37**	50.1653	55.3172	**75**	101.6863	121.6977	**165**	223.7099			
28.0274	**38**	51.5211	56.0547	**76**	103.0422	125.3856	**170**	230.4890			

Approximate Equivalent Hardness Numbers and Tensile Strengths for Vickers Hardness Numbers for Steel (a)

Vickers hardness	Brinell hardness, 3000-kg load, 10-mm ball		Rockwell hardness				Rockwell superficial hardness, superficial Brale indenter			Knoop hardness, 500-g load and greater	Sclero-scope hardness	Tensile strength (approx), ksi	Vickers hardness
	Standard ball	Tungsten carbide ball	A scale, 60-kg load, Brale indenter	B scale, 100-kg load, 1/16-in. diam ball	C scale, 150-kg load, Brale indenter	D scale, 100-kg load, Brale indenter	15N scale, 15-kg load	30N scale, 30-kg load	45N scale, 45-kg load				
940	· · ·	· · ·	85.6	· · ·	68.0	76.9	93.2	84.4	75.4	920	97	· · ·	940
920	· · ·	· · ·	85.3	· · ·	67.5	76.5	93.0	84.0	74.8	908	96	· · ·	920
900	· · ·	· · ·	85.0	· · ·	67.0	76.1	92.9	83.6	74.2	895	95	· · ·	900
880	· · ·	(767)	84.7	· · ·	66.4	75.7	92.7	83.1	73.6	882	93	· · ·	880
860	· · ·	(757)	84.4	· · ·	65.9	75.3	92.5	82.7	73.1	867	92	· · ·	860
840	· · ·	(745)	84.1	· · ·	65.3	74.8	92.3	82.2	72.2	852	91	· · ·	840
820	· · ·	(733)	83.8	· · ·	64.7	74.3	92.1	81.7	71.8	837	90	· · ·	820
800	· · ·	(722)	83.4	· · ·	64.0	73.8	91.8	81.1	71.0	822	88	· · ·	800
780	· · ·	(710)	83.0	· · ·	63.3	73.3	91.5	80.4	70.2	806	87	· · ·	780
760	· · ·	(698)	82.6	· · ·	62.5	72.6	91.2	79.7	69.4	788	86	· · ·	760
740	· · ·	(684)	82.2	· · ·	61.8	72.1	91.0	79.1	68.6	772	84	· · ·	740
720	· · ·	(670)	81.8	· · ·	61.0	71.5	90.7	78.4	67.7	754	83	· · ·	720
700	· · ·	(656)	81.3	· · ·	60.1	70.8	90.3	77.6	66.7	735	81	· · ·	700
690	· · ·	(647)	81.1	· · ·	59.7	70.5	90.1	77.2	66.2	725	· · ·	· · ·	690
680	· · ·	(638)	80.8	· · ·	59.2	70.1	89.8	76.8	65.7	716	80	355	680
670	· · ·	(630)	80.6	· · ·	58.8	69.8	89.7	76.4	65.3	706	· · ·	348	670
660	· · ·	620	80.3	· · ·	58.3	69.4	89.5	75.9	64.7	697	79	342	660
650	· · ·	611	80.0	· · ·	57.8	69.0	89.2	75.5	64.1	687	78	336	650
640	· · ·	601	79.8	· · ·	57.3	68.7	89.0	75.1	63.5	677	77	328	640
630	· · ·	591	79.5	· · ·	56.8	68.3	88.8	74.6	63.0	667	76	323	630
620	· · ·	582	79.2	· · ·	56.3	67.9	88.5	74.2	62.4	657	75	317	620
610	· · ·	573	78.9	· · ·	55.7	67.5	88.2	73.6	61.7	646	· · ·	310	610
600	· · ·	564	78.6	· · ·	55.2	67.0	88.0	73.2	61.2	636	74	303	600
590	· · ·	554	78.4	· · ·	54.7	66.7	87.8	72.7	60.5	625	73	298	590
580	· · ·	545	78.0	· · ·	54.1	66.2	87.5	72.1	59.9	615	72	293	580
570	· · ·	535	77.8	· · ·	53.6	65.8	87.2	71.7	59.3	604	· · ·	288	570
560	· · ·	525	77.4	· · ·	53.0	65.4	86.9	71.2	58.6	594	71	283	560
550	(505)	517	77.0	· · ·	52.3	64.8	86.6	70.5	57.8	583	70	276	550
540	(496)	507	76.7	· · ·	51.7	64.4	86.3	70.0	57.0	572	69	270	540
530	(488)	497	76.4	· · ·	51.1	63.9	86.0	69.5	56.2	561	68	265	530
520	(480)	488	76.1	· · ·	50.5	63.5	85.7	69.0	55.6	550	67	260	520
510	(473)	479	75.7	· · ·	49.8	62.9	85.4	68.3	54.7	539	· · ·	254	510
500	(465)	471	75.3	· · ·	49.1	62.2	85.0	67.7	53.9	528	66	247	500
490	(456)	460	74.9	· · ·	48.4	61.6	84.7	67.1	53.1	517	65	241	490
480	(448)	452	74.5	· · ·	47.7	61.3	84.3	66.4	52.2	505	64	235	480
470	441	442	74.1	· · ·	46.9	60.7	83.9	65.7	51.3	494	· · ·	228	470
460	433	433	73.6	· · ·	46.1	60.1	83.6	64.9	50.4	482	62	223	460
450	425	425	73.3	· · ·	45.3	59.4	83.2	64.3	49.4	471	· · ·	217	450
440	415	415	72.8	· · ·	44.5	58.8	82.8	63.5	48.4	459	59	212	440
430	405	405	72.3	· · ·	43.6	58.2	82.3	62.7	47.4	447	58	205	430

Approximate Equivalent Hardness Numbers and Tensile Strengths for Vickers Hardness Numbers for Steel (a) (continued)

Vickers hardness	Brinell hardness, 3000-kg load, 10-mm ball		Rockwell hardness				Rockwell superficial hardness, superficial Brale indenter			Knoop hardness, 500-g load and greater	Scleroscope hardness	Tensile strength (approx), ksi	Vickers hardness
	Standard ball	Tungsten carbide ball	A scale, 60-kg load, Brale indenter	B scale, 100-kg load, 1/16-in. diam ball	C scale, 150-kg load, Brale indenter	D scale, 100-kg load, Brale indenter	15N scale, 15-kg load	30N scale, 30-kg load	45N scale, 45-kg load				
420	397	397	71.8	· · ·	42.7	57.5	81.8	61.9	46.4	435	57	199	420
410	388	388	71.4	· · ·	41.8	56.8	81.4	61.1	45.3	423	56	193	410
400	379	379	70.8	· · ·	40.8	56.0	80.8	60.2	44.1	412	55	187	400
390	369	369	70.3	· · ·	39.8	55.2	80.3	59.3	42.9	400	· · ·	181	390
380	360	360	69.8	(110.0)	38.8	54.4	79.8	58.4	41.7	389	52	175	380
370	350	350	69.2	· · ·	37.7	53.6	79.2	57.4	40.4	378	51	170	370
360	341	341	68.7	(109.0)	36.6	52.8	78.6	56.4	39.1	367	50	164	360
350	331	331	68.1	· · ·	35.5	51.9	78.0	55.4	37.8	356	48	159	350
340	322	322	67.6	(108.0)	34.4	51.1	77.4	54.4	36.5	346	47	155	340
330	313	313	67.0	· · ·	33.3	50.2	76.8	53.6	35.2	337	46	150	330
320	303	303	66.4	(107.0)	32.2	49.4	76.2	52.3	33.9	328	45	146	320
310	294	294	65.8	· · ·	31.0	48.4	75.6	51.3	32.5	318	· · ·	142	310
300	284	284	65.2	(105.5)	29.8	47.5	74.9	50.2	31.1	309	42	138	300
295	280	280	64.8	· · ·	29.2	47.1	74.6	49.7	30.4	305	· · ·	136	295
290	275	275	64.5	(104.5)	28.5	46.5	74.2	49.0	29.5	300	41	133	290
285	270	270	64.2	· · ·	27.8	46.0	73.8	48.4	28.7	296	· · ·	131	285
280	265	265	63.8	(103.5)	27.1	45.3	73.4	47.8	27.9	291	40	129	280
275	261	261	63.5	· · ·	26.4	44.9	73.0	47.2	27.1	286	39	127	275
270	256	256	63.1	(102.0)	25.6	44.3	72.6	46.4	26.2	282	38	124	270
265	252	252	62.7	· · ·	24.8	43.7	72.1	45.7	25.2	277	· · ·	122	265
260	247	247	62.4	(101.0)	24.0	43.1	71.6	45.0	24.3	272	37	120	260
255	243	243	62.0	· · ·	23.1	42.2	71.1	44.2	23.2	267	· · ·	117	255
250	238	238	61.6	99.5	22.2	41.7	70.6	43.4	22.2	262	36	115	250
245	233	233	61.2	· · ·	21.3	41.1	70.1	42.5	21.1	258	35	113	245
240	228	228	60.7	98.1	20.3	40.3	69.6	41.7	19.9	253	34	111	240
230	219	219	· · ·	96.7	(18.0)	· · ·	· · ·	· · ·	· · ·	243	33	106	230
220	209	209	· · ·	95.0	(15.7)	· · ·	· · ·	· · ·	· · ·	234	32	101	220
210	200	200	· · ·	93.4	(13.4)	· · ·	· · ·	· · ·	· · ·	226	30	97	210
200	190	190	· · ·	91.5	(11.0)	· · ·	· · ·	· · ·	· · ·	216	29	92	200
190	181	181	· · ·	89.5	(8.5)	· · ·	· · ·	· · ·	· · ·	206	28	88	190
180	171	171	· · ·	87.1	(6.0)	· · ·	· · ·	· · ·	· · ·	196	26	84	180
170	162	162	· · ·	85.0	(3.0)	· · ·	· · ·	· · ·	· · ·	185	25	79	170
160	152	152	· · ·	81.7	(0.0)	· · ·	· · ·	· · ·	· · ·	175	23	75	160
150	143	143	· · ·	78.7	· · ·	· · ·	· · ·	· · ·	· · ·	164	22	71	150
140	133	133	· · ·	75.0	· · ·	· · ·	· · ·	· · ·	· · ·	154	21	66	140
130	124	124	· · ·	71.2	· · ·	· · ·	· · ·	· · ·	· · ·	143	20	62	130
120	114	114	· · ·	66.7	· · ·	· · ·	· · ·	· · ·	· · ·	133	18	57	120
110	105	105	· · ·	62.3	· · ·	· · ·	· · ·	· · ·	· · ·	123	· · ·	· · ·	110
100	95	95	· · ·	56.2	· · ·	· · ·	· · ·	· · ·	· · ·	112	· · ·	· · ·	100
95	90	90	· · ·	52.0	· · ·	· · ·	· · ·	· · ·	· · ·	107	· · ·	· · ·	95
90	86	86	· · ·	48.0	· · ·	· · ·	· · ·	· · ·	· · ·	102	· · ·	· · ·	90
85	81	81	· · ·	41.0	· · ·	· · ·	· · ·	· · ·	· · ·	97	· · ·	· · ·	85

(a) For carbon and alloy steels in the annealed, normalized, and quenched-and-tempered conditions: less accurate for cold worked condition and for austenitic steels. The values in bold-faced type correspond to the values in the joint SAE-ASM-ASTM hardness conversions as printed in ASTM E140, Table 1. The values in parentheses are beyond normal range and are given for information only.

Approximate Equivalent Hardness Numbers and Tensile Strengths for Brinell Hardness Numbers for Steel(a)

Brinell indentation diam, mm	Brinell hardness (b), 3000-kg load, 10-mm ball		Vickers hardness	Rockwell hardness				Rockwell superficial hardness, superficial Brale indenter			Knoop hardness, 500-g load and greater	Sclero-scope hardness	Tensile strength (approx), ksi	Brinell indentation diam, mm
	Standard ball	Tungsten carbide ball		A scale, 60-kg load, Brale indenter	B scale, 100-kg load, 1/16-in.-diam ball	C scale, 150-kg load, Brale indenter	D scale, 100-kg load, Brale indenter	15N scale, 15-kg load	30N scale, 30-kg load	45N scale, 45-kg load				
2.25	· · ·	(745)	840	84.1	· · ·	65.3	74.8	92.3	82.2	72.2	852	91	· · ·	2.25
2.30	· · ·	(712)	783	83.1	· · ·	63.4	73.4	91.6	80.5	70.4	808	· · ·	· · ·	2.30
2.35	· · ·	(682)	737	82.2	· · ·	61.7	72.0	91.0	79.0	68.5	768	84	· · ·	2.35
2.40	· · ·	(653)	697	81.2	· · ·	60.0	70.7	90.2	77.5	66.5	732	81	· · ·	2.40
2.45	· · ·	627	667	80.5	· · ·	58.7	69.7	89.6	76.3	65.1	703	79	347	2.45
2.50	· · ·	601	640	79.8	· · ·	57.3	68.7	89.0	75.1	63.5	677	77	328	2.50
2.55	· · ·	578	615	79.1	· · ·	56.0	67.7	88.4	73.9	62.1	652	75	313	2.55
2.60	· · ·	555	591	78.4	· · ·	54.7	66.7	87.8	72.7	60.6	626	73	298	2.60
2.65	· · ·	534	569	77.8	· · ·	53.5	65.8	87.2	71.6	59.2	604	71	288	2.65
2.70	· · ·	514	547	76.9	· · ·	52.1	64.7	86.5	70.3	57.6	579	70	273	2.70
2.75	(495)	· · ·	539	76.7	· · ·	51.6	64.3	86.3	69.9	56.9	571	· · ·	269	2.75
	· · ·	495	528	76.3	· · ·	51.0	63.8	85.9	69.4	56.1	558	68	263	
2.80	(477)	· · ·	516	75.9	· · ·	50.3	63.2	85.6	68.7	55.2	545	· · ·	257	2.80
	· · ·	477	508	75.6	· · ·	49.6	62.7	85.3	68.2	54.5	537	66	252	
2.85	(461)	· · ·	495	75.1	· · ·	48.8	61.9	84.9	67.4	53.5	523	· · ·	244	2.85
	· · ·	461	491	74.9	· · ·	48.5	61.7	84.7	67.2	53.2	518	65	242	
2.90	444	· · ·	474	74.3	· · ·	47.2	61.0	84.1	66.0	51.7	499	· · ·	231	2.90
	· · ·	444	472	74.2	· · ·	47.1	60.8	84.0	65.8	51.5	496	63	229	
2.95	429	429	455	73.4	· · ·	45.7	59.7	83.4	64.6	49.9	476	61	220	2.95
3.00	415	415	440	72.8	· · ·	44.5	58.8	82.8	63.5	48.4	459	59	212	3.00
3.05	401	401	425	72.0	· · ·	43.1	57.8	82.0	62.3	46.9	441	58	202	3.05
3.10	388	388	410	71.4	· · ·	41.8	56.8	81.4	61.1	45.3	423	56	193	3.10
3.15	375	375	396	70.6	· · ·	40.4	55.7	80.6	59.9	43.6	407	54	184	3.15
3.20	363	363	383	70.0	· · ·	39.1	54.6	80.0	58.7	42.0	392	52	177	3.20
3.25	352	352	372	69.3	(110.0)	37.9	53.8	79.3	57.6	40.5	379	51	172	3.25
3.30	341	341	360	68.7	(109.0)	36.6	52.8	78.6	56.4	39.1	367	50	164	3.30
3.35	331	331	350	68.1	(108.5)	35.5	51.9	78.0	55.4	37.8	356	48	159	3.35
3.40	321	321	339	67.5	(108.0)	34.3	51.0	77.3	54.3	36.4	345	47	154	3.40
3.45	311	311	328	66.9	(107.5)	33.1	50.0	76.7	53.3	34.4	336	46	149	3.45
3.50	302	302	319	66.3	(107.0)	32.1	49.3	76.1	52.2	33.8	327	45	146	3.50
3.55	293	293	309	65.7	(106.0)	30.9	48.3	75.5	51.2	32.4	318	43	142	3.55
3.60	285	285	301	65.3	(105.5)	29.9	47.6	75.0	50.3	31.2	310	42	138	3.60
3.65	277	277	292	64.6	(104.5)	28.8	46.7	74.4	49.3	29.9	302	41	134	3.65
3.70	269	269	284	64.1	(104.0)	27.6	45.9	73.7	48.3	28.5	294	40	131	3.70
3.75	262	262	276	63.6	(103.0)	26.6	45.0	73.1	47.3	27.3	286	39	127	3.75
3.80	255	255	269	63.0	(102.0)	25.4	44.2	72.5	46.2	26.0	279	38	123	3.80

Approximate Equivalent Hardness Numbers and Tensile Strengths for Brinell Hardness Numbers for Steel(a) (continued)

| Brinell indentation diam, mm | Brinell hardness (b), 3000-kg load, 10-mm ball | | Vickers hardness | Rockwell hardness | | | | Rockwell superficial hardness, superficial Brale indenter | | | Knoop hardness, 500-g load and greater | Sclero-scope hardness | Tensile strength (approx), ksi | Brinell indentation diam, mm |
	Standard ball	Tungsten carbide ball		A scale, 60-kg load, Brale indenter	B scale, 100-kg load, 1/16-in.-diam ball	C scale, 150-kg load, Brale indenter	D scale, 100-kg load, Brale indenter	15N scale, 15-kg load	30N scale, 30-kg load	45N scale, 45-kg load				
3.85	248	248	261	62.5	(101.0)	24.2	43.2	71.7	45.1	24.5	272	37	120	3.85
3.90	241	241	253	61.8	100.0	22.8	42.0	70.9	43.9	22.8	265	36	116	3.90
3.95	235	235	247	61.4	99.0	21.7	41.4	70.3	42.9	21.5	259	35	114	3.95
4.00	229	229	241	60.8	98.2	20.5	40.5	69.7	41.9	20.1	253	34	111	4.00
4.05	223	223	234	· · ·	97.3	(19.0)	· · ·	· · ·	· · ·	· · ·	247	· · ·	107	4.05
4.10	217	217	228	· · ·	96.4	(17.7)	· · ·	· · ·	· · ·	· · ·	242	33	105	4.10
4.15	212	212	222	· · ·	95.5	(16.4)	· · ·	· · ·	· · ·	· · ·	237	32	102	4.15
4.20	207	207	218	· · ·	94.6	(15.2)	· · ·	· · ·	· · ·	· · ·	232	31	100	4.20
4.25	201	201	212	· · ·	93.7	(13.8)	· · ·	· · ·	· · ·	· · ·	227	· · ·	98	4.25
4.30	197	197	207	· · ·	92.8	(12.7)	· · ·	· · ·	· · ·	· · ·	222	30	95	4.30
4.35	192	192	202	· · ·	91.9	(11.5)	· · ·	· · ·	· · ·	· · ·	217	29	93	4.35
4.40	187	187	196	· · ·	90.9	(10.2)	· · ·	· · ·	· · ·	· · ·	212	· · ·	90	4.40
4.45	183	183	192	· · ·	90.0	(9.0)	· · ·	· · ·	· · ·	· · ·	207	28	89	4.45
4.50	179	179	188	· · ·	89.0	(8.0)	· · ·	· · ·	· · ·	· · ·	202	27	87	4.50
4.55	174	174	182	· · ·	88.0	(6.7)	· · ·	· · ·	· · ·	· · ·	198	· · ·	85	4.55
4.60	170	170	178	· · ·	87.0	(5.4)	· · ·	· · ·	· · ·	· · ·	194	26	83	4.60
4.65	167	167	175	· · ·	86.0	(4.4)	· · ·	· · ·	· · ·	· · ·	190	· · ·	81	4.65
4.70	163	163	171	· · ·	85.0	(3.3)	· · ·	· · ·	· · ·	· · ·	186	25	79	4.70
4.75	159	159	167	· · ·	83.9	(2.0)	· · ·	· · ·	· · ·	· · ·	182	· · ·	78	4.75
4.80	156	156	163	· · ·	82.9	(0.9)	· · ·	· · ·	· · ·	· · ·	178	24	76	4.80
4.85	152	152	159	· · ·	81.9	· · ·	· · ·	· · ·	· · ·	· · ·	174	· · ·	75	4.85
4.90	149	149	156	· · ·	80.8	· · ·	· · ·	· · ·	· · ·	· · ·	170	23	73	4.90
4.95	146	146	153	· · ·	79.7	· · ·	· · ·	· · ·	· · ·	· · ·	166	· · ·	72	4.95
5.00	143	143	150	· · ·	78.6	· · ·	· · ·	· · ·	· · ·	· · ·	163	22	71	5.00
5.10	137	137	143	· · ·	76.4	· · ·	· · ·	· · ·	· · ·	· · ·	157	21	67	5.10
5.20	131	131	137	· · ·	74.2	· · ·	· · ·	· · ·	· · ·	· · ·	151	· · ·	65	5.20
5.30	126	126	132	· · ·	72.0	· · ·	· · ·	· · ·	· · ·	· · ·	145	20	63	5.30
5.40	121	121	127	· · ·	69.8	· · ·	· · ·	· · ·	· · ·	· · ·	140	19	60	5.40
5.50	116	116	122	· · ·	67.6	· · ·	· · ·	· · ·	· · ·	· · ·	135	18	58	5.50
5.60	111	111	117	· · ·	65.4	· · ·	· · ·	· · ·	· · ·	· · ·	131	17	56	5.60

(a) For carbon and alloy steels in the annealed, normalized, and quenched-and-tempered conditions; less accurate for cold worked condition and for austenitic steels. Values in bold-faced type correspond to the values in the joint SAE-ASM-ASTM hardness conversions as printed in ASTM E140, Table 3. Values in parentheses are beyond normal range and are given for information only. (b) Brinell numbers are based on the diameter of impressed indentation. If the ball distorts (flattens) during test, Brinell numbers will vary in accordance with the degree of such distortion when related to hardnesses determined with a Vickers diamond pyramid, Rockwell Brale, or other indenter that does not sensibly distort. At high hardnesses, therefore, the relationship between Brinell and Vickers or Rockwell scales is affected by the type of ball used. Standard steel balls tend to flatten slightly more than tungsten carbide balls, resulting in a larger indentation and a lower Brinell number than shown by a tungsten carbide ball. Thus, on a specimen of about 539 to 547 HV, a standard ball will leave a 2.75-mm indentation (495 HB), and a tungsten carbide ball a 2.70-mm indentation (514 HB). Conversely, identical indentation diameters for both types of ball will correspond to different Vickers and Rockwell values. Thus, if indentations in two different specimens both are 2.75 mm in diameter (495 HB), the specimen tested with a standard ball has a Vickers hardness of 539, whereas the specimen tested with a tungsten carbide ball has a Vickers hardness of 528.

Approximate Equivalent Hardness Numbers and Tensile Strengths for Rockwell C Hardness Numbers for Steel (a)

Rockwell C-scale hardness	Vickers hardness	Brinell hardness, 3000-kg load, 10-mm ball		Rockwell hardness			Rockwell superficial hardness, superficial Brale indenter			Knoop hardness, 500-g load and greater	Scleroscope hardness	Tensile strength (approx), ksi	Rockwell C-scale hardness
		Standard ball	Tungsten carbide ball	A scale, 60-kg load, Brale indenter	B scale, 100-kg load, 1/16-in.-diam ball	D scale, 100-kg load, Brale indenter	15N scale, 15-kg load	30N scale, 30-kg load	45N scale, 45-kg load				
68	940	· · ·	· · ·	85.6	· · ·	76.9	93.2	84.4	75.4	920	97	· · ·	68
67	900	· · ·	· · ·	85.0	· · ·	76.1	92.9	83.6	74.2	895	95	· · ·	67
66	865	· · ·	· · ·	84.5	· · ·	75.4	92.5	82.8	73.3	870	92	· · ·	66
65	832	· · ·	(739)	83.9	· · ·	74.5	92.2	81.9	72.0	846	91	· · ·	65
64	800	· · ·	(722)	83.4	· · ·	73.8	91.8	81.1	71.0	822	88	· · ·	64
63	772	· · ·	(705)	82.8	· · ·	73.0	91.4	80.1	69.9	799	87	· · ·	63
62	746	· · ·	(688)	82.3	· · ·	72.2	91.1	79.3	68.8	776	85	· · ·	62
61	720	· · ·	(670)	81.8	· · ·	71.5	90.7	78.4	67.7	754	83	· · ·	61
60	697	· · ·	(654)	81.2	· · ·	70.7	90.2	77.5	66.6	732	81	· · ·	60
59	674	· · ·	(634)	80.7	· · ·	69.9	89.8	76.6	65.5	710	80	351	59
58	653	· · ·	615	80.1	· · ·	69.2	89.3	75.7	64.3	690	78	338	58
57	633	· · ·	595	79.6	· · ·	68.5	88.9	74.8	63.2	670	76	325	57
56	613	· · ·	577	79.0	· · ·	67.7	88.3	73.9	62.0	650	75	313	56
55	595	· · ·	560	78.5	· · ·	66.9	87.9	73.0	60.9	630	74	301	55
54	577	· · ·	543	78.0	· · ·	66.1	87.4	72.0	59.8	612	72	292	54
53	560	· · ·	525	77.4	· · ·	65.4	86.9	71.2	58.6	594	71	283	53
52	544	(500)	512	76.8	· · ·	64.6	86.4	70.2	57.4	576	69	273	52
51	528	(487)	496	76.3	· · ·	63.8	85.9	69.4	56.1	558	68	264	51
50	513	(475)	481	75.9	· · ·	63.1	85.5	68.5	55.0	542	67	255	50
49	498	(464)	469	75.2	· · ·	62.1	85.0	67.6	53.8	526	66	246	49
48	484	(451)	455	74.7	· · ·	61.4	84.5	66.7	52.5	510	64	238	48
47	471	442	443	74.1	· · ·	60.8	83.9	65.8	51.4	495	63	229	47
46	458	432	432	73.6	· · ·	60.0	83.5	64.8	50.3	480	62	221	46
45	446	421	421	73.1	· · ·	59.2	83.0	64.0	49.0	466	60	215	45
44	434	409	409	72.5	· · ·	58.5	82.5	63.1	47.8	452	58	208	44
43	423	400	400	72.0	· · ·	57.7	82.0	62.2	46.7	438	57	201	43
42	412	390	390	71.5	· · ·	56.9	81.5	61.3	45.5	426	56	194	42
41	402	381	381	70.9	· · ·	56.2	80.9	60.4	44.3	414	55	188	41
40	392	371	371	70.4	· · ·	55.4	80.4	59.5	43.1	402	54	182	40
39	382	362	362	69.9	· · ·	54.6	79.9	58.6	41.9	391	52	177	39
38	372	353	353	69.4	· · ·	53.8	79.4	57.7	40.8	380	51	171	38
37	363	344	344	68.9	· · ·	53.1	78.8	56.8	39.6	370	50	166	37
36	354	336	336	68.4	(109.0)	52.3	78.3	55.9	38.4	360	49	161	36
35	345	327	327	67.9	(108.5)	51.5	77.7	55.0	37.2	351	48	157	35
34	336	319	319	67.4	(108.0)	50.8	77.2	54.2	36.1	342	47	153	34
33	327	311	311	66.8	(107.5)	50.0	76.6	53.3	34.9	334	46	149	33
32	318	301	301	66.3	(107.0)	49.2	76.1	52.1	33.7	326	44	145	32
31	310	294	294	65.8	(106.0)	48.4	75.6	51.3	32.5	318	43	141	31
30	302	286	286	65.3	(105.5)	47.7	75.0	50.4	31.3	311	42	138	30
29	294	279	279	64.7	(104.5)	47.0	74.5	49.5	30.1	304	41	135	29
28	286	271	271	64.3	(104.0)	46.1	73.9	48.6	28.9	297	40	131	28
27	279	264	264	63.8	(103.0)	45.2	73.3	47.7	27.8	290	39	128	27
26	272	258	258	63.3	(102.5)	44.6	72.8	46.8	26.7	284	38	125	26
25	266	253	253	62.8	(101.5)	43.8	72.2	45.9	25.5	278	38	122	25
24	260	247	247	62.4	(101.0)	43.1	71.6	45.0	24.3	272	37	119	24
23	254	243	243	62.0	100.0	42.1	71.0	44.0	23.1	266	36	117	23
22	248	237	237	61.5	99.0	41.6	70.5	43.2	22.0	261	35	114	22
21	243	231	231	61.0	98.5	40.9	69.9	42.3	20.7	256	35	112	21

Approximate Equivalent Hardness Numbers and Tensile Strengths for Rockwell B Hardness Numbers for Steel (a)

Rockwell B-scale hardness	Vickers hardness	Brinell hardness, 10-mm-diam ball		Rockwell hardness			Rockwell superficial hardness, 1/16-in.-diam ball			Knoop hardness, 500-g load and greater	Scleroscope hardness	Tensile strength (approx), ksi	Rockwell B-scale hardness
		500-kg load	3000-kg load	A scale, 60-kg load, Brale indenter	C scale, 150-kg load, Brale indenter	F scale, 60-kg load, 1/16-in.-diam ball	15T scale, 15-kg load	30T scale, 30-kg load	45T scale, 45-kg load				
98	228	189	228	60.2	(19.9)	· · ·	92.5	81.8	70.9	241	34	107	98
97	222	184	222	59.5	(18.6)	· · ·	92.1	81.1	69.9	236	33	104	97
96	216	179	216	58.9	(17.2)	· · ·	91.8	80.4	68.9	231	32	102	96
95	210	175	210	58.3	(15.7)	· · ·	91.5	79.8	67.9	226	· · ·	99	95
94	205	171	205	57.6	(14.3)	· · ·	91.2	79.1	66.9	221	31	97	94
93	200	167	200	57.0	(13.0)	· · ·	90.8	78.4	65.9	216	30	94	93
92	195	163	195	56.4	(11.7)	· · ·	90.5	77.8	64.8	211	· · ·	92	92
91	190	160	190	55.8	(10.4)	· · ·	90.2	77.1	63.8	206	29	90	91
90	185	157	185	55.2	(9.2)	· · ·	89.9	76.4	62.8	201	28	88	90
89	180	154	180	54.6	(8.0)	· · ·	89.5	75.8	61.8	196	27	86	89
88	176	151	176	54.0	(6.9)	· · ·	89.2	75.1	60.8	192	· · ·	84	88
87	172	148	172	53.4	(5.8)	· · ·	88.9	74.4	59.8	188	26	82	87
86	169	145	169	52.8	(4.7)	· · ·	88.6	73.8	58.8	184	26	81	86
85	165	142	165	52.3	(3.6)	· · ·	88.2	73.1	57.8	180	25	79	85
84	162	140	162	51.7	(2.5)	· · ·	87.9	72.4	56.8	176	· · ·	78	84
83	159	137	159	51.1	(1.4)	· · ·	87.6	71.8	55.8	173	24	76	83
82	156	135	156	50.6	(0.3)	· · ·	87.3	71.1	54.8	170	24	75	82
81	153	133	153	50.0	· · ·	· · ·	86.9	70.4	53.8	167	· · ·	73	81
80	150	130	150	49.5	· · ·	· · ·	86.6	69.7	52.8	164	23	72	80
79	147	128	147	48.9	· · ·	· · ·	86.3	69.1	51.8	161	· · ·	70	79
78	144	126	144	48.4	· · ·	· · ·	86.0	68.4	50.8	158	22	69	78
77	141	124	141	47.9	· · ·	· · ·	85.6	67.7	49.8	155	22	68	77
76	139	122	139	47.3	· · ·	· · ·	85.3	67.1	48.8	152	· · ·	67	76
75	137	120	137	46.8	· · ·	99.6	85.0	66.4	47.8	150	21	66	75
74	135	118	135	46.3	· · ·	99.1	84.7	65.7	46.8	148	21	65	74
73	132	116	132	45.8	· · ·	98.5	84.3	65.1	45.8	145	· · ·	64	73
72	130	114	130	45.3	· · ·	98.0	84.0	64.4	44.8	143	20	63	72
71	127	112	127	44.8	· · ·	97.4	83.7	63.7	43.8	141	20	62	71
70	125	110	125	44.3	· · ·	96.8	83.4	63.1	42.8	139	· · ·	61	70
69	123	109	123	43.8	· · ·	96.2	83.0	62.4	41.8	137	19	60	69
68	121	107	121	43.3	· · ·	95.6	82.7	61.7	40.8	135	19	59	68
67	119	106	119	42.8	· · ·	95.1	82.4	61.0	39.8	133	19	58	67
66	117	104	117	42.3	· · ·	94.5	82.1	60.4	38.7	131	· · ·	57	66
65	116	102	116	41.8	· · ·	93.9	81.8	59.7	37.7	129	18	56	65
64	114	101	114	41.4	· · ·	93.4	81.4	59.0	36.7	127	18	· · ·	64
63	112	99	112	40.9	· · ·	92.8	81.1	58.4	35.7	125	18	· · ·	63
62	110	98	110	40.4	· · ·	92.2	80.8	57.7	34.7	124	· · ·	· · ·	62
61	108	96	108	40.0	· · ·	91.7	80.5	57.0	33.7	122	17	· · ·	61
60	107	95	107	39.5	· · ·	91.1	80.1	56.4	32.7	120	· · ·	· · ·	60
59	106	94	106	39.0	· · ·	90.5	79.8	55.7	31.7	118	· · ·	· · ·	59
58	104	92	104	38.6	· · ·	90.0	79.5	55.0	30.7	117	· · ·	· · ·	58
57	103	91	103	38.1	· · ·	89.4	79.2	54.4	29.7	115	· · ·	· · ·	57
56	101	90	101	37.7	· · ·	88.8	78.8	53.7	28.7	114	· · ·	· · ·	56
55	100	89	100	37.2	· · ·	88.2	78.5	53.0	27.7	112	· · ·	· · ·	55

(a) For carbon and alloy steels in the annealed, normalized, and quenched-and-tempered conditions; less accurate for cold worked condition and for austenitic steels. The values in bold-faced type correspond to the values in the joint SAE-ASM-ASTM hardness conversions as printed in ASTM E140, Table 2. The values in parentheses are beyond normal range and are given for information only.

Section 14________________________

Source: *Style Manual*, Periodical Publications, American Society for Metals, Metals Park, Ohio 44073, 1968.

Selections—in order of appearance:

    Spelling
    Codes and Specifications
    Preferred Usages
    Marks for Copy Preparation
       and Proofreading

19. Spelling

19-1. Many words end in ise, ize, or yze. The letter **l** is followed by **yze** if the word expresses an idea of loosening or separating, as **analyze.** With the exception of words ending in **wise** and those on the following list, all other words in this class end in **ize.**

advertise	compromise	excise	prise (to force)
advise	demise	exercise	prize (to value)
affranchise	despise	exorcise	reprise
apprise (to inform)	devise	franchise	revise
apprize (to appraise)	disenfranchise	improvise	rise
arise	disguise	merchandise	supervise
chastise	emprise	misadvise	surmise
circumcise	enfranchise	mortise	surprise
comprise	enterprise	premise	

19-2. Only one word ends in **sede (supersede);** only three end in **ceed (exceed, proceed, succeed);** all other words in this class end in **cede (precede, secede).**

19-3. Nouns ending in **o** preceded by a vowel take an **s** to form the plural (**portfolios);** with the exception of words in the following list, nouns ending in **o** preceded by a consonant take an **es** to form the plural (**tomatoes).** Exceptions: **dynamos, halos, magnetos, mementos, provisos, quartos, salvos, solos, tobaccos, twos, tyros, virtuosos, zeros.**

19-4. In forming plurals of compound terms, the significant word takes the plural form (**aides-de-camp, adjutants general, postmasters general, assistant attorneys, deputy chiefs of staff, general counsels, deputy judges, lieutenant colonels).**

19-5. When a preposition is hyphened with a noun, the plural is formed on the noun **(goings-on, lookers-on, makers-up, passers-by).**

19-6. When neither word in a compound is a noun, the plural is formed on the last word **(also-rans, come-ons, run-offs, tie-ins, write-ups).**

19-7. In nouns ending with **ful,** form the plural by adding an **s (five bucketfuls of the mixture).**

19-8. Plural forms of the following words often cause trouble:

agendum, agenda	matrix, matrices
addendum, addenda	medium, mediums
antenna, antennas	memorandum, memorandums
appendix, appendixes	minutia, minutiae
axis, axes	nucleus, nuclei
basis, bases	parenthesis, parentheses
chassis (singular and plural)	phenomenon, phenomena
Co., Cos.	radius, radii
crisis, crises	stimulus, stimuli
datum, data	stratum, strata
equilibrium, equilibriums	syllabus, syllabi
erratum, errata	synopsis, synopses
formula, formulas	terminus, termini
hypothesis, hypotheses	thesis, theses
index, indexes	

19-9. Joint possession is indicated by placing an apostrophe on the last element in a series **(Brown & Nelson's store).** Indicate individual possession in a series with an apostrophe on each element **(the Army's and the Navy's work).**

19-10. A single consonant following a single vowel and ending a monosyllable or a final accented syllable is doubled before a suffix beginning with a vowel **(bag, baggage; get, getting; red, reddish; rob, robbing; allot, allotted, concur, concurred).**

19-11. Use indefinite article **a** before a consonant and an aspirated **h.** Use the indefinite article **an** before a silent **h** and all vowels. Exceptions: **u (pronounced yu)** and **o (as in one).** Write: **a historical event, an honor.**

19-12. When it is possible, avoid Latin plurals **(formulas, not formulae; indexes, not indices).** Use Latin forms when the English versions are awkward **(nuclei, not nucleuses; radii, not radiuses).**

19-13. Use **er,** not **re,** endings for words like **center, liter, fiber, miter, meter, theater.** Exception: retain **re** endings in official names of businesses, trade names, organizations, and places **(Chicago Amphitheatre).**

19-14. Avoid **our** endings in words like **labor** and **honor.** They are British usages.

19-15. Retain British spellings of words like aluminum in official names of businesses, organizations, and places **(Aluminium Ltd.).**

19-16. Do not tack a final **s** on **toward.**

19-17. Approved spellings:

abscissas
acknowledgment
adapter
adaption
adoption
adviser
aeronautical
afterward
aging
aileron
airborne
aircrew
airfield
airframe
airplane
airtight
align
alkalis
all-round
all-out
all right
all-time
already (adv.)
aluminum
anybody
anyone
anyplace
anything
any time
anyway
anywhere
austenitizing
ax

babbitt
backfire
backlash
backstage
back-up (adj.)
back up (v.)
bandwagon
bedplate
benchboard
benchmark
benefited
beside
beveling
bird's-eye
blowhole
blow off
blow out
blue-annealed
blueprint
bluecollar
boulder
boxcar
brake shoe
brazeable, -ility
breakdown
break-in (n., adj.)
breakthrough
briefcase
Brinell
briquet
bucketful
buildup (n.)
build up (v.)
built-in (adj.)
built in (v.)
built-up (adj.)
built up (v.)
bull's-eye
buses

businessman
butt weld
bylaw
byline
bypass
byproduct

caliber
calk
camshaft
cancel
canceled
cancellation
cannot
canyon
cap screw
carat
carbuilder
carburetor
carburize (to impregnate with
    carbon, rather than car-
    bonize, which means to
    char or graphitize)
carload
carry over (n.)
carry over (v.)
case harden
catalog
catalyst
catchall
centerline
chamfer
changeover (n.)
checklist
chemical milling
chromating (treatment of sur-
    face with chromic acid)
chrome (in sense of refractories
    or of coloring, as chrome
    brick, or chrome-orange)
chromite (ore)
chromium
chromizing (cementation with
    chromium)
cipher
circuit breaker
cleanup (n.)
clean up (v.)
clearcut
clockwise
cloverleaf
clue
clipsheet
closedown (n.)
close down (v.)
close-up (n.)
close up (v.)
cold-roll (v.)
columbium (not niobium)
common sense (n.)
common-sense (v.)
computer
consensus
continuously cast
controlled
conveyor
cooperate
coordinate
coreblower
corebox
coremaker
corrodent

corrodible
countercurrent
crankcase
crankpin
crisscross
criticism
criticize
cross-sectional (adj.)
cross section (n.)
cross-section (v.)
cross slide
crosswise
crystalline
crystallize
cutaway (n., adj.)
cut away (v.)
cutback (n.)
cutoff (n., adj.)
cut off (v.)
cutthroat

damping
daywork
deadline
debugging
defense
degassing
desalination
desirable
dew point
diagrammed
dialog
die-cast (adj.)
die cast (v.)
die casting (n.)
diesinking
disc
dispatch
distributor
downcomer
downflow
downtime
downward
draft
drafting room
draftsman
dragline
drainpipe
drawbar
drawing board
drier (adj.)
dripproof
drop out (v.)
dropout (n., adj.)
drought
dryer
dusttight
dyeing

earthmoving
electric weld
embed
employ
employee
enameling
encase
enclose
endorse
endwise
enforce
enroll
en route
entrust

equalizing
equalized
erosible (not **erodible**)
escalator
everybody
everyday
everyone
everything
every time
everywhere
eyebar
eyebolt

facsimile
falsework
farsighted
feedback (n., **adj.**)
feed back (**v.**)
feldspar
fiber
fiberboard
fiber glass
fingertip
firebox
firebrick
fire clay
firedamp
firsthand (adv., **adj.**)
first hand (n.)
fishplate
flashboard
flier
flywheel
focusing
followthrough (n.)
follow through (v.)
followup (n., adj.)
footboard
footnote
forego
forecast
foreword
forgeable
formulas
forward
freehand
fulfill
funneled
fuse

gage
gases
gasoline
gassing
gearbox
gear cutter
gearmotor
gear wheel
glycerin
goodby
gray
gray iron
grill (cooking utensil)
grille (auto part)

hacksaw
halftone
halfway
handbook
handful
handhole
handmade
handrail
handsaw
handwheel
hardfacing

hardsurfacing
hassle
head gate
headstock
hollow ware
homemade
hydroelectric
hygroscopic
hypotenuse

I-beam
incase
inclinable
incrust
indenter
indexes
infrared
in-line (adj.)
inoculation
input
inquire
insure
interoffice
inward
ironclad
ironware

jackshaft
journal box
judgment

karat (for alloys)
kerosine
keyboard
keyway
keyword
kilogram
know-how

lag screw
landmark
lawnmower
layaways (n.)
layoff (n.)
lay off (v.)
lead screw
leadtime
leakproof
lefthand (adj.)
left hand (n.)
lengthwise
lifetime
lightweight (adj.)
likable
liter
locknut
longeron
lookout (n., adj.)
look out (v.)
loudspeaker
LP-gas
Lüders' lines

Mach
machinable
machine shop
magnetism, magnetize
makeready
makeshift
makeup (n.)
make up (v.)
malleablize
malleabilizing
malleabilization

manhole
manhour
manpower
marketplace
markup
material (not **materiél**)
material handling
medalist
media
metallize (not **metalize**)
midweek
miter box
Mohs' scale
mold
molybdenum
monkey wrench
mortise
movable
multispeed

nacelle
nameplate
naphtha
naphthalene, naphthol
nationwide
nickel
nitroglycerin
nodular iron
nonferrous
nonnitrogenous
normalize
noticeable

occurred
offense
offhand
offset
offshore
oilless
oil pan
oil ring
oiltight
O.K. (n., adj., v., adv.)
O.K.'d (past part.)
O.K.'ing (pres. part.)
Olsen cup test
on-line (adj.)
ore bed
ore body
outboard
outbound
output
outward
over-all
overflow
overload
overnight
overrunning
overshot
overtime
overturn
oxyacetylene

paleo
paneling
paperwork
parceling
patternmaker
pattern shop
payoff (n.)
peacetime
per cent
percentage

petcock
phosphating (not
    phosphatizing)
phosphorus
pickle
pickup (adj., n.)
piggyback
pipeline
pipe wrench
piston ring
plastic (adj.)
plastics
plug-in (adj.)
pocketbook
polyphase
post mortem
postwar
powder metal
powerhouse
powerplant
preceding
presstime
preventive (not
    preventative)
prewar
principal
principle
proceeding
procedure
program
programmed
programming
propellent
pro rata
prorate
proration
proved
pushbutton

quarreling
quick-freeze

raveling
readymade
reconnaissance
reduction in area
removable
replaceable
reservoir
re-use (n., v.)
right of way
roadbuilding
roundup
rundown (n., adj.)
run down (v.)
runway

salable
salvageable
sandblast
sand cast (v.)
sandpaper
screw-down (adj.)
screw down (v.)
self-contained
semiannual
semifinished
semi-independent
Sendzimir
servoamplifier
servomechanism
servomotor
set-asides
setscrew
setup (adj., n.)
shakeout (n.)

shake out (v.)
short circuit
shortcut
shotblasting
shutdown (n.)
shut down (v.)
sidewise
siliceous
silicon
silicone
single-phase (adj.)
siphon
sizable
skillful
sliderule
slip fit
slowdown (n.)
slow down (v.)
sometimes
sparkplug
speedup (n.)
speed up (v.)
spheroidized
spoonfuls
spotcheck
spotface
squirrel-cage (adj.)
squirrel cage (n.)
standby (n., adj.)
stand by (v.)
station wagon
steelworks
stepdown (n., adj.)
step down (v.)
stepup (n.)
step up (v.)
stockpile
stockyards
stopgap
stress raiser
stretchout (n.)
stretch out (v.)
strong-arm (adj., v.)
subassembly
subbituminous
subboundary
subgrade
subpoena
sulfocyanide
sulfur (n.)
sulfuric (adj.)
superheat
synchro

tailormade
taillight
tailrace
tailstock
takehome (n., adj.)
takeoff (n.)
tearsheet
telltale
template
testpiece
textbook
theater
thermit
through hardened (not
    thorough hardened)
throwaway
thumbscrew
tiebolt
tie-in (adj., n.)
tierod
tie-up (n.)

tie up (v.)
toluol
toolholder
topheavy
topnotch (adj., n.)
totaled
toward
trade-in (n.)
trade in (v.)
trademark
trade name
trucktrailer
truing
tryout (n.)
try out (v.)
turk's-head
turnover (n., adj.)
turn over (v.)
turntable
tuyere

ultraviolet
underframe
underway
update
up-to-date (adj.)
upturn
upward
usable
usage
usefulness

vapor
vendor

wartime
washout (n., adj.)
wash out (v.)
watershed
waterworks
waveform
waveguide
weekday
weekend
well-being
westward
wheelbase
whitecollar
widespread
Widmanstätten
willful
workbench
workforce
workload
workman
workmonth
workpart
workpiece
workplace
workstation
worktable
workweek
worthwhile
wraparound
writeoff (n., adj.)
write off (v.)

X-ray
xylol

yearend
year-round

zinciferous

20. Codes and Specifications

20-1. In designating an alloy by content, try to state elements in order of decreasing quantity. Exceptions: when a minor element must be mentioned first by virtue of its importance in the context of the article, or when industry usage lists the minor element first (as in **beryllium copper).**

Use the per cent sign with single elements only: **18% Ni.** In writing a composition, use this form: **54 Cr, 5 Mn, 2 Fe.** If the composition listed serves as a designation for the alloy or is a binary alloy, hyphens may be used: **18Cr-8Ni, Ti-6Al-4V, 60Cu-40Zn, 38-18-1.**

20-2. Here are some specifying organizations and their codes:

Society of Automotive Engineers—SAE 4130

American Iron & Steel Institute — AISI 3220, X1112, E4640, type 446

American Society for Testing & Materials — ASTM B17-34T, A59-49; 1330-H

Aeronautical Materials Specifications — AMS 5063

Federal Specifications — QQ-S-624(-2); QQ-B-726c; WW-T-791(-1)

    Type, grade or class — FS-310; C1041; gr 446, gr A

Military Specifications — MIL-T-1223A; MIL-T-8504(-2)ASG; MIL-16301(BuOrd)

    Type, grade or class — comp 5, gr IV, cl a

Alloy Casting Institute — ACI CA-15; CD-4MCu; HA; CF-8C

20-3. Here are some representative alloy designations:

Aluminum Alloys	Copper, Brass and Bronze
B195	CDA No. 102
A355	
2025-T	Hafnium Alloys
2024-T81	Hf-27 Ta
7075-T6	
3003-O	High Strength Steels
	A242
Beryllium Alloys	AW-441
Lockalloy	AWX-45
	AWX-50
Cobalt Alloys	AWX-55
S816	Char-Pac
Elgiloy	Con-Pac
Haynes Alloy No. 25	Cor-Ten
L605	Dofascoloy
Rex 78	Ex-Ten 42
MAR-M302	GLX-45-W
UMCo 50	Hi-Yield 42
WI-52	IH 50
	IHX-45
Columbium Alloys	INX-42
Haynes Alloy Cb-751	Jalloy-AR 360
Columbium B-33	Jalloy-S-340
D43	Jalten #3
D40M	Kaisaloy 45 FG
F48 Alloy	Man-Ten
Su-16	Mayari R

ML-F
N-A-XTRA 90
NAX- Finegrain
HY80
Par-Ten ⪦1
Pitt-Ten ⧣1
PX SKA 45
SSS-100
"T-1"
"T-1" type A
Tri-Ten
X-A-R-30 Abrasion
   Resistant
V45
YSW-42
Yoloy HSX
Yoloy S
300M
Hy-Tuf
HS220
Vasco Matrix II
Lescalloy 270
HP9-4-25

High Temperature Alloys
HTS1100
Unitemp 14HV
H11
5CrMoV
Vascojet 1000
H13
D6A, D6AC
Pyromet X-12
Haynes Alloy No. 506
20Cb-3
Unitemp 1416MV
Lapelloy
Greek Ascoloy
N155
Tenelon
19-9DL
16-25-6
V57
S590

Magnesium Alloys
AZ63A
AM100A
ZK61A
EZ33A
HK31A
K1A
QE22A

Maraging Steels
Unimar 25
NiMark II
18 NiCoMo 250
VascoMax 300 CVM
RSM 200

Molybdenum Alloys
Mo-0.5Ti
TZM

Nickel Alloys
Incoloy Alloy 901
Monel Alloy 400

Inconel Alloy 600
Inconel Alloy X-750
René 41
Haynes Alloy No. R41
Hastelloy Alloy B
Hastelloy Alloy C
Permanickel Alloy 300
Duranickel Alloy 300
Duranickel Alloy 301
Nickelvac W
Ni-Span C
Nickel 201
TD Nickel: 6, 28, 50
Nichrome V
GMR235
Waspaloy
Allvac 500
M252
Udimet 1753
RA333
D979
MAR-M200
DCM
Udimet 520

Stainless Steels
type 201
type 303Se
type 430FSe
17-7 PH
PH 15-7 Mo
type 410
AM350
AM355
NS355
A286
NS-A286

Tantalum Alloys
GE-473
Haynes Alloy Ta-78Z

Titanium Alloys
Ti-5Al-2.5Sn
Ti-8Al-1Mo-1V
Ti-7Al-2Cb-1Ta

Tool Steels
H11
M2
T1
O1

Tungsten Alloys
Elkonite 100W
Gyromet 1000
77 Metal
C33
G-E WN100
Mallory 2000

Zirconium Alloys
RA1
R1
Zircalloy 2

20-4. Follow copy exactly when it contains designations of crystal structure. For example, retain brackets, parentheses, braces, carets, and overbars above numbers: **(100) ($\bar{1}$12) {100} [100] <100>**

20-5. Use standard symbols in figures and text when referring to areas or points on a transformation diagram (**M_s, M_f, Ae_1, Ae_3**).

20-6. Anglicized versions of Greek letters are used to designate metallurgical products, such as **alpha iron,** and such terms as **gamma ray** and **sigma phase.** In phase diagrams, use Greek forms.

20-7. Use this style to state the composition of a thermocouple. If it is **platinum vs platinum-rhodium one, containing 15% Rh,** write: **Pt/Pt-15Rh.**

20-8. State hardness numbers this way. Brinell: **Bhn 435;** Knoop: **Knoop (10 kg) 200;** Rockwell: **Rc 42, Rb 37, R15N 85;** diamond pyramid hardness (or Vickers): **dph (30 kg) 860.**

20-9. Use atomic weights as superscripts (**U^{235}**) and atomic numbers as subscripts (**$_{92}U^{235}$**). If the element is spelled out, use this form for atomic weights: **uranium-235.**

20-10. In converting foreign weights and measures, use this form: **212 F (100 C).**

21. Preferred Usages (in alphabetical order)

21-1. **Actual facts:** all facts are actual. Drop actual.

21-2. **Affect, effect:** affect (always a verb) means to have an influence on, the verb **effect** means to cause, to produce, to result in. (The salt environment **affected** corrosion resistance. The company **effected** several changes in its price structure.) In many instances, affect and effect are weak verbs, and we can do a better job of expressing ourselves by finding more precise language. (The salt environment **speeded up** corrosion.)

21-3. **Allusion** is a casual reference; **illusion** is a false impression.

21-4. **Alternative:** many people insist this is a "one of two" situation. However, "one of any number" is a legitimate usage, such as "there are three alternatives."

21-5. **Among:** use for a group of three or more jointly involved; **between:** use for two only, or for bilateral action in a group **(among** departments in the company . . . or **between** the shop and accounting departments).

21-6. **And/or:** this usage should be avoided. Say **or . . . or both.** Sometimes **and** alone is enough. (Failure may result from fatigue **or** poor resistance to sudden overstress **or both.)**

21-7. **As:** this word is not a substitute for **because** or **since.** **Wrong: As** the alloy is expensive, it will not be used.

21-8. **Beside** means "next to." **Besides** means **"in addition to."**

21-9. **Capital, capitol: capital** refers to a city or money assets. **Capitol** refers to a building.

21-10. **Christian names** are not abbreviated. Official spellings of company names are an exception **(Geo.** F. Marchant Co.).

21-11. **Comparisons:** only like things can be compared. Also, complete all comparisons — usually with a pronoun. **Wrong: Properties** of high strength steels **are similar to stainless steels.** **Right: Properties** of high strength steels **are usually similar to those of** stainless steels. In the first example "properties" and "stainless steels" are compared. They are unlike things, and we can't rely on the implied comparison of "properties" and "properties."

21-12. **Compared with, compared to:** use **with** to show differences; use **to** to show likenesses. (Sales were $1 billion this year **compared with** $1.1 billion last year. Gross sales, when **compared to** last year's, are about the same.)

21-13. **Continual, continuous:** if it is continual, it recurs regularly, but is not constant. If it is continuous, it is constant. Avoid the common industry practice of using the adjective form instead of the correct adverbial construction. **Wrong: continuous cast** ingot. **Right: continuously cast** ingot.

21-14. **Data:** this is the plural form of datum. It takes a plural verb and a plural pronoun. **(These data are in the table.)**

21-15. **Different from:** use instead of **different than.**

21-16. **Farther, further: farther** means distance. **Further** means "in addition to."

21-17. **Got** is preferred over **gotten.**

21-18. **Had** implies volition when it is used in this sense: **he had his arm broken.**

21-19. **Just:** when this word is an adverb, it has an exact meaning: precisely, closely, or precisely at the time referred to, now, or a moment ago. It is overused as a synonym for recently.

21-20. **Latter, former:** always used with two persons or things. When the number is more than two, use first and last. However, **latter, former,** and the related **respectively** are not encouraged. They tend to force a reader to back up and find out which is the former and which is the latter.

21-21. **Locate:** a building is located when its site is chosen. Thereafter, it is situated.

21-22. **Microphotograph** means a small photograph. A **photomicrograph** is a photograph of a microscopic object.

21-23. **Past and last:** last is favored in the sense of "last month."

21-24. **Plurals** that are singular: treat all sums as singular in number. Companies and organizations take singular verbs and singular, impersonal pronouns. (R. B. Smith & Associates **is . . . it is** a fine organization. The Business & Defense Administration **is . . .**) The British use plural, personal pronouns in this area. They look upon a company as an organization made up of people. In the United States, we regard a company as a legal entity with an identity of its own.

21-25. **Position of adverbs:** they are usually placed after the first element in a compound verb, not before the compound. **Wrong:** He **usually will take** the opposite side. **Right:** He **will usually take** the opposite side.

21-26. **Preventive** is favored over **preventative,** which is an irregular formation of preventive.

21-27. **Principal, principle: principal** means highest in rank, authority, or importance; chief; main. **Principle** generally involves a fundamental truth or basic law.

21-28. **Proved** is the past tense of prove. **Proven** is the adjective form. (It was **proved** beyond doubt. He has **proven** ability.)

21-29. **Provided that** is preferred over **providing that.**

21-30. **Subjunctive and past tense:** both forms should be avoided in reporting present facts. **Wrong:** Dr. Baker **said** that the tanks, now half finished, **were** 22 by 50 ft. **Right:** Dr. Baker **said** that the tanks, now half finished, **are** 22 by 50 ft. If Dr. Baker is making his report in the story, we should use **says** not **said.**

21-31. Stating ranges: write **between** 25 **and** 40 F. Ranging **from** 25 **to** 40 F.

21-32. **Toward: towards** is British usage.

21-33. **Try to:** don't write **try and.**

21-34. **Virtually:** in essence or effect, but not in fact, is meant. It is not a synonym for practically or nearly.

21-35. **While:** do not use in the sense of whereas, although, and, or but. It is used in reference to the simultaneity of two or more events. (Nero fiddled **while** Rome burned.)

22. Marks for Copy Preparation and Proofreading

Style

wf	Wrong font (size or style)
lc	Use a Lower Case LETTER
caps	SET IN caps
c&lc	Set in Caps and Lower Case
sc	SET IN small caps
sc	SET IN CAPS AND SMALL CAPS
rom	Set in roman
ital	Set in italic
lf	Set in lightface
bf	**Set in boldface**
bfc	**Set in boldface caps**
	Set Superior figure[3]
	Set inferior figure[3]
	Begin paragraph
	Run in.
no ¶	No paragraph

Position

	Move to the right
	Move to the left
	Lower material marked
	Raise material marked
	Straighten type horizontally
	Align type vertically
tr	Transpose material ringed
tr	Transpose order words of or letters

tr	Rearrange in numbers order of
< >	Center
	Close up completely
less #	Less space between words
eq #	Equalize space between words
#	Insert space
	Em quad space
the	Insert marginal addition
	Delete. Take out
stet	Leave as set (all material above or inside dots)
×	Replace broken type
	Turn over (upside down)
	Push down printing space
sp out	Spell out 20 words
?/lies/?	Please answer query

Punctuation

	Insert period
	Insert comma
	Insert colon
	Insert semicolon
	Insert apostrophe
	Insert quotation marks
	Insert hyphen
EN	En dash
	One-em dash
	Insert parentheses

Section 15

Source: *ASM Stylebook*, American Society for Metals, Metals Park, Ohio
          44073, 1979.

Selections—in order of appearance:

Personal Titles
Geographical Terms
Expressions of Time
Numerals
Abbreviations
Abbreviations and Symbols for
    Technical Terms
Mechanical and Physical
    Properties
Engineering Quantities
Mathematical Symbols
Coordinate Systems
Chemical Reactions
Symbols for the Chemical
    Elements

Crystal Structure
Transformations
Atomic Information
Designating an Alloy by
    Content
Specifications of Various
    Agencies
Aluminum Alloys
Copper, Brass and Bronze
Ferrous Materials
Heat-Resisting Alloys
Magnesium Alloys
Nickel Alloys
Titanium Alloys
Heat Treatments

5. Personal Titles

5-1. Civil, government, military, professional, and hereditary titles are capitalized, and many are abbreviated when they precede the full name (Sen. E. L. Bartlett).

5-2. Titles are generally capitalized when they precede the full name (President John Smith of ABC Co., Executive Vice President John Smith of ABC Assn., Chairman John Smith of the ABC Standards Committee).

5-3. Do not capitalize such titles as engineer, metallurgist, and researcher when they precede the full name.

5-4. Generally, a title preceding a last name is spelled out (Governor Rhodes). Exceptions include Mr., Mrs., and Dr.

5-5. Abbreviations for civil, governmental, military, and professional titles, and college degrees:

Civil Titles

Miss
Misses (plural for Miss)
Mr.
Mrs.
Messrs. (plural for Mr.)
Mlle. (mademoiselle)
Mme. (madame)
Mmes. (plural for Mrs.)
M. (Monsieur)
Ms.

Government Titles

Gov.
Lt. Gov.
Hon.
Right Hon.
Sen. (senator)
Rep. (representative)

Army, Air Force, and Marine Corps Titles

Gen.
Lt. Gen.
Maj. Gen.
Brig. Gen.
Adj. Gen.
 (adjutant general)
Paymaster Gen.
Quartermaster Gen.
Col.
Lt. Col.
Maj.
Capt.
1st Lt.
2nd Lt.
W.O. (warrant officer)

Navy Titles

Adm.
Vice Adm.
Rear Adm.
Capt.
Cdr. (commander)
Ltcdr.
Lt.
Lt. (jg)
Ens.
W.O. (warrant officer)
P.O. (petty officer)

College Degrees

Bachelor of Arts	AB or BA
Bachelor of Business Administration	BBA
Bachelor of Science	BS or BSe
Master of Arts	MA or AM
Master of Business Administration	MBA
Master of Science	MS or MSc
Doctor of Philosphy	PhD
Doctor of Science	DSc

Professional Titles

Dr.
Rev.
Msgr. (monsignor)
Prof.

6. Geographical Terms

6-1. The names of cities, states, countries, territories, provinces, islands, and other geographical units are capitalized (Chicago, Maryland, England, Ontario, Prince Edward Island).

6-2. The names of political subdivisions are capitalized (Lake County, Mentor Township, Fifth Ward). Such terms as state, county, township, and ward are not capitalized when they are used alone.

6-3. Names commonly used to identify a group of related states (South, Midwest, West Coast, Gulf Coast) are capitalized.

6-4. Names commonly used to identify areas of the world (Iron Curtain, Free World, the Orient) are capitalized.

6-5. Do not capitalize areas that are identified by compass direction alone (northeast Ohio). The designation is not specific.

6-6. Generally, the state is abbreviated when it is used in combination with the city (Waco, Tex.).

6-7. The names of the following widely known cities in the United States are normally used without the state abbreviation:

Akron (Ohio)	Des Moines (Iowa)	New Orleans (La.)
Atlanta (Ga.)	Detroit (Mich.)	New York (N.Y.)
Atlantic City (N.J.)	Duluth (Minn.)	Oklahoma City (Okla.)
Baltimore (Md.)	Ft. Worth (Tex.)	Omaha (Nebr.)
Baton Rouge (La.)	Honolulu (Hawaii)	Philadelphia (Pa.)
Birmingham (Ala.)	Houston (Tex.)	Pittsburgh (Pa.)
Boston (Mass.)	Indianapolis (Ind.)	St. Louis (Mo.)
Brooklyn (N.Y.)	Los Angeles (Calif.)	St. Paul (Minn.)
Buffalo (N.Y.)	Louisville (Ky.)	Salt Lake City (Utah)
Chicago (Ill.)	Memphis (Tenn.)	San Diego (Calif.)
Cincinnati (Ohio)	Miami (Fla.)	San Francisco (Calif.)
Cleveland (Ohio)	Milwaukee (Wis.)	Seattle (Wash.)
Dallas (Tex.)	Minneapolis (Minn.)	Washington (D.C.)
Denver (Colo.)	Nashville (Tenn.)	Youngstown (Ohio)

6-8. Use both city and state in Postal addresses.

6-9. Except in postal addresses (see 6-12) use these abbreviations for states, posses-
sions (Puerto Rico and Virgin Islands) and Canadian provinces. Alaska, Hawaii,
Idaho, Iowa, Maine, Ohio, and Utah are never abbreviated.

Ala.	Alabama	N. Dak.	North Dakota
Ariz.	Arizona	Okla.	Oklahoma
Ark.	Arkansas	Oreg.	Oregon
Calif.	California	Pa.	Pennsylvania
Colo.	Colorado	R.I.	Rhode Island
Conn.	Connecticut	S.C.	South Carolina
Del.	Delaware	S. Dak.	South Dakota
D.C.	District of Columbia	Tenn.	Tennessee
Fla.	Florida	Tex.	Texas
Ga.	Georgia	Vt.	Vermont
Ill.	Illinois	Va.	Virginia
Ind.	Indiana	Wash.	Washington
Kans.	Kansas	W. Va.	West Virginia
Ky.	Kentucky	Wis.	Wisconsin
La.	Louisiana	Wyo.	Wyoming
Md.	Maryland	P.R.	Puerto Rico
Mass.	Massachusetts	V.I.	Virgin Islands
Mich.	Michigan	Alta.	Alberta
Minn.	Minnesota	B.C.	British Columbia
Miss.	Mississippi	Man.	Manitoba
Mo.	Missouri	N.B.	New Brunswick
Mont.	Montana	N.F.	Newfoundland
Nebr.	Nebraska	N.S.	Nova Scotia
Nev.	Nevada	Ont.	Ontario
N.H.	New Hampshire	P.E.I.	Prince Edward Island
N.J.	New Jersey	Que.	Quebec
N. Mex.	New Mexico	Sask.	Saskatchewan
N.Y.	New York	Yukon	Yukon Territory
N.C.	North Carolina		

6-10. Check Webster's New World Dictionary, the Pronouncing Gazeteer in the
back of Webster's New Collegiate Dictionary, the U.S. Postal Guide, or The
World Almanac for spellings not covered in this section.

6-11. Saint, Mount, and Fort in names of cities and military posts are abbreviated
(St. Louis; Mt. Vernon, N.Y.; Ft. Wayne, Ind.; Ft. Dix, N.J.). Do not abbreviate
parts of official names which are compass directions (South St. Paul, Minn.;
East Chicago, Ind.).

6-12. In postal addresses, follow these ZIP code practices:

Alaska	AK		Montana	MT
Alabama	AL		Nebraska	NB
Arizona	AZ		Nevada	NV
Arkansas	AR		New Hampshire	NH
California	CA		New Jersey	NJ
Canal Zone	CZ		New Mexico	NM
Colorado	CO		New York	NY
Connecticut	CT		North Carolina	NC
Delaware	DE		North Dakota	ND
District of Columbia	DC		Ohio	OH
Florida	FL		Oklahoma	OK
Georgia	GA		Oregon	OR
Guam	GU		Pennsylvania	PA
Hawaii	HI		Puerto Rico	PR
Idaho	ID		Rhode Island	RI
Illinois	IL		South Carolina	SC
Indiana	IN		South Dakota	SD
Iowa	IA		Tennessee	TN
Kansas	KS		Texas	TX
Kentucky	KY		Utah	UT
Louisiana	LA		Vermont	VT
Maine	ME		Virginia	VA
Maryland	MD		Virgin Islands	VI
Massachusetts	MA		Washington	WA
Michigan	MI		West Virginia	WV
Minnesota	MN		Wisconsin	WI
Mississippi	MS		Wyoming	WY
Missouri	MO			

7. Expressions of Time

7-1. Do not abbreviate May, June or July.

7-2. Nine months may be abbreviated when they are used with a day, such as 25 Dec.
They are:

Jan.............January	Sept........... September
Feb. ....February	Oct...............October
Mar. ....March	Nov. ....November
Apr. .... April	Dec. ....December
Aug. .... August	

7-3. In expressing two dates within the same century, write: 1901-55. If two dates are
in different centuries, write: 1890-1950.

7-4. Spell out expressions like "the soaring sixties" or use figures, the early 1960's.

7-5. Days of the week can be abbreviated when space is at a premium, as in programs
or tabular matter:

Sun. .... Sunday	Thurs........... Thursday
Mon. .... Monday	Fri. ....Friday
Tues............. Tuesday	Sat..............Saturday
Wed...........Wednesday	

7-6. In expressing clock time, use a colon to separate hours and minutes (2:30 p.m.).
If minutes are not expressed, do not add ciphers after the hour; make it 2 p.m.

7-7. In writing clock time use a.m. and p.m. You may use 12 a.m. or 12 noon, 12 p.m.
or 12 midnight.

8. Numerals

8-1. Generally, whole numbers running one through ten are written out (one, ten). Generally, whole numbers over ten are used as numerals.

8-2. When a number is specific, use numerals and proper abbreviations to denote weight, measure, distance, clock time (seconds, minutes, hours), money, percentage, degrees, votes, ratios, proportions and age (1 lb, $43, 6%). If a specific number is not given, as in several pounds, several dollars, an abbreviation is not used.

8-3. In a series, cardinal numbers may be written as numerals (1,2,3) or spelled out (one, two, three). Ordinal numbers are spelled out (first, second, third). Use a consistent style within a series.

    If cardinal numbers are used as numerals, follow this style: Introduce series with colon: 1. Place period after each number. 2. Period at end of each statement. 3. Start each new statement with a capital. 4. Do not use "and" before the last number in the series.

    If ordinal numbers are used, or cardinal numbers are spelled out, follow this style. Introduce series with a colon: First, (or One,) put a comma behind the ordinal or cardinal number. Second, (or Two,) lower case the word following the number. Third (or Three,) end each statement with a period. Fourth, (or Four,) do not use "and" before the last number in the series.

8-4. If numbers in a series are over and under 11, express all as figures (1 battleship, 7 cruisers, 12 destroyers).

8-5. When two numerals are in apposition (11 12-room houses), spell out the first number (eleven 12-room houses). If the first number is clumsy when it is spelled out, use numerals to express it and write out the second number (263 eleven-room houses).

8-6. Use figures and capitalize references like Fig. 1, No. 1, Part 1, Class 1.

8-7. Whole numbers and fractions that begin a sentence are spelled out and capitalized (One man was there. One-third of the year is gone). Do not start a sentence with a whole number and a fraction, or a number which is awkward when spelled out (267 000, for example).

8-8. Fractions may be written out (one-fourth), used as figures (¼), or expressed as decimals (0.25). Point off decimals with period and put a zero in front of the decimal if there is no whole number (0.21). Use decimal system when you want to be precise.

8-9. Two forms may be used to express ratios and proportions (1 to 4 and 1:4).

8-10. Do not use zeros in expressing whole sums of money ($2, not $2.00).

8-11. Do not use comma to point off numbers. Follow metric practice. Where there are four digits, write solid (4000). Where there are five or six digits, use spaces in this manner: 10 000 and 100 000. Where numbers have seven or more digits, drop zeros. Use this style: 1 million, 2 billion. Treat terms like million and billion as singular.

8-12. In stating square and cubic measurements, follow this order: length, width, height. Use "by" between dimensions. In tabular matter and exponential numbers, use multiplication symbol (8 x 12 in., 9 x 10^6).

8-13. Use Arabic numerals for figures (Fig. 1) and Roman numerals for tables (Table I) in periodicals. Use Arabic numerals in reference publications to identify both figures and tables.

8-14. Use the apostrophe to form the plural of figures (8's).

8-15. Do not use a comma to separate elements in expressions like 6 ft 2 in. wide.

8-16. Use round numbers (100, 250) to express magnifications.

10. Abbreviations

10-1. Capitalize an abbreviation if the word it stands for is capitalized when it is spelled out (U.S. for United States).

10-2. Use singular abbreviations for singular and plural words (1 lb, 10 lb, p. 1, p. 6-10). Also make it Fig. 1, 2, not Figs. 1, 2; or No. 1, 3, 7, not Nos. 1, 3, 7.

10-3. Generally, if an abbreviation is a word in itself, a period is used to avoid confusion (in., not in, for example). There are exceptions, such as Mr. and Mrs.

10-4. The general rules in this section may be disregarded where space is at a minimum, as in headlines and tabular matter; but great care is advised. If a reader does not known an abbreviation, it means nothing to him.

10-5. Symbols used in equations should be defined below the equation. For a list of standard symbols, together with rules for the use of symbols, see section 10-11.

10-6. Spacing is generally standard when an abbreviation is used (100 lb).

10-7. Do not use the degree sign in expressing temperature. Write 100 F, not 100° F.

10-8. Names of countries may be abbreviated when the abbreviation is well known (U.S., USSR, UK).

10-9 Do not start a sentence with an abbreviation. Write: Figure 1 shows, not Fig. 1 shows. Exceptions include Mr., Mrs., Dr.

10-10. Abbreviations and Symbols for Technical Terms

In the following list, asterisks indicate that an abbreviation should be used only in tabular matter.

acre ....spell out
Aerospace Material Specification
   (of SAE).......................................AMS
air cooled...AC*
alternating current (noun or adj.) ....ac
Alloy Casting Institute.......................ACI
Aluminum Association ....AA
American Chemical Society...............ACS
American Foundrymen's Society ....AFS
American Institute of Mining,
   Metallurgical and Petroleum
   Engineers..................................AIME
American Iron and Steel Institute ..... AISI
American National Standards
   Institute...................................ANSI
American Petroleum Institute.............API
American Society for Metals.............ASM
American Society of Mechanical
   Engineers................................ ASME
American Society for Quality
   Control...................................ASQC
American Society for Steel
   Treating..................................ASST
American Society for Testing
   and Materials .... ASTM
American Standards Association
   (superseded by ANSI).................... ASA
   (use only for film-speed ratings)
American Welding Society ....AWS
American wire gage ....Awg
ampere... A
ampere hour...................................... Ah
amperes per square foot.....................A/ft^2
analyzed reagent ....AR
angstrom (unit)...............................avoid
                (convert to NM)
annealed...ann*
ante meridiem .... a.m.
antilogarithm .... antilog
approximately.............................. approx
Association ....Assn.
Associates .... Assoc.
atmosphere (pressure) .... atm
atomic percent ....at. %
atomic weight................................. at. wt
audio frequency (adjective only)........... AF
average...avg*

back rake (angle) ....BR
barn .... b
Baume.. Be
becquerel..Bq
Birmingham wire gage .... Bwg
body-centered cubic ....bcc
Brinell hardness number ....HB
British thermal unit............................Btu
Brown & Sharpe (gage) ....B&S

calorie ....cal
candela...cd
Celsius..C
Cemented Carbide Producers
   Association...................................CCPA
cent (money) .... ¢
centerline average
   (surface roughness)...........................cla
centigram ....avoid
                (convert to mg)
centilitre...avoid
                (convert to mL)
centimetre ....cm
centimetre-gram-second (system) .... cgs
centipoise...cP
centistokes..cSt
chemically pure..................................CP
close-packed hexagonal.......................cph
coefficient...coef*
cold rolled..CR*
cologarithm.......................................colog
commercial quality ....CQ*
concentrated..................................... conc*
conductivity ....cond*
Copper Development
   Association....................................CDA
coulomb..C
cubic centimetre ....cm^3
cubic feet per hour.............................ft^3/h
cubic feet per minute ....ft^3/min
cubic feet per second ....ft^3/s
cubic foot...ft^3
cubic inch...$in.^3$
cubic metre ....m^3
cubic millimetre ....mm^3
cubic yard ....yd^3
curie..Ci

Abbreviations and Symbols for Technical Terms

current density ....spell out
cycles ....spell out
cycles per second (hertz) ....Hz

day.. d
decibel...dB
decimetre ....avoid
(convert to mm or cm)
degree (angular) .... °
Defense Metals Information
    Center.................................DMIC
diameter (dimensional) .... diam
direct current (noun or adj.) ....dc
dollar .... $
dozen ....doz
drawing quality..............................DQ*

e (base of natural logarithms)..................e
electrolytic tough pitch (copper) ....ETP
electromotive force .... emf
electron volt.......................................eV
elongation (%)...............................elong*
end cutting edge angle ....ECEA
end rake (angle) .... ER
equation...Eq
estimated..est
extreme-pressure (lubricant)................ EP

face-centered cubic..............................fcc
Fahrenheit......................................F
farad ....F
feet per minute...............................ft/min
feet per second.................................ft/s
feet per year ....ft/yr
fluid ounce....................................fl oz
foot...ft
foot pound ....ft lb
foot-pound-second (system) ....fps

gallon..gal
gallons per minute ....gal/min
gallons per second.......................... gal/s
gram... g
grams per cubic centimetre.............. g/cm³
gravity (acceleration) .... g
gray (absorbed dose of radiation)........Gy
half hard ....1/2 hard*
hectare .... ha
henry..H
hertz..Hz
hexagonal, close-packed (see close-
    packed hexagonal)

horsepower .... hp
horsepower-hour...............................hp-h
hot rolled ....HR*
hour .... h
hydrogen-ion activity, log of...............pH

inch .... in.
inch(es) per minute .... in./min
inches per revolution ....in./rev
inch(es) per second ....in./s
inches per tooth ....spell out
inches per year (penetration) ....in./yr
inch pound..............................in.-lb
inside diameter (when preceded
    by dimensional value, or in a
    table ....ID
International Annealed Copper
    Standard .... IACS

joule.. J

kelvin .... K
kilocalorie ....kcal
kilocycle ....spell out
kilocyles per second (kilohertz)...........kHz
kilogram... kg
kilograms per cubic metre................ kg/m³
kilograms per second .... kg/s
kilograms per square millimetre........avoid
kilohertz..kHz
kilolitre ....avoid
(convert to m³)
kilometre.. km
kilometres per second .... km/s
kilovolt ....kV
kilovolt ampere.................................kVA
kilowatt...kW
kilowatt hour .... kW-h
Knoop hardness number..................... HK

left-hand (thread)............................LH
liquid .... liq*
litre L
logarithm (base 10)............................ log
(base e)..............................ln
lumen...lm
lux..lx

man-hour...............................spell out
manufacturer ....spell out
maximum.. max
megacycle..............................spell out

Abbreviations and Symbols for Technical Terms

megacycles per second (megahertz)	MHz
mega electron volt	MeV
megahertz	MHz
megajoule	MJ
megapascal	MPa
megavolt	MV
megohm	$M\Omega$
metre	m
microampere	μA
microfarad	μF
micro-inch	micro-in.
micron (micrometre)	μm
microhm	$\mu\Omega$
microsecond	μs
microvolt	μV
microwatt	μW
miles per hour	mph*
milliampere	mA
milliampere second	mA-s
milligram	mg
milligrams per sq decimetre per day	avoid
	(convert to mg/m^2 d)
millilitre	mL
millimetre	mm
millimicron (nanometre)	nm
millisecond	ms
millivolt	mV
mils per year	Mils/yr
minimum	min
minute (time)	min
minute (angular)	'
molar (solution)	*M*
molecular weight	mol wt
mol percent	mol %
mole	mol
nanometre	nm
nanosecond	ns
National coarse (thread)	NC
National fine (thread)	NF
National Formulary	NF
National pipe straight (thread)	NPS
National pipe taper (thread)	NPT
negative (—)	neg*
newton	N
newton/m² (pascal)	Pa
normal (solution)	N
normalized	norm*
nose radius	NR
number	No.
ohm	Ω
oil quenched	OQ*
open-back inclinable (press)	OBI
ounce	oz
ounces per sq ft	oz/ft²
outside diameter (when preceded by dimensional value, or in a table)	OD
oxygen-free high-conductivity (copper)	OFHC
parts per billion	ppb
parts per million	ppm
pascal	Pa
per cent	%
pieces per hour	pcs/h*
pitch diameter	PD
poise	P
positive (+)	pos*
post meridiem	p.m.
pound	lb
pound foot	lb-ft
pound inch	lb-in.
pound molecular weight	lb-mol
pounds per cubic foot	lb/ft³
pounds per square inch	psi
1000 psi	ksi
pounds per square inch, absolute	psia
pounds per square inch, gage	psig
power factor	spell out
quart	qt
quarter hard	1/4 hard*
quenched and tempered	Q&T*
rad (absorbed dose)	rd
radian	rad
radio frequency (adjective only)	RF
Rankine	R
reduction in area (%)	RA*
Reference	Ref
remainder	rem*
resistance-capacitance (circuit)	R-C*
revolutions per minute	rpm
revolutions per second	rps
right-hand (thread)	RH
rise per tooth	rpt
Rockwell A hardness	HRA
Rockwell B hardness	HRB
Rockwell C hardness	HRC

Abbreviations and Symbols for Technical Terms

roentgen . . . . .R
roentgens per hour at one metre . . . . R hm
roentgens per minute at one
    metre . . . . R mm
room temperature . . . . RT*
root mean square . . . . rms
roughness-height rating . . . . rhr
round-corner square . . . . rcs*

Saybolt universal seconds . . . . SUS
second (time) . . . . .s
second (angular) . . . . ″
side cutting-edge (angle) . . . . SCEA
side rake (angle) . . . . SR
side relief (angle) . . . . SRF
siemens (formerly mho) . . . . .S
Society of Automotive Engineers . . . . SAE
specific gravity . . . . sp gr
specific heat . . . . .sp ht*
square centimetre . . . . cm^2
square foot . . . . ft^2
square inch . . . . $in.^2$
square metre . . . . m^2
square root of mean square . . . . rms
square millimetre . . . . mm^2
square yard . . . . yd^2
standard temperature and pressure . . . . . STP
steradian . . . . sr
stoke . . . . .St

technically pure . . . . TP
temperature . . . . .temp*
tesla . . . . .T

thousand pounds per square inch . . . . .ksi
three-quarters hard . . . . .3/4 hard*
tonne (metric ton, or 1 Mg) . . . . .t
tons per square inch . . . . avoid
                      (convert to ksi)
total indicator reading .TIR
total indicator variation . . . . TIV
transformation-induced plasticity
    (steels) . . . . TRIP
typical . . . . typ*

Unified coarse (thread) . . . . UNC
Unified extra fine (thread) . . . . UNEF
Unified fine (thread) . . . . UNF
Unified special (thread) . . . . UNS
United States Pharmacopeia . . . . .USP

Vickers hardness number . . . . .HV
volt v
volt ampere . . . . vA
Volume (in Ref lists) . . . . .Vol
volume percent . . . . vol %

water quenched . . . . WQ*
watt . . . . W
watt-hour . . . . W-h
weber . . . . .Wb
weight percent . . . . wt %
    (use % alone, except to avoid
    confusion with atomic per cent
    or volume per cent)

yard . . . . yd
year . . . . yr*

10-11. The following symbols represent ASM preference. In equations, all letters used
to represent terms of the equation should be set in italic. When these symbols
are used in text, they should be used in expressions derived from equations --
for example: $2t$ (meaning two times the thickness). Numbers (including subscripts and superscripts), the symbol for the chemical elements, letter designations for north and south magnetic poles, mathematical operators such as log,
exp, and ln, and trigonometric functions such as sin, cos, and tan should be set
in roman. (Exception: the mathematical operators for derivative, d, and
function, f, should be set in italic -- for example, dy/dx or $f/(x)$.) Boldface
roman letters should be used for vectors in both equations and text.

Wherever possible, lower-case letters should be used for linear dimensions
such as length, depth, width, height, or thickness. An exception is the use of
capital D and R for outside diameter and radius when inside diameter and
radius are cited in the same context. Lower-case Greek letters are usually
preferred for those quantities commonly represented by Greek letters.

10-12a. Mechanical and Physical Properties

coefficient of linear
thermal expansion .... α

density .... ρ

strength .... S

    ultimate (tensile) strength ...... S_u

    yield strength .... S_y

    shear strength .... S_s

electrical resistivity .... ρ

Poisson's ratio .... μ

plane-strain fracture
toughness .... K_{Ic}

modulus
    elastic (Young's) .... E

    shear .... G

    bulk .... K

surface tension .... γ

thermal conductivity .... κ

viscosity .... η

temperature
    Curie .... T_C

    melting .... T_M

10-12b. Engineering Quantities

acceleration ....a

angle.........$\theta, \alpha, \beta, \gamma, \phi, \psi$

arc length ....s

area ....A

circumference c

current (electrical)
   direct or average (rms). I
   instantaneous...............i

depth ....d

diameter
   general or inside ....d
   outside ....D

distance ....d

frequency......................f

friction coefficient ....ν

height ....h

impedance (acoustic or
   electrical) ....Z

inertia
   moment of I
   polar moment of.........J

length l

load (force).................... L

magnetic north ....N

magnetic south .... S

notch factor..................K

pressure .... P

radian frequency
   $(2\pi f)$.......................ω

radius
   general or inside ....r
   outside ....R

reactance X

resistance (electrical).......R

strain ....ϵ

strain-energy-release
   rate ....G

strain rate.....................ϵ

stress
   normal.....................τ
   shear........................τ

stress-intensity factor......K
   critical plane-strain,
   or plane-strain fracture
   toughness.................K_{Ic}

temperature; torque........ T

thickness; time............... t

velocity
   angular.....................ω
   linear.......................v

voltage
   (electrical potential) .... E

volume.........................V

wavelength....................λ

weight......................... W

width ....w, b

10-12c. Mathematical Symbols

plus　+

minus　−　　　　plus or minus　±

times　×, or juxtaposition: $2ab$, or period centered: a·b

divided by　÷, or fractional notation (virgule or horizontal bar)

equals　=

is identical with　≡

does not equal　≠

equals approximately　≈, or ≅, or ≃

less than　<

much less than　≪

greater than　>

much greater than　≫

less than or equal to　≤　or　⩽

greater than or equal to　≥　or　⩾

varies with　∝

ratio　:

proportion　::

per cent　%

aggregation:　( ) parenthesis, [ ] bracket,　brace　(in order of use)

power　2^2, 2^{-2}, 2

square root　√　　　cube root　√

infinity　∞

variable number　n

unspecified number　N

constant　C, k, K

difference, increment　△ (u.c. delta, one size smaller than
　　　accompanying type)

derivative　d　　　partial derivative　∂

integral　∫　　　definite integral　$_a\int^h$

summation　Σ　　　definite summation　Σ

mean, or average, on χ　　　χ

standard deviation　σ

variance δ

parallel $\|$

perpendicular $\perp$

angular measure: degree $^{\circ}$, minute $'$, second $''$

trigonometric functions: sin, cos, tan, sec, csc, cot

hyperbolic functions: sinh, cosh, tanh, sech, csch, coth

inverse functions: $\sin^{-1}$, $\cos^{-1}$, $\sinh^{-1}$, etc. (arcsin, arccos, etc.)

base of natural (Napierian) logarithms   e   (equal to 2.71828)

pi π (equal to 3.14159)

natural logarithm   ln

common logarithm   log
    (base 10)

antilogarithm $\ln^{-1}$ or $\log^{-1}$

exponential function   exp $[A \exp (x\text{-}2) \equiv Ae^{K-2}]$

10-12d. Coordinate Systems

Cartesian (rectilinear)   axes labeled x, y, z

Polar r, θ, z

Spherical r, θ, ϕ

10-12e. Chemical Reactions

forms by chemical reaction $\rightarrow$

forms by reversible chemical reaction*

substance passes off as a gas $\uparrow$

substance precipitates from solution $\downarrow$

chemically combined with $\cdot$ (period centered, as in $Fe_2O_3 \cdot 5H_2O$)

10-12f. Symbols for the Chemical Elements

Symbols for elements are used in alloy compositions where weight percentages are given in numbers. (Example: 0.20 C, 0.80 Mn, 0.15 Si, rem Fe; where two or more elements are given, the per cent sign is omitted.)

Actinium	Ac	Gadolinium	Gd
Aluminum	Al	Gallium	Ga
Americium	Am	Germanium	Ge
Antimony	Sb	Gold	Au
Argon	Ar		
Arsenic	As	Hafnium	Hf
Astatine	At	Helium	He
		Holmium	Ho
Barium	Ba	Hydrogen	H
Berkelium	Bk		
Beryllium	Be	Indium	In
Bismuth	Bi	Iodine	I
Boron	B	Iridium	Ir
Bromine	Br	Iron	Fe
Cadmium	Cd	Krypton	Kr
Calcium	Ca		
Californium	Cf	Lanthanum	La
Carbon	C	Lawrencium	Lr
Cerium	Ce	Lead	Pb
Cesium	Cs	Lithium	Li
Chlorine	Cl	Lutetium	Lu
Chromium	Cr		
Cobalt	Co	Magnesium	Mg
Columbium		Manganese	Mn
(Niobium)	Cb	Mendelevium	Md
Copper	Cu	Mercury	Hg
Curium	Cm	Molybdenum	Mo
Dysprosium	Dy	Neodymium	Nd
		Neon	Ne
Einsteinium	Es	Neptunium	Np
Erbium	Er	Nickel	Ni
Europium	Eu	Niobium (Columbium)	Nb
		Nitrogen	N
Fermium	Fm	Nobelium	No
Fluorine	F		
Francium	Fr	Osmium	Os
		Oxygen	O

10-12f. Symbols for the Chemical Elements (*continued*)

Palladium	Pd	Strontium	Sr
Phosphorus	P	Sulfur	S
Platinum	Pt		
Plutonium	Pu	Tantalum	Ta
Polonium	Po	Technetium	Tc
Potassium	K	Tellurium	Te
Praseodymium	Pr	Terbium	Tb
Promethium	Pm	Thallium	Tl
Protactinium	Pa	Thorium	Th
		Thulium	Tm
Radium	Ra	Tin	Sn
Radon	Rn	Titanium	Ti
Rhenium	Re	Tungsten	W
Rhodium	Rh		
Rubidium	Rb	Uranium	U
Ruthenium	Ru		
		Vanadium	V
Samarium	Sm		
Scandium	Sc	Xenon	Xe
Selenium	Se		
Silicon	Si	Ytterbium	Yb
Silver	Ag	Yttrium	Y
Sodium	Na		
		Zinc	Zn
		Zirconium	Zr

10-12g. Crystal Structure

Follow copy exactly in editing manuscripts containing designations of crystal structures because the brackets, parentheses, braces and carets, and the placement of the minus sign above numbers, have special significance. The following examples show some of the forms used. See pages 234-235 of Vol 8 of the 8th Edition of Metals Handbook for an explanation of the meaning of the different expressions.

(111)   (112) $\{100\}$ [100] $\langle 100 \rangle$

10-12h. Transformations

Transformations in iron and steel are expressed by certain standard symbols that are used in figures and in text when referring to the areas or points on the transformation diagram. For a list and definition of these symbols, see "transformation temperature" in Glossary of Metallurgical Terms and Engineering Tables (ASM 1979).

10-12j. Atomic Information

Normally, atomic information will be presented in the following format:

$$\begin{array}{lll} \text{Mass number}\dots\dots\dots14 & 4+ & \text{Valence state} \\ \qquad\qquad\qquad\qquad\text{C} & & \\ \text{Atomic number}\dots\dots\ 6 & 2 & \text{Number of atoms} \\ & & \text{per molecule} \end{array}$$

Not all of these indicators will be needed at the same time. For instance, only in very rare instances will the atomic number be used at all; the chemical symbol implies the same information given by the atomic number. The mass number will be used only when it is necessary to differentiate between isotopes (such as to distinguish fissionable ^{235}U from nonfissionable ^{238}U).

10-12k. Designating an Alloy by Content

Arrange the elements in order of decreasing importance: Thus a copper alloy containing 6% Si would be generally spoken of as a "copper-silicon alloy," except in such a definite expression as a "6% Si copper alloy."

10-12l. Specifications of Various Agencies

Society of Automotive Engineers: SAE 4130
Aerospace Material Specifications (of SAE): AMS 5063
American Iron and Steel Institute: AISI 3220; X1112; E4640; type 446
American Society for Testing and Materials: ASTM B17
Federal Specifications: QQ-S-624 (-2); QQ-B-726; WW-%-791 (-1); type, grade or
    class: FS-310; C1041; gr 446; grA
Military Specifications: MIL-C-3383; MIL-T-8504 (-2) ASG; MIL-16301 (BuOrd);
    type, grade or class: comp 5; gr IV; cl a
General Motors Corp: GM 2315
Alloy Casting Institute: ACI CA-15; CD-4MCu
American Society of Mechanical Engineers: ASME SA-213, grade T11
Because SAE, AISI and ACI designations are well known, the "SAE," "AISI," and
"ACI" are usually omitted.

ASTM and ASME specifications for ferrous and nonferrous alloys used in boilers,
heat exchangers, and pressure vessels have the same numerical designation but different prefixes -- for instance, ASTM A334 is equivalent to ASME SA-334. The agency
should be listed for these materials so that any minor differences in the specifications
would be accurately cited.

10-12m. Aluminum Alloys

Wrought: Aluminum Association designations are used (without "AA"): 2024; 7075;
    3003
Cast: Aluminum Association designations are used (without "AA") except that only
    the first three digits are shown when the reference is general -- to the alloy, not the
    form: 356; 413.0; 295.2
Powder Metals: SAP 865; XAP004

10-12n. Copper, Brass and Bronze

Wrought and Cast: Unified Numbering Systems (UNS) designations are used for
    both wrought and cast alloys (without "UNS"); C10200, C26000, C67500, C86200
Casting Ingot: ASTM ingot identification numbers are preferred, but Brass and
    Bronze Ingot Institute designations may be used.

10-12o. Ferrous Materials

Wrought steel: AISI and SAE designations: 1010; 4130; 5160; 8620H; 50B50
Cast steel:
  ASTM: ASTM A27, grade 60-30
  SAE: SAE 090; SAE 0105

Gray iron:
    ASTM: class 20; class 35
    SAE: SAE G1800; SAE G2500; SAE G3000
Malleable iron: ASTM: grade 35018; grade 32510. SAE: M3210; M5503; M8501
Ductile iron: (Say "ductile", not "nodular" or "spherulitic".)
    ASTM: class 80-60-03; class 100-70-03. SAE: D4018; D5506; DQ&T
Wrought stainless steel: AISI: type 310 (rather than SAE 30310). UNS: S31000
    Other: 17-7 PH; PH 15-7 Mo; AM-350
Alloy castings: ACI: CA-15; CD-4MCu

10-12p. Heat-Resisting Alloys

Industry names: S-590; K-42-B; 16-25-6; A-286; M-308; D-979; R-235; 19-9 DL;
Hastelloy B; N-155; Nimonic 80A; U-212 and so on. See ASTM DS45 for cross-
reference of trade names and product specifications.

10-12q. Magnesium Alloys

ASTM designations are used for both wrought and cast alloys (without "ASTM"):
AZ81A; EK41A; HM21A

10-12r. Nickel Alloys

Use industry designations: Nickel 200, Nickel 270, Monel 400; Inconel X-750; Incoloy
800.

10-12s. Titanium Alloys

Ti grade 1; Ti code 12; Ti-641-4V; Ti-5Al-2.5 Sn; Ti-8Mn; Ti-2Fe-2C-2Mo; Ti-7Al-
4Mo

10-12t. Heat Treatments

Aluminum: -0; -76; -W (suffixes)
Magnesium: -T6; -0; -H31 (suffixes)
Precipitation-hardening alloys: RH 950; TH 1050

Index